INVERSE HEAT TRANSFER

FUNDAMENTALS AND APPLICATIONS

INVERSE HEAT TRANSFER

FUNDAMENTALS AND APPLICATIONS

M. Necati Özisik
Department of Mechanical and Aerospace Engineering
North Carolina State University
Raleigh, North Carolina

Helcio R. B. Orlande
Department of Mechanical Engineering, EE/COPPE
Federal University of Rio de Janeiro
Rio de Janeiro, Brazil

CRC Press is an imprint of the
Taylor & Francis Group, an **informa** business

CRC Press
Taylor & Francis Group
6000 Broken Sound Parkway NW, Suite 300
Boca Raton, FL 33487-2742

First issued in paperback 2019

ISBN-13: 978-0-367-44739-7 (pbk)
ISBN-13: 978-1-56032-838-4 (hbk)

Visit the Taylor & Francis Web site at
http://www.taylorandfrancis.com

and the CRC Press Web site at
http://www.crcpress.com

To Sevgi, Hakan and to the memory of Gul
(M.N.O.)

To Fernanda and Arthur José
(H.R.B.O.)

INVERSE HEAT TRANSFER

FUNDAMENTALS AND APPLICATIONS

Contents

Preface

Inverse Heat Transfer Problems (IHTP) rely on temperature and/or heat flux measurements for the estimation of unknown quantities appearing in the analysis of physical problems in thermal engineering. As an example, inverse problems dealing with heat conduction have been generally associated with the estimation of an unknown boundary heat flux, by using temperature measurements taken below the boundary surface. Therefore, while in the classical direct heat conduction problem the cause (boundary heat flux) is given and the effect (temperature field in the body) is determined, the inverse problem involves the estimation of the cause from the knowledge of the effect. An advantage of IHTP is that it enables a much closer collaboration between experimental and theoretical researchers, in order to obtain the maximum of information regarding the physical problem under study.

Difficulties encountered in the solution of IHTP should be recognized. IHTP are mathematically classified as ill-posed in a general sense, because their solutions may become unstable, as a result of the errors inherent to the measurements used in the analysis. Inverse problems were initially taken as not of physical interest, due to their ill-posedness. However, some heuristic methods of solution for inverse problems, which were based more on pure intuition than on mathematical formality, were developed in the 50's. Later in the 60's and 70's, most of the methods, which are in common use nowadays, were formalized in terms of their capabilities to treat ill-posed unstable problems. The basis of such formal methods resides on the idea of reformulating the inverse problem in terms of an approximate well-posed problem, by utilizing some kind of regularization (stabilization) technique. In this sense, it is recognized here the pioneering works of scientists who found different forms of overcoming the instabilities of inverse problems, including A. N. Tikhonov, O. M. Alifanov and J. V. Beck.

The field of inverse heat transfer is wide open and diversified. Therefore, an orderly and systematic presentation of the scientific material is essential for the understanding of the subject. This principle has been the basic guideline in the preparation of this book.

This book is intended for graduate and advanced undergraduate levels of teaching, as well as to become a reference for scientists and practicing engineers. We have been motivated by the desire to make an application-oriented book, in order to address the needs of readers seeking solutions of IHTP, without going through detailed mathematical proofs.

The main objectives of the book can be summarized as follows:

- Introduce the fundamental concepts regarding IHTP;
- Present in detail the basic steps of four techniques of solution of IHTP, as a parameter estimation approach and as a function estimation approach;
- Present the application of such techniques to the solution of IHTP of practical engineering interest, involving conduction, convection and radiation; and
- Introduce a formulation based on generalized coordinates for the solution of inverse heat conduction problems in two-dimensional regions.

The book consists of six chapters.

Chapter 1 introduces the reader to the basic concepts of IHTP.

Chapter 2 is concerned with the description of four techniques of solution for inverse problems. The four techniques considered in this book include:

Technique I: The Levenberg-Marquardt Method for Parameter Estimation
Technique II: The Conjugate Gradient Method for Parameter Estimation
Technique III: The Conjugate Gradient Method with Adjoint Problem for Parameter Estimation
Technique IV: The Conjugate Gradient Method with Adjoint Problem for Function Estimation

These techniques were chosen for use in this book because, based on the authors' experience, they are sufficiently general, versatile, straightforward and powerful to overcome the difficulties associated with the solution of IHTP.

In Chapter 2 the four techniques are introduced to the reader in a systematic manner, as applied to the solution of a simple, but illustrative, one-dimensional inverse test-problem, involving the estimation of the transient strength of a plane heat-source in a slab. The basic steps of each technique, including the iterative procedure, stopping criterion and computational algorithm, are described in detail in this chapter. Results obtained by using simulated measurements, as applied to the solution of the test-problem, are discussed. The mathematical and physical significances of sensitivity coefficients are also discussed in Chapter 2 and three different methods are presented for their computation. Therefore, in Chapter 2 the reader is exposed to a full inverse analysis involving a simple test-problem, by using the four techniques referred to above, which will be applied later in the book to more involved physical situations, including Conduction Heat Transfer in **Chapter 3**, Convection Heat Transfer in **Chapter 4** and Radiation Heat Transfer in **Chapter 5**.

Chapter 6 is concerned with the solution of inverse heat conduction problems of estimating the transient heat flux applied on part of the boundary of irregular two-dimensional regions, by using Technique IV. The irregular region in the physical domain (x,y) is transformed into a rectangle in the computational domain (ξ,η). Different quantities required for the solution are formulated in terms of the generalized coordinates (ξ,η). Therefore, the present formulation is general and can be applied to the solution of boundary inverse heat conduction problems over any region that can be mapped into a rectangle. The present approach is illustrated with an inverse problem of practical engineering interest, involving the cooling of electronic components.

The pertinent **References** and sets of **Problems** are included at the end of each chapter. The proposed problems expose the reader to practical situations in a gradual level of increasing complexity, so that he(she) can put into practice the general concepts introduced in the book.

We would like to acknowledge the financial support provided by CNPq, CAPES and FAPERJ, agencies for science promotion of the Brazilian and Rio de Janeiro State governments, as well as by NSF-USA, for the visits of M. N. Özisik to the Federal University of Rio de Janeiro (UFRJ) and of H. R. B. Orlande to the North Carolina State University (NCSU). The hospitality of the Mechanical Engineering Departments at both institutions is greatly appreciated. This text was mainly typed by M. M. Barreto, who has demonstrated extreme dedication to the work and patience in understanding our handwriting in the original manuscript. The works of collaborators of the authors, acknowledged throughout the text, were essential for transforming an idea for a book into a reality. We would like to thank Prof. M. D. Mikhailov for invaluable suggestions regarding the contents of Chapter 2 and Prof. R. M. Cotta for introducing us to the editorial vice president of Taylor & Francis. We are indebted to several students from the Department of Mechanical Engineering of the Federal University of Rio de Janeiro, who helped us at different points during the preparation of the book. They include E. N. Macedo, R. N. Carvalho, M. J. Colaço, M. M. Mejias, L. M. Pereira, L. B. Dantas, H. A. Machado, L. F. Saker, L. A. Sphaier, L. S. B. Alves, L. R. S. Vieira and C. F. T. Matt. H. R. B. Orlande is thankful for the kind hospitality of several friends during his visits to Raleigh, who certainly made the preparation of this book more pleasant and joyful. They include the Ferreiras, the Gonzalezes and the Özisiks. Finally, we would like to express our deep appreciation for the love, prayers and support of our families.

Nomenclature

C	volumetric heat capacity
$C_j(t)$	trial function, $j = 1, ..., N$
$\mathrm{cov}(X,Y)$	covariance of X and Y
c_p	specific heat
d	direction of descent
I	number of transient measurements per sensor
$\mathbf{J}$	sensitivity matrix
J_{ij}	sensitivity coefficient
k	thermal conductivity
L	length
M	number of sensors
N	number of unknown parameters
$\mathbf{P}$	vector of unknown parameters
P_j	j^{th} unknown parameter, $j = 1, ..., N$
q	heat flux
r, θ, φ	polar spatial coordinates
r, θ, z	cylindrical spatial coordinates
S	objective function or objective functional
t	time
t_f	final time
$\mathbf{T}$	vector of estimated temperatures
T_i	estimated temperature at time t_i, $i = 1, ..., I$
$\mathbf{V}$	covariance matrix of estimated parameters
x,y,z	cartesian spatial coordinates
$\mathbf{Y}$	vector with measured temperatures
Y_i	measured temperature at time t_i, $i = 1, ..., I$

GREEK SYMBOLS

α	thermal diffusivity
β	search step size
γ	conjugation coefficient
$\delta(\cdot)$	Dirac delta function
Δ	variation
ΔT	sensitivity function
ε	tolerance
Ω	diagonal matrix for the iterative procedure of Technique I
μ	damping parameter for the iterative procedure of Technique I
λ	Lagrange Multiplier satisfying the adjoint problem
ρ	density
σ	standard deviation of the measurements
χ_N^2	chi-square distribution with N degrees of freedom
$\nabla S(\cdot)$	gradient direction

SUBSCRIPTS

est	estimated
ex	exact
f	final
i	i^{th} measurement
j	j^{th} parameter
m	sensor number
meas	measurement location for a single sensor

SUPERSCRIPTS

k	iteration number
T	transpose

Part ONE

FUNDAMENTALS

Chapter 1

BASIC CONCEPTS

In recent years interest has grown in the theory and application of Inverse Heat Transfer Problems (IHTP); it is encountered in almost every branch of science and engineering. Mechanical, aerospace, chemical and nuclear engineers, mathematicians, astrophysicists and statisticians are all interested in this subject, each group with different applications in mind.

The space program has played a significant role in the advancement of solution techniques for the IHTP in late 50's and early 60's. For example, aerodynamic heating of space vehicles is so high during reentry in the atmosphere that the surface temperature of the thermal shield cannot be measured directly with temperature sensors. Therefore, temperature sensors are placed beneath the hot surface of the shield and the surface temperature is recovered by inverse analysis. Inverse analysis can also be used in the *estimation* of thermophysical properties of the shield during operating conditions at such high temperatures.

Direct measurement of heat flux at the surface of a wall subjected to fire by using conventional methods is a difficult matter; but it can readily be estimated by an inverse analysis utilizing transient temperature recordings taken at a specified location beneath the heated surface.

In situations when the well established classical methods for property estimation cannot provide the desired degree of accuracy or become inapplicable, the IHTP technique can be used.

Difficulties associated with the solution of IHTP should also be recognized. Mathematically, inverse heat transfer problems belong to a class called *Ill-posed* [1-6], whereas standard heat transfer problems are ***Well-posed.*** The concept of a well-posed problem, originally introduced by Hadamard [2], requires that its solution should satisfy the following three conditions:

- The solution must exist;
- The solution must be unique;
- The solution must be stable under small changes to the input data (i.e., stability condition)

The existence of a solution for an inverse heat transfer problem may be assured by physical reasoning; for example, if there is a change in the values of the measured temperature in a transient problem, there exists a causal characteristic, say, a boundary heat flux, to be estimated. On the other hand, the uniqueness of the solution of inverse problems can be mathematically proved only for some special cases [5,6]. Also, the inverse problem is very sensitive to random errors in the measured input data, thus requiring special techniques for its solution in order to satisfy the stability condition.

For a long time it was thought that, if any of the conditions required for well-posedness were violated, the problem would be unsolvable or the results obtained from such a solution would be meaningless, hence would have no practical importance. As a result, interest waned by the mathematicians, physicists and engineers in the solution of inverse problems [5]. It was *Tikhonov's regularization procedure* [3,7-9], *Alifanov's iterative regularization techniques* [1,5,10-24] and *Beck's function estimation approach* [6,25] that revitalized the interest in the solution of inverse heat transfer problems. A successful solution of an inverse problem generally involves its reformulation as an approximate well-posed problem. In most methods, the solution of inverse heat transfer problems are obtained in the least squares sense. Tikhonov's *regularization procedure* modifies the least squares equation by adding smoothing terms in order to reduce the unstable effects of the measurement errors. In the *iterative regularization principle*, a sequential improvement of the solution takes place. The stopping criterion for such iterative procedure is chosen so that the final solution is stabilized with respect to errors in the input data.

As a result of such new solution techniques and the availability of high speed, large capacity computers, successful solution of inverse heat transfer problems has now become feasible. The past three decades have been most active in the advancement of solution techniques for the IHTP. One of the earliest discussion of thermal inverse problems is due to Giedt [26] who examined the heat transfer at the inner surface of a gun barrel. Stolz [27] presented a procedure for estimating surface temperature and heat flux from the temperature measurements taken within a body being quenched. Several other works on the theory and application of inverse heat transfer problems can be found in references [28-110] and a number of books are also available on the subject [3-6,21,111-120].

In this chapter, we present a general discussion of inverse heat transfer problems, including basic concepts, application areas, classification, an overview of various solution techniques and difficulties involved in such solutions.

1–1 INVERSE HEAT TRANSFER PROBLEM CONCEPT

The physical significance of the inverse heat transfer problem concept is better envisioned by referring to the following standard, one-dimensional transient heat conduction problem in a slab of thickness L. The temperature distribution in the slab is initially $F(x)$. For times $t > 0$, a transient heat flux $f(t)$ is applied on the boundary $x = 0$, while the boundary $x = L$ is maintained at the constant temperature T_L. The mathematical formulation of this problem is given by:

$$\frac{\partial}{\partial x}\left(k\frac{\partial T}{\partial x}\right)=\rho c_p \frac{\partial T}{\partial t} \qquad \text{in} \quad 0<x<L, \qquad \text{for} \quad t>0 \qquad (1.1.1.a)$$

$$-k\frac{\partial T}{\partial x}=f(t) \qquad \text{at} \quad x=0, \qquad \text{for} \quad t>0 \qquad (1.1.1.b)$$

$$T=T_L \qquad \text{at} \quad x=L, \qquad \text{for} \quad t>0 \qquad (1.1.1.c)$$

$$T=F(x) \qquad \text{for} \quad t=0, \qquad \text{in} \quad 0<x<L \qquad (1.1.1.d)$$

For the case where the boundary conditions $f(t)$ and T_L, the initial condition $F(x)$, and the thermophysical properties ρ, c_p and k are all specified, the problem given by equations (1.1.1) is concerned with the determination of the temperature distribution $T(x, t)$ in the interior region of the solid, as a function of time and position. This is called the *Direct Problem*.

We now consider a problem similar to that given by equations (1.1.1), but the boundary condition function $f(t)$ at the surface $x = 0$ is unknown, while all the other quantities appearing in equations (1.1.1), such as T_L, $F(x)$, k, ρ and c_p, are known. We then wish to determine the unknown boundary condition $f(t)$. To compensate for the lack of information on the boundary condition, measured temperatures $T(x_{meas}, t_i) \equiv Y_i$ are given at an interior point x_{meas} at different times t_i ($i = 1, 2, ..., I$), over a specified time interval $0 < t \leq t_f$, where t_f is the final time. This is an *Inverse Problem* because it is concerned with the *estimation* of the unknown surface condition $f(t)$. Here the terminology *estimation* is used in place of determination. The reason is that the measured temperature data used in the inverse analysis contain measurement errors. As a result, the quantity recovered by the inverse analysis. (i.e., the boundary condition $f(t)$ in the example above) is not exact, but it is only an estimate within the measurement errors.

Then, the mathematical formulation of this *Inverse Problem* is given by

$$\frac{\partial}{\partial x}\left(k\frac{\partial T}{\partial x}\right)=\rho c_p \frac{\partial T}{\partial t} \qquad \text{in} \quad 0<x<L, \qquad \text{for } 0<t\leq t_f \qquad (1.1.2.a)$$

$$-k\frac{\partial T}{\partial x}=f(t) = \text{? (unknown)} \quad \text{at} \quad x=0, \qquad \text{for } 0<t\leq t_f \tag{1.1.2.b}$$

$$T=T_L \quad \text{at} \quad x=L, \qquad \text{for } 0<t\leq t_f \tag{1.1.2.c}$$

$$T=F(x) \quad \text{for } t=0, \qquad \text{in } 0<x<L \tag{1.1.2.d}$$

and temperature measurements at an interior location x_{meas} at different times t_i are given by

$$T(x_{meas}, t_i) \equiv Y_i \quad \text{at} \quad x=x_{meas}, \quad \text{for} \quad t=t_i \, (i=1,2,...,I) \tag{1.1.3}$$

The main objective of the direct problem is to recover the temperature field $T(x, t)$ in the solid, when all the *causal characteristics* (i.e., boundary conditions and their parameters, initial condition, thermophysical properties of the medium and energy generation term, if there is any) are specified. On the other hand, the objective of the inverse problem is to estimate one or more of such unknown causal characteristics, from the knowledge of the measured temperature (*the effect*) at some specified section of the medium. *In the direct problem the causes are given, the effect is determined; whereas in the inverse problem the effect is given, the cause (or causes) is estimated.*

In the inverse problem given above, the boundary surface function $f(t)$ is unknown. Hence, the problem is referred to as a *boundary inverse heat transfer problem*. Analogously, one envisions inverse heat transfer problems of unknown initial condition, energy generation, thermophysical properties, and so on. So far we considered an inverse heat transfer problem of conduction; similarly, we can have inverse problems of convection, body or surface radiation, mixed modes of heat transfer and numerous others.

Inverse problems can be solved either as a *parameter estimation* or as a *function estimation* approach. If some information is available on the functional form of the unknown quantity, the inverse problem is reduced to the estimation of few unknown parameters. Let us consider the boundary inverse problem given by equations (1.1.2, 1.1.3) and assume that the unknown function $f(t)$ can be represented as a polynomial in time in the form

$$f(t)=P_1+P_2t+P_3t^2+...+P_Nt^{N-1} \tag{1.1.4.a}$$

or in the more general linear form as

$$f(t)=\sum_{j=1}^{N} P_j C_j(t) \tag{1.1.4.b}$$

where P_j , j=1,...,N , are unknown constants and $C_j(t)$ are known trial functions. Therefore, the inverse problem of estimating the unknown function $f(t)$ is reduced to the problem of *estimating a finite number of parameters P_j*, where the number N of parameters is supposed to be chosen in advance. Another example of parameter estimation is the recovering of unknown constant thermophysical properties, such as the thermal conductivity k or the volumetric heat capacity ρc_p, appearing in equations (1.1.2). If no prior information is available on the functional form of the unknown, the inverse problem can be regarded as a *function estimation approach in an infinite dimensional space of functions.* Techniques for the solution of inverse problems as a parameter estimation, as well as a function estimation approach, will be presented in the following chapter.

1-2 APPLICATION AREAS OF INVERSE HEAT TRANSFER

With the advent of modern complex materials having thermophysical properties strongly varying with temperature and position, the use of conventional methods for determining thermophysical properties has become unsatisfactory. Similarly, the operation of modern industrial concerns is becoming more and more sophisticated, and an accurate *in situ* estimation of thermophysical properties under actual operating conditions is becoming necessary. The inverse heat transfer problem approach can provide satisfactory answers for such situations.

The principal advantage of the IHTP is that it enables to conduct experiments as close to the real conditions as possible. Practical applications of IHTP techniques include, among others, the following specific areas:

- Estimation of thermophysical properties of materials [4,6,20-23,68,96,103,110]. For example, properties of heat shield material during its reentry into the earth's atmosphere, and estimation of temperature dependence of thermal conductivity of a cooled ingot during steel tempering.
- Estimation of bulk radiation properties and boundary conditions in absorbing, emitting and scattering semi-transparent materials [73-79,87].
- Control of the motion of the solid-liquid interface during solidification [89-91].
- Estimation of inlet condition and boundary heat flux in forced convection inside ducts [72,80,81,108,109].
- Estimation of timewise varying unknown interface conductance between metal solidification and metal mold during casting [82,85].
- Estimation of interface conductance between periodically contacting surfaces [83].
- Monitoring radiation properties of reflecting surfaces of heaters and cryogenic panels [5].
- Estimation of heat release during friction of two solids [5].

- Estimation of reaction function [84,97].
- Control and optimization of the curing process of rubber [98,99].
- Estimation of the boundary shapes of bodies [93,94,112].

The estimation of such quantities with conventional techniques is an extremely difficult or impossible matter. However, with the application of the inverse heat transfer analysis, such problems not only can be handled, but the information value of the studies is enhanced and the experimental work is accelerated.

1–3 CLASSIFICATION OF INVERSE HEAT TRANSFER PROBLEMS

Most of the early works on the solution of inverse heat transfer problems have been concerned with heat conduction in one-dimensional geometries. The application of inverse analysis techniques to multi-dimensional problems, as well as to problems involving convection and radiation, is more recent.

Inverse heat transfer problems can be classified in accordance with the nature of the heat transfer process, such as:

IHTP of conduction
IHTP of convection (forced or natural)
IHTP of surface radiation
IHTP of radiation in participating medium
IHTP of simultaneous conduction and radiation
IHTP of simultaneous conduction and convection
IHTP of phase change (melting or solidification)

Another classification can be one based on the type of causal characteristic to be estimated. For example:

IHTP of boundary conditions
IHTP of thermophysical properties
IHTP of initial condition
IHTP of source term
IHTP of geometric characteristics of a heated body

Inverse heat transfer problems can be one-, two- or three-dimensional. Also, IHTP can be linear or nonlinear. The factors affecting the linearity will be apparent in the following chapters.

1-4 DIFFICULTIES IN THE SOLUTION OF INVERSE HEAT TRANSFER PROBLEMS

To illustrate the inherent difficulties in the solution of inverse heat transfer problems, we consider a semi-infinite solid ($0 < x < \infty$) initially at zero temperature. For times $t > 0$, the boundary surface at $x = 0$ is subjected to a periodically varying heat flux in the form

$$q(t) = q_0 \cos \omega t$$

where q_0 and ω are the amplitude and frequency of oscillations for the heat flux, respectively, and t is the time variable. After the transients have passed, the quasi-stationary temperature distribution in the solid is given by [113,121]

$$T(x,t) = \frac{q_0}{k}\sqrt{\frac{\alpha}{\omega}} \exp\left(-x\sqrt{\frac{\omega}{2\alpha}}\right) \cos\left(\omega t - x\sqrt{\frac{\omega}{2\alpha}} - \frac{\pi}{4}\right) \tag{1.4.1.a}$$

where α is the thermal diffusivity and k is the thermal conductivity of the solid.

Equation (1.4.1.a) shows that the temperature response is lagged with respect to the heat flux excitation at the surface of the body, and such lagging is more pronounced for points located deeper inside the body. The temperature lagging indicates the need for measurements taken after the moment that the heat flux is applied, if such heat flux is to be estimated.

The amplitude for the temperature oscillation at any location, $|\Delta T(x)|$, is obtained by setting $\cos(\cdot) = 1$ in equation (1.4.1.a). Hence,

$$|\Delta T(x)| = \frac{q_0}{k}\sqrt{\frac{\alpha}{\omega}} \exp\left(-x\sqrt{\frac{\omega}{2\alpha}}\right) \tag{1.4.1.b}$$

Equation (1.4.1.b) shows that $|\Delta T(x)|$ attenuates exponentially with increasing depth below the surface and with increasing frequency ω. On the other hand, if the amplitude of the surface heat flux, q_0, is to be estimated by utilizing directly the measured temperatures at an interior point, any measurement error on $|\Delta T(x)|$ will be magnified exponentially with the depth x and with the frequency ω, as shown below in equation (1.4.1.c).

$$q_0 = k|\Delta T(x)|\sqrt{\frac{\omega}{\alpha}} \exp\left(x\sqrt{\frac{\omega}{2\alpha}}\right) \tag{1.4.1.c}$$

It is easy to notice that, in order to be able to estimate the boundary heat flux, a sensor must be located within a depth below the surface where the amplitude of the temperature oscillation is much greater than the measurement errors. Otherwise, it is impossible to distinguish if the measured temperature oscillation is due to changes in the boundary heat flux or due to measurement errors, thus resulting in the non-uniqueness of the inverse problem solution.

The foregoing discussion reveals that, depending on the location of the sensor and the frequency of oscillations, the solution of the inverse problem may become very sensitive to measurement errors in the input data. Since the accuracy of the solution obtained by an inverse analysis is affected by the errors involved in temperature measurements, it is instructive to present the eight standard assumptions proposed by Beck [4,6,86], regarding the *statistical description* of such errors. They are:

1. The errors are additive, that is

$$Y_i = T_i + \varepsilon_i$$

   where Y_i is the measured temperature, T_i is the actual temperature and ε_i is the random error.

2. The temperature errors ε_i have a zero mean, that is,

$$E(\varepsilon_i) = 0$$

   where $E(\cdot)$ is the expected value operator. The errors are then said to be unbiased.

3. The errors have constant variance, that is,

$$\sigma_i^2 = E\{[Y_i - E(Y_i)]^2\} = \sigma^2 = \text{constant}$$

   which means that the variance of Y_i is independent of the measurement.

4. The errors associated with different measurements are uncorrelated. Two measurement errors ε_i and ε_j, where $i \neq j$, are uncorrelated if the covariance of ε_i and ε_j is zero, that is,

$$\text{cov}(\varepsilon_i, \varepsilon_j) \equiv E\{[\varepsilon_i - E(\varepsilon_i)][\varepsilon_j - E(\varepsilon_j)]\} = 0 \qquad \text{for } i \neq j$$

   Such is the case if the errors ε_i and ε_j have no effect on or relationship to the other.

5. The measurement errors have a normal (Gaussian) distribution. By taking into consideration the assumptions 2, 3 and 4 above, the probability distribution function of ε_i is given by

$$f(\varepsilon_i) = \frac{1}{\sigma\sqrt{2\pi}} \exp\left(\frac{-\varepsilon_i^2}{2\sigma^2}\right)$$

6. The statistical parameters describing ε_i, such as σ, are known.

7. The only variables that contain random errors are the measured temperatures. The measurement times, measurement positions, dimensions of the heated body, and all other quantities appearing in the formulation of the inverse problem are all *accurately* known.

8. There is no prior information regarding the quantities to be estimated, which can be either parameters or functions. If such information exists, it can be utilized to obtain improved estimates.

All of the eight assumptions above rarely apply in actual experiments. For example, if the magnitudes of the measurement errors are quite unequal, the standard deviations σ_i are likely to be different. However, such assumptions are assumed to be valid throughout this book. They permit the verification of the applicability of a method of solution to a specific inverse problem, as well as of the stability of the inverse problem solution with respect to measurement errors, number of sensors, sensor locations, experiment duration, etc, by using simulated measurements in the inverse analysis. Such type of measurements will be described latter in Chapter 2. We have included in NOTE 1 at the end of this chapter a brief review of statistical concepts.

1-5 AN OVERVIEW OF SOLUTION TECHNIQUES FOR INVERSE HEAT TRANSFER PROBLEMS

We present below various techniques used for the solution of IHTP. Such techniques generally require the solution of the associated direct problem. Therefore, it is difficult to present the techniques of solving inverse problems without referring to those associated with the solution of direct problems. Such techniques can be loosely classified under the following groups:

1. Integral equation approach [26-30,32-35].
2. Integral transform techniques [36,37, 39-45,106,110].
3. Series solution approach [46-49].
4. Polynomial approach [50-52].
5. Hyperbolization of the heat conduction equation [53-55].
6. Numerical methods such as finite differences [56-62,68-70,80-85,105,107-109], finite elements [31,63-67,89-91] and boundary elements [92-95,103,112].

7. Space marching techniques together with filtering of the noisy input data, such as in the mollification method [38,69,116].
8. Iterative filtering techniques [88].
9. Steady-state techniques [101-103,112].
10. Beck's sequential function specification method[6,25,30,31,56,59,86,118].
11. Levenberg-Marquardt method for the minimization of the least-squares norm [4,73,75,79,96,104,110,113,122-125].
12. Tikhonov's regularization approach [3,5-9,116-118,126-128].
13. Iterative regularization methods for parameter and function estimations [1,5,10-24,68,72,76-85,89-94,97-100,104-109,113].
14. Genetic algorithms [111].

The *time domain* over which measurements are used in the inverse analysis may be another way to classify the methods of solution [6]. Consider, as an example, the estimation of the boundary heat flux $f(t)$ in the time domain $0<t\le t_f$, as discussed in section 1-1, equations (1.1.2, 1.1.3). Three different possible time domains for the measurements used in the estimation of the heat flux component $f(t_i)$ at time $t_i < t_f$ include:

a. up to time $t_i < t_f$ [6,27].
b. up to the time $t_i < t_f$ plus few time steps [6,25,30,31,56,59,86,118].
c. the whole time domain $0<t\le t_f$ [1,4-6,10-24,68,72-85,88-94,97-100,104-109,113].

Methods based on the time domains (a) and (b) are *sequential* in nature. Methods based on measurements up to time t_i (a) permit the exact matching of estimated and measured temperatures, if a single sensor is used in the analysis [6,27]. Although apparently attractive, they have the disadvantage that the solution algorithms are extremely sensitive to measurement errors. The use of measurements up to time t_i plus few time steps, originally proposed by Beck [6,25,30,31,56,59,86,118], improves the stability of the sequential algorithms. Such an approach is based on the fact that the temperature response is lagged with respect to the excitation, as discussed in section 1-4. We note, however, that sequential methods based on the time domains (a) and (b) generally become unstable as small time steps are used in the analysis [6]. The *whole time domain* approach (c) is very powerful because very small time steps can be taken for the solution. This is quite important in order to estimate, with good resolution, time dependent unknown functions, such as the boundary heat flux of the example. However, methods based on the whole time domain are not as computationally efficient as the sequential ones.

From the foregoing review of the methods, it is apparent that a variety of techniques has been used to solve inverse heat transfer problems. Therefore, it is useful to list some criteria proposed for the evaluation of IHTP solution procedures [5,6,86]:

1. The predicted quantity should be accurate if the measured data are of high accuracy.
2. The method should be stable with respect to measurement errors.
3. The method should have a statistical basis and permit various statistical assumptions for the measurement errors.
4. The method should not require the input data to be *a priori* smoothed.
5. The method should be stable for small time steps or intervals. This permits a better resolution of the time variation of the unknown quantity than is permitted by large time steps.
6. Temperature measurements from one or more sensors should be permitted.
7. The method should not require continuous first derivatives of unknown functions. Furthermore, the method should be able to recover functions containing jump discontinuities.
8. Knowledge of the precise starting time of the application of an unknown surface heat flux or source term should not be required.
9. The method should not be restricted to any fixed number of measurements.
10. The method should be able to treat complex physical situations, including, among others, composite solids, moving boundaries, temperature dependent properties, convective and radiative heat transfer, combined modes of heat transfer, multi-dimensional problems and irregular geometries.
11. The method should be easy for computer programming.
12. The computer cost should be moderate.
13. The user should not have to be highly skilled in mathematics in order to use the method.
14. The method should permit extension to more than one unknown.

Generally, inverse problems are solved by minimizing an *objective function* with some stabilization technique used in the estimation procedure. If all of the eight statistical assumptions stated above in section 1-4 are valid, the *objective function*, S, that provides minimum variance estimates is the *ordinary least squares norm* [4,123] (i.e., the sum of the squared residuals) defined as

$$S=(\mathbf{Y}-\mathbf{T})^T(\mathbf{Y}-\mathbf{T}) \tag{1.5.1}$$

where $\mathbf{Y}$ and $\mathbf{T}$ are the vectors containing the measured and estimated temperatures, respectively, and the superscript T indicates the transpose of the vector. The estimated temperatures are obtained from the solution of the direct problem with estimates for the unknown quantities. We consider the following three particular cases:

a. When the transient readings Y_i taken at times t_i, $i=1,...I$ of a *single sensor* are used in the inverse analysis, the *transpose vector of the residuals*, $(\mathbf{Y}-\mathbf{T})^T$, is given by

$$(\mathbf{Y}-\mathbf{T})^T=(Y_1-T_1\,,\,Y_2-T_2\,,\dots,Y_I-T_I) \tag{1.5.2.a}$$

and the least squares norm, equation (1.5.1), can be written as

$$S = (\mathbf{Y}-\mathbf{T})^T (\mathbf{Y}-\mathbf{T}) = \sum_{i=1}^{I} (Y_i - T_i)^2 \qquad (1.5.2.b)$$

b. When the transient readings of *multiple sensors* are used in the inverse analysis, the *transpose vector of the residuals* is then given by

$$(\mathbf{Y}-\mathbf{T})^T = (\vec{Y}_1 - \vec{T}_1, \vec{Y}_2 - \vec{T}_2, \ldots, \vec{Y}_I - \vec{T}_I) \qquad (1.5.3.a)$$

where, for time t_i, $(\vec{Y}_i - \vec{T}_i)$ is a row vector of length equal to the *number of sensors*, *M*, that is,

$$(\vec{Y}_i - \vec{T}_i) = (Y_{i1} - T_{i1}, Y_{i2} - T_{i2}, \ldots, Y_{iM} - T_{iM}) \qquad (1.5.3.b)$$

In equation (1.5.3.b), the first subscript refers to time t_i and the second subscript refers to the sensor number. Thus, the ordinary least squares norm, equation (1.5.1), can be written as

$$S = (\mathbf{Y}-\mathbf{T})^T (\mathbf{Y}-\mathbf{T}) = \sum_{m=1}^{M} \sum_{i=1}^{I} (Y_{im} - T_{im})^2 \qquad (1.5.3.c)$$

c. If the values of the standard deviations of the measurements are quite different, the ordinary least squares method does not yield minimum variance estimates[4, 123]. In such a case, the *objective function* is given by the *weighted least squares norm*, S_w, defined as

$$S_w = (\mathbf{Y}-\mathbf{T})^T \mathbf{W} (\mathbf{Y}-\mathbf{T}) \qquad (1.5.4)$$

where **W** is a *diagonal weighting matrix*. Such matrix is usually taken as the inverse of the covariance matrix of the measurement errors, in cases where the other statistical hypotheses presented in section 1-4 remain valid [4,123]. By assuming available the measurements of a single sensor, the weighting matrix **W** is then given by:

$$\mathbf{W} = \begin{bmatrix} 1/\sigma_1^2 & & & 0 \\ & 1/\sigma_2^2 & & \\ & & \ddots & \\ 0 & & & 1/\sigma_I^2 \end{bmatrix} \qquad (1.5.5)$$

and S_W given by Equation (1.5.4) can be written in explicit form as:

$$S_W = \sum_{i=1}^{I} \frac{(Y_i - T_i)^2}{\sigma_i^2} \tag{1.5.6.a}$$

where σ_i is the standard deviation of the measurement Y_i at time t_i.

Similarly, for cases involving M sensors equation (1.5.4) can be written as

$$S_W = \sum_{m=1}^{M} \sum_{i=1}^{I} \frac{(Y_{im} - T_{im})^2}{\sigma_{im}^2} \tag{1.5.6.b}$$

where σ_{im} is the standard deviation of the measurement Y_{im} of sensor m at time t_i.

If the inverse heat transfer problem involves the estimation of only few unknown parameters, such as the estimation of a thermal conductivity value from the transient temperature measurements in a solid, the use of the *ordinary least squares norm* given by equations (1.5.2.b) or (1.5.3.c) can be stable. However, if the inverse problem involves the estimation of a large number of parameters, such as the recovery of the unknown transient heat flux components $f(t_i) \equiv f_i$ at times t_i, i=1,...,I, in equations (1.1.2, 1.1.3), excursion and oscillation of the solution may occur. One approach to reduce such instabilities is to use the procedure called *Tikhonov's regularization* [3,5-9,116-118,126-128], which modifies the least squares norm by the addition of a term such as

$$S[f(t)] = \sum_{i=1}^{I} (Y_i - T_i)^2 + \alpha^* \sum_{i=1}^{I} f_i^2 \tag{1.5.7}$$

where α^* (> 0) is the *regularization parameter* and the second summation on the right is the *whole-domain zeroth-order regularization term.* In equation (1.5.7), f_i is the heat flux at time t_i, which is supposed to be constant in the interval $t_i - \Delta t/2 < t < t_i + \Delta t/2$, where Δt is the time interval between two consecutive measurements. The values chosen for the regularization parameter α^* influence the stability of the solution as the minimization of equation (1.5.7) is performed. As $\alpha^* \to 0$ the solution may exhibit oscillatory behavior and become unstable, since the summation of f_i^2 terms may attain very large values and the estimated temperatures tend to match those measured. On the other hand, with large values of α^* the solution is damped and deviates from the exact result.

The *whole-domain first-order regularization* procedure for a single sensor involves the minimization of the following modified least squares norm

$$S[f(t)] = \sum_{i=1}^{I} (Y_i - T_i)^2 + \alpha^* \sum_{i=1}^{I-1} (f_{i+1} - f_i)^2 \tag{1.5.8}$$

For $\alpha^* \rightarrow 0$, exact matching between estimated and measured temperatures is obtained as the minimization of $S[f(t)]$ is performed and the inverse problem solution becomes unstable. For large values of α^*, when the second summation in equation (1.5.8) is dominant, the heat flux components f_i tend to become constant for $i = 1, 2, ..., I$, that is, the first derivative of $f(t)$ tends to zero.

Instabilities on the solution can be alleviated by proper selection of the value of α^*, as discussed in references [3,5-9,116-118,126-128]. Tikhonov [3] suggested that α^* should be selected so that the minimum value of the objective function would be equal to the sum of the squares of the errors expected for the measurements. The cross-validation approach introduced in references [126-128] can also be used to determine the optimum value of α^*. Fortunately, in several cases a relatively wide range of values for α^* can be used. For example, the values of α^* ranged from 10^{-1} to 10^{-4} in reference [126].

The regularization method described above can be related to damped least squares methods [4,6], such as the one due to Levenberg [124] and Marquardt [125]. The so-called *Levenberg-Marquardt Method* is a powerful iterative technique for nonlinear parameter estimation, which has been applied to the solution of various inverse heat transfer problems [4,73,75,79,96,104,110,113,122-125].

An alternative approach for the regularization scheme described above is the use *Alifanov's Iterative Regularization Methods* [1,5,10-24,68,72,76-85,89-94,97-100,104-109,113]. In these methods, the number of iterations plays the role of the regularization parameter α^* and the stopping criterion is so chosen that reasonably stable solutions are obtained. Therefore, there is no need to modify the original objective function, as opposed to Tikhonov's approach. The iterative regularization approach is sufficiently general and can be applied to both parameter and function estimations, as well as to linear and non-linear inverse problems.

In this book we focus our attention on the application of *Levenberg-Marquardt's method of parameter estimation and Alifanov's method of iterative regularization for both parameter and function estimations*. These methods are quite stable, powerful and straightforward and can be applied to the solution of a large variety of inverse heat transfer problems. They meet the majority of criteria enumerated above in this section regarding the evaluation of inverse problems solution procedures. In the following chapters of this book, we shall use these methods based on the whole time domain approach.

Chapter 2 is concerned with the detailed solution of a model inverse heat conduction problem by using the Levenberg-Marquardt Method and the Conjugate Gradient Method. The Conjugate Gradient Method with a suitable stopping criterion belongs to the class of iterative regularization techniques. The subsequent Chapters 3-6 are devoted to the application of these methods for the solution of a wide class of inverse heat transfer problems, involving conduction, convection and radiation.

PROBLEMS

1-1 **Derive the analytical solution of the direct heat conduction problem given by equations (1.1.1).**

1-2 **Use the analytical solution derived above in problem 1-1 to plot the transient temperatures at different locations inside a steel slab [ρ = 7753 kg/m^3, c_p = 0.486 kJ/(kgK) and k = 36 W/(mK)] of thickness L = 5 cm, initially at the uniform temperature of 200 °C. The boundary at x = 0 cm is kept insulated while the boundary at x = 5 cm is maintained at the constant temperature of 20 °C.**

1-3 **Repeat problem 1-2 for a slab made of brick [ρ = 1600 kg/m^3, c_p = 0.84 kJ/(kgK) and k = 0.69 W/(mK)] instead of steel. Compare the temperature variations in the steel and brick slabs at selected positions, say, x = 0, 2 and 4 cm.**

1-4 **Consider a physical problem involving one-dimensional heat conduction in a slab of thickness L, with initial temperature distribution $F(x)$. Assume constant thermophysical properties. A time-dependent heat flux $f(t)$ is supplied at the surface x=0, while the surface at x=L is kept insulated. Energy is generated in the medium at a rate $g(x,t)$ per unit time and per unit volume. What is the mathematical formulation of this heat conduction problem?**

1-5 **Derive the analytical solution of the direct problem associated with the above heat conduction problem 1-4.**

1-6 **Use the solution developed in problem 1-5 to plot the transient temperatures at several locations inside an aluminum slab [ρ = 2707 kg/m^3, c_p = 0.896 kJ/(kgK) and k = 204 W/(mK)] of thickness L = 3 cm, initially at the uniform temperature of 20 °C. No heat is generated inside the medium and a constant heat flux of 8000 W/m^2 is supplied at the surface x = 0 cm.**

1-7 **Consider a physical problem involving one-dimensional heat conduction in a slab of thickness 2L, with initial temperature distribution $F(x)$. Assume constant thermophysical properties. Heat is lost by convection to an ambient at the temperature T_∞ with a heat transfer coefficient h, at the surfaces x=-L and x=L. Energy is generated in the medium at a rate $g(x,t)$ per unit time and per unit volume. What is the mathematical formulation of this heat conduction problem?**

1-8 **Derive the analytical solution of the direct problem associated with the above heat conduction problem 1-7.**

1-9 **Use the solution developed in problem 1-8 to plot the transient temperatures at several locations inside an iron slab [ρ = 7850 kg/m^3, c_p = 0.460 kJ/(kgK) and k = 60 W/(mK)] of thickness 2L = 5 cm, initially at the uniform temperature of 250 °C. No heat is generated inside the slab and the ambient temperature is 25 °C. The heat transfer coefficient at both slab surfaces is 500 W/(m^{2}K).**

1-10 Consider a physical problem involving one-dimensional heat conduction in a solid cylinder of radius b, with initial temperature distribution $F(r)$. Assume constant thermophysical properties. Heat is lost by convection to an ambient at the temperature T_∞ with a heat transfer coefficient h, at the surface $r=b$. Energy is generated in the medium at a rate $g(r,t)$ per unit time and per unit volume. What is the mathematical formulation of this heat conduction problem?

1-11 Derive the analytical solution of the direct problem associated with the above heat conduction problem 1-10.

1-12 Use the solution developed in problem 1-11 to plot the transient temperatures at several locations inside an iron cylinder [$\rho = 7850$ kg/m^3, $c_p = 0.460$ kJ/(kgK) and $k = 60$ W/(mK)] of radius $b = 2.5$ cm, initially at the uniform temperature of 250 °C. No heat is generated inside the cylinder and the ambient temperature is 25 °C. The heat transfer coefficient at the cylinder surface is 500 W/(m^{2}K).

1-13 Repeat problems 1-10, 1-11 and 1-12, for a solid sphere of radius $r = b$, instead of a solid cylinder.

1-14 Compare the transient temperature variations at $x = r = 0$ cm, in problems 1-9, 1-12 and 1-13, for a slab, cylinder and sphere, respectively.

1-15 Consider a physical problem involving two-dimensional heat conduction in a plate of width a and height b, with initial temperature distribution $F(x,y)$. Assume constant thermophysical properties. Heat is lost by convection to an ambient at the temperature T_∞ with a heat transfer coefficient h, at all plate surfaces. Energy is generated in the medium at a rate $g(x,y,t)$ per unit time and per unit volume. What is the mathematical formulation of this heat conduction problem?

1-16 Derive the analytical solution of the direct problem associated with the above heat conduction problem 1-15.

1-17 Use the solution developed in problem 1-16 to plot the transient temperatures for the central point in a square iron plate [$\rho = 7850$ kg/m^3, $c_p = 0.460$ kJ/(kgK) and $k = 60$ W/(mK)] with sides $a = b = 5$ cm, initially at the uniform temperature of 250 °C. No heat is generated inside the plate and the ambient temperature is 25 °C. The heat transfer coefficient at the plate surfaces is 500 W/(m^{2}K).

1-18 Consider a physical problem involving a plate with width a and thickness b, which moves horizontally along the x direction with a constant velocity u. Assume constant physical properties. Also, suppose the plate to be infinitely long in the axial (x) direction. The plate looses heat by convection through its lateral surfaces, at $y = 0$ and $y = a$, to an ambient at temperature T_∞ with a heat transfer coefficient h. The bottom surface at $z = 0$ is supposed insulated, while a transient heat flux with distribution $q(x,y,t)$ is supplied at the top surface $z = b$, in the region $x > 0$. The initial temperature in the medium is $F(x,y,z)$ and the plate enters into the heated zone ($x > 0$) with a uniform temperature T_0 at $x = 0$. Heat is generated in the medium at rate $g(x,y,z,t)$ per unit time and per unit volume. What is the mathematical formulation of this problem?

1-19 Simplify the formulation developed in problem 1-18, for the steady-state heat transfer problem in a plate with negligible lateral heat losses and no heat generation. The heat flux at $z = b$ is a function of x only, say, $q(x)$.

1-20 Derive the analytical solution for the direct problem formulated in problem 1-19.

1-21 By using the analytical solution derived in problem 1-20, find the temperature field in a steel plate [$\rho = 7753$ kg/m^3, $c_p = 0.486$ kJ/(kgK) and $k = 36$ W/(mK)] of thickness $b = 2.5$ cm, moving with a velocity 0.15 m/s, for $T_0 = 20$ °C, $q(x) = 50\text{x}10^4$ W/m^2 in $1 < x < 2$ cm, and $q(x) = 0$ W/m^2 outside this region.

1-22 Review, in basic Heat Transfer books, the physics and formulation of heat transfer by radiation in non-participating and participating media.

1-23 For the heat transfer problems formulated above in problems 1-4, 1-7,1-10, 1-13, 1-15 and 1-18, devise inverse problems of:

(i) Boundary condition;
(ii) Initial condition;
(iii) Energy source-term;
(iv) Thermophysical Properties.

How would you address the solution of such inverse problems? In terms of parameter or of function estimation?

1-24 Plot the temperature variation given by equation (1.4.1.a) in a steel semi-infinite solid [$\rho = 7753$ kg/m^3, $c_p = 0.486$ kJ/(kgK) and $k = 36$ W/(mK)], at different locations below the surface, for a heat flux with amplitude $q_0=10^4$ W/m^2 and frequency: (i) $\omega=1$ rad/s, (ii) $\omega=10$ rad/s and (iii) $\omega=100$ rad/s. What would be the maximum depth that a temperature sensor could be located for the recovery of q_0 in such cases?

1-25 Use equation (1.4.1.c) in order to recover the amplitude of the heat flux, q_0, by using the readings of a sensor located within the maximum depths obtained in problem 1-24. Perturb the maximum amplitude of the temperature variation with different levels of random errors. What are the effects of the sensor location and random errors on the estimated quantity?

1-26 Repeat problems 1-24 and 1-25 for brick [$\rho = 1600$ kg/m^3, $c_p = 0.84$ kJ/(kgK) and $k = 0.69$ W/(mK)] instead of steel. Compare the results of maximum depths and estimated values for q_0 , obtained with these two materials.

1-27 Derive equation (1.5.2.b).

1-28 Derive equation (1.5.3.c).

1-29 Derive equation (1.5.6.a).

1-30 Derive equation (1.5.6.b).

REFERENCES

1. Alifanov, O. M., "Determination of Heat Loads from a Solution of the Nonlinear Inverse Problem", *High Temperature,* **15**(3), 498-504, 1977.
2. Hadamard, J., *Lectures on Cauchy's Problem in Linear Differential Equations*, Yale University Press, New Haven, CT, 1923.
3. Tikhonov, A. N. and Arsenin, V. Y., *Solution of Ill-Posed Problems*, Winston & Sons, Washington, DC, 1977.
4. Beck, J. V. and Arnold, K. J., *Parameter Estimation in Engineering and Science*, Wiley Interscience, New York, 1977.
5. Alifanov, O. M., *Inverse Heat Transfer Problems*, Springer-Verlag, New York, 1994.
6. Beck, J. V., Blackwell, B. and St. Clair, C. R., *Inverse Heat Conduction: Ill-Posed Problems*, Wiley Interscience, New York, 1985.
7. Tikhonov, A. N., "Solution of Incorrectly Formulated Problems and the Regularization Method", *Soviet Math. Dokl.*, **4**(4), 1035-1038, 1963.
8. Tikhonov, A. N., "Regularization of Incorrectly Posed Problems", *Soviet Math. Dokl.*, **4**(6), 1624-1627, 1963.
9. Tikhonov, A. N., "Inverse Problems in Heat Conduction", *J. Eng. Phys.*, **29**(1), 816-820, 1975.
10. Alifanov, O. M., "Solution of an Inverse Problem of Heat-Conduction by Iterative Methods", *J. Eng. Phys.*, **26**(4), 471-476, 1974.
11. Artyukhin, E.A. and Nenarokomov, A.V., "Coefficient Inverse Heat Conduction Problem", *J. Eng. Phys.*, **53**, 1085-1090, 1988.
12. Alifanov, O. M. and Kerov, N. V., "Determination of External Thermal Load Parameters by Solving the Two-Dimensional Inverse Heat-Conduction Problem", *J. Eng. Phys.*, **41**(4), 1049-1053, 1981.
13. Alifanov, O. M. and Klibanov, M. V., "Uniqueness Conditions and Method of Solution of the Coefficient Inverse Problem of Thermal Conductivity", *J. Eng. Phys.*, **48**(6), 730-735, 1985.
14. Alifanov, O. M. and Mikhailov, V. V., "Solution of the Nonlinear Inverse Thermal Conductivity Problem by the Iteration Method", *J. Eng. Phys.*, **35**(6), 1501-1506, 1978.
15. Alifanov, O. M. and Mikhailov, V. V., "Determining Thermal Loads from the Data of Temperature Measurements in a Solid", *High Temperature*, **21**(5), 724-730, 1983.
16. Alifanov, O. M. and Mikhailov, V. V., "Solution of the Overdetermined Inverse Problem of Thermal Conductivity Involving Inaccurate Data", *High Temperature*, **23**(1), 112-117, 1985.
17. Alifanov, O. M. and Rumyantsev, S. V., "One Method of Solving Incorrectly Stated Problems", *J. Eng. Phys.*, **34**(2), 223-226, 1978.
18. Alifanov, O. M. and Rumyantsev, S. V., "On the Stability of Iterative Methods for the Solution of Linear Ill-Posed Problems", *Soviet Math. Dokl.*, **20**(5) 1133-1136, 1979.

19. Alifanov, O. M. and Rumyantsev, S. V., "Regularizing Gradient Algorithms for Inverse Thermal-Conduction Problems", *J. Eng. Phys.*, **39**(2), 858-861, 1980.
20. Alifanov, O. M. and Tryanin, A. P., "Determination of the Coefficient of Internal Heat Exchange and the Effective Thermal Conductivity of a Porous Solid on the Basis of a Nonstationary Experiment", *J. Eng. Phys.*, **48**(3), 356-365, 1985.
21. Alifanov, O. M., Artyukhin, E. and Rumyantsev, A., *Extreme Methods for Solving Ill-Posed Problems with Applications to Inverse Heat Transfer Problems*, Begell House, New York, 1995.
22. Artyukhin, E. A., "Reconstruction of the Thermal Conductivity Coefficient from the Solution of the Nonlinear Inverse Problem", *J. Eng. Phys.*, **41**(4), 1054-1058, 1981.
23. Artyukhin, E. A., "Iterative Algorithms for Estimating Temperature-Dependent Thermophysical Characteristics", *1st International Conference on Inverse Problems in Engineering – Proceedings*, 101-108, Palm Coast, Fl, 1993.
24. Artyukhin, E. A. and Rumyantsev, S. V., "Descent Steps in Gradient Methods of Solution of Inverse Heat Conduction Problems", *J. Eng. Phys.*, **39**, 865-868, 1981.
25. Beck, J. V., "Calculation of Surface Heat Flux from an Internal Temperature History", *ASME Paper 62-HT-46*, 1962.
26. Giedt, W. H., "The Determination of Transient Temperatures and Heat Transfer at a Gas-Metal Interface Applied to a 40-mm Gun Barrel", *Jet Propulsion*, **25**, 158-162, 1955.
27. Stolz, G., "Numerical Solutions to an Inverse Problem of Heat Conduction for Simple Shapes", *ASME J. Heat Transfer*, **82**, 20-26, Feb. 1960.
28. Masket, A. V. and Vastano, A. C., "Interior Value Problems of Mathematics Physics, Part II. Heat Conduction", *American J. of Physics*, **30**, 796-803, 1962.
29. Sabherwal, K. C., "An Inverse Problem of Transient Heat Conduction", *Indian J. of Pure and Applied Physics*, **3**, 397-398, 1965.
30. Beck, J. V., "Surface Heat Flux Determination Using an Integral Method", *Nuclear Engineering and Design*, **7**, 170-178, 1968.
31. Osman, A. M., Dowding, K. J. and Beck, J. V., "Numerical Solution of the General Two-dimensional Inverse Heat Conduction Problem", *ASME J. Heat Transfer*, **119**, 38-44, 1997.
32. Temkin, A. G., "Integral Solutions of Inverse Heat Conduction Problems", *Heat Transfer Sov. Res.*, **10**, 20-32, 1978.
33. Abramovich, B. G. and Trofimov, V. S., "Integral Form of the Inverse Heat Conduction Problem for a Hollow Cylinder", *High Temp. (USSR)*, **17**, 552-554, 1979.
34. Hills, R. G. and Mulholland, G. P., "The Accuracy and Resolving Power of One-Dimensional Transient Inverse Heat Conduction Theory as Applied to Discrete and Inaccurate Measurements", *Int. J. Heat Mass Transfer*, **22**, 1221-1229, 1979.

35. Hills, R. G., Mulholland, G. P., and Matthews, L. K., "The Application of the Backus-Gilbert Method to the Inverse Heat Conduction Problem in Composite Media", *ASME Paper No. 82-HT-26*, 1982.
36. Chen, C. J. and Thomsen, D. M., "On Transient Cylindrical Surface Heat Flux Predicted from Interior Temperature Response", *AIAA Journal*, **13**, 697-699, 1975.
37. Chen, C. J. and Thomsen, D. M., "On Determination of Transient Surface Temperature and Heat Flux by Imbedded Thermocouple in a Hollow Cylinder", *Tech. Rept.*, Rock Island Arsenal, Rock Island, IU., March 1974.
38. Murio, D. A., "The Mollification Method and the Numerical Solution of an Inverse Heat Conduction Problem", *SIAM J. Sci. Stat. Comput.*, **2**, 17-34, 1981.
39. Sparrow, E. M., Haji-Sheikh, A. and Lundgren, T. S., "The Inverse Problem in Transient Heat Conduction", *J. Appl. Mech.*, **86E**, 369-375, 1964.
40. Deverall, L. I. and Channapragda, R. S., "A New Integral Equation for Heat Flux in Inverse Heat Conduction", *J. of Heat Transfer*, **88**, 327-328, 1966.
41. Imber, M. and Khan, J., "Prediction of Transient Temperature Distributions with Embedded Thermocouples", *AIAA Journal*, **10**, 784-789, 1972.
42. Khan, J., *A New Analytical Solution of the Inverse Heat Conduction Problem*, Ph.D. Thesis, Polytechnic Inst. of Brooklyn, New York, 1972.
43. Imber, M., "Temperature Extrapolation Mechanism for Two-Dimensional Heat Flow", *AIAA J.*, **12**, 1089-1093, 1974.
44. Imber, M., "Two-Dimensional Inverse Conduction Problem - Further Observations", *AIAA J.*, **13**, 114-115, 1975.
45. Woo, K. C. and Chow, L. C., "Inverse Heat Conduction by Direct Inverse Laplace Transform", *Numerical Heat Transfer*, **4**, 499-504, 1981.
46. Burggraf, O. R., "An Exact Solution of the Inverse Problem in Heat Conduction Theory and Applications", *J. Heat Transfer*, **86C**, 373-382, 1964.
47. Kover'yano, V. A., "Inverse Problem of Non-Steady Thermal Conductivity", *Teplofizika Vysokikh Temp.*, 5 141-148, 1967.
48. Makhin, J. A. and Shmukin, A. A., "Inverse Problems of Unsteady Heat Conduction", *Heat Transfer - Sov. Res.*, **5**, 160-165, 1973.
49. Langford, D., "New Analytical Solutions of the One-Dimensional Heat Equation for Temperature and Heat Flow Rate Both Prescribed at the Same Fixed Boundary (with Applications to the Phase Change Problem)", *Quarterly of Applied Mathematics*, **24**, 315-322, 1976.
50. Frank, I., "An Application of Least Squares Method to the Solution of the Inverse Problem of Heat Conduction", *J. Heat Transfer*, **85C**, 378-379, 1963.
51. Mulholland, G. P. and San Martin, R. L., "Indirect Thermal Sensing in Composite Media", *Int. J. Heat Mass Transfer*, **16**, 1056-1060, 1973.

52. Mulholland, G. P., Gupta, B. P., and San Martin, R. L., "Inverse Problem of Heat Conduction in Composite Media", *ASME Paper No. 75-WA/HT-83*, 1975.
53. Novikov, N. A., "Hyperbolic Equation of Thermal Conductivity: Solution of the Direct and Inverse Problems for a Semi-Infinite Bar", *J. Eng. Phys.*, **35**, 1253-1257, 1978.
54. Novikov, A., "Solution of the Linear One-Dimensional Inverse Heat-Conduction Problem on the Basis of a Hyperbolic Equation", *J. Eng. Phys.*, **40**(6), 1093-1098, 1981.
55. Weber, C. F., "Analysis and Solution of the Ill-posed Inverse Heat Conduction Problem", *Int. J. Heat Mass Transfer*, **24**, 1783-1792, 1981.
56. Beck, J. V. and Wolf, H., "The Nonlinear Inverse Heat Conduction Problem", *ASME Paper No. 65-HT-40*, 1965.
57. Powell, W. B. and Price, T. W., "A Method for the Determination of Local Heat Flux Transient Temperature Measurements", *ISA Transactions*, **3**(3), 246-254, 1964.
58. Fidelle, T. P. and Zinsmeister, G. E., "A Semi-Discrete Approximate Solution of the Inverse Problem of Transient Heat Conduction", *ASME Paper No. 68-WA/HT-26*, 1968.
59. Beck, J. V., "Nonlinear Estimation Applied to the Nonlinear Inverse Heat Conduction Problem", *Int. J. Heat Mass Transfer*, **13**, 703-716, 1970.
60. D'Souza, N., "Numerical Solution of One-Dimensional Inverse Transient Heat Conduction by Finite Difference Method", *ASME Paper No. 75-WA/HT-81*, 1975.
61. Garifo, L., Schrock, V. E., and Spedicato, E., "On the Solution of the Inverse Heat Conduction Problem by Finite Differences", *Energia Nucleare*, **22**, 452-464, 1975.
62. Randall, J. D., "Finite Difference Solution of the Inverse Heat Conduction Problem and Ablation", *Technical Report*, Johns Hopkins University, Laurel, MD, 1976.
63. Blackwell, B. F., "An Efficient Technique for the Solution of the One-Dimensional Inverse Problem of Heat Conduction", *Technical Report*, Sandia Laboratories, Thermal Test and Analysis Division, Albuquerque, NM, 1980.
64. Muzzy, R. J., Avila, J. H., and Root, R. E., "Determination of Transient Heat Transfer Coefficients and the Resultant Surface Heat Flux from Internal Temperature Measurements", *General Electric, GEAP-20731*, 1975.
65. Hore, P. S., Krutz, G. W., and Schoenhals, R. J., "Application of the Finite Element Method to the Inverse Heat Conduction Problem", *ASME Paper No. 77-WA/TM-4*, 1977.
66. Bass, B. R., "Incor: A Finite Element Program for One-Dimensional Nonlinear Inverse Heat Conduction Analysis", *Technical Report RNC/NUREG/CSD/TM-8*, Oak Ridge National Lab., 1979.

67. Bass, B. R. and Ott, L. J., "A Finite Element Formulation of the Two-Dimensional Nonlinear Inverse Heat Conduction Problem", *Advances in Computer Technology*, **2**, 238-248, 1980.
68. Dantas, L. and Orlande, H. R. B., "A Function Estimation Approach for Determining Temperature-Dependent Thermophysical Properties", *Inverse Problems in Engineering*, **3**, 261-279, 1996.
69. Murio, D. A., "The Mollification Method and the Numerical Solution of the Inverse Heat Conduction Problem by Finite Difference", *Comput. Math. Appl.*, **17**, 1385-1896, 1989.
70. Raynaud, M. and J. Bransier, "A New Finite Difference Method for the Nonlinear Heat Conduction Problem", *Numer. Heat Transfer*, **9**, 27-42, 1986.
71. Jarny, Y., Özisik, M. N. and Bardon, J. P., "A General Optimization Method Using Adjoint Equation for Solving Muldimensional Inverse Heat Conduction", *Int. J. Heat Mass Transfer*, **34**, 2911-2929,1991.
72. Huang, C. H. and Özisik, M. N., "Inverse Problem of Determining Unknown Wall Heat Flux in Laminar Flow Through a Parallel Plate Duct", *Numerical Heat Transfer Part A*, **21**, 55-70, 1992.
73. Silva Neto, A. J. and Özisik, M. N., "An Inverse Analysis of Simultaneously Estimating Phase Function, Albedo and Optical Thickness", *ASME/AIChe Conference*, San Diego, CA, Aug. 9-12, 1992.
74. Li, H. Y. and Özisik, M. N., "Identification of the Temperature Profile in an Absorbing, Emitting, and Isotropically Scattering Medium by Inverse Analysis", *J. Heat Transfer*, **114**, 1060-1063, 1992.
75. Ho, C. H. and Özisik, M. N., "An Inverse Radiation Problem", *Int. J. Heat and Mass Transfer*, **32**, 335-341, 1989.
76. Li, H. Y. and Özisik, M. N., "Inverse Radiation Problem for Simultaneous Estimation of Temperature Profile and Surface Reflectivity", *J. Thermophysics and Heat Transfer*, **7**, 88-93, 1993.
77. Li, H. Y. and Özisik, M. N. "Estimation of the Radiation Source Term with a Conjugate-Gradient Method of Inverse Analysis", *J. Quantitative Spectroscopy and Radiative Transfer*, **48**, 237-244, 1992.
78. Ho, C. H. and Özisik, M. N., "Inverse Radiation Problem in Inhomogeneous Media", *J. Quantitative Spectroscopy and Radiative Transfer*, **40**, 553-560, 1988.
79. Bokar, J. C. and Özisik, M. N., "An Inverse Problem for the Estimation of Radiation Temperature Source Term in a Sphere", *Inverse Problems in Engineering*, **1**, 191-205, 1995.
80. Machado, H. A., Orlande, H. R. B., "Inverse Analysis of Estimating the Timewise and Spacewise Variation of the Wall Heat Flux in a Parallel Plate Channel", *Int. J. Numer. Meth. Heat and Fluid Flow*, **7**, 696-710, 1997.
81. Bokar, J. C. and Özisik, M. N., "An Inverse Analysis for Estimating Time Varying Inlet Temperature in Laminar Flow Inside a Parallel Plate Duct", *Int. J. Heat Mass Transfer*, **38**, 39-45, 1995.

82. Huang, C. H., Özisik, M. N., and Sawaf, B., "Conjugate Gradient Method for Determining Unknown Contact Conductance During Metal Casting", *Int. J. Heat Mass Transfer*, **35**, 1779-1789, 1992.
83. Orlande, H. R. B. and Özisik, M. N., "Inverse Problem of Estimating Interface Conductance Between Periodically Contacting Surfaces", *J. Thermophysics and Heat Transfer*, **7**, 319-325, 1993.
84. Orlande, H. R. B. and Özisik, M. N., "Determination of the Reaction Function in a Reaction-Diffusion Parabolic Problem," *J. Heat Transfer*, **116**, 1041-1044, 1994.
85. Özisik, M. N., Orlande, H. R. B., Hector, L.G. and Anyalebechi, P.N., "Inverse Problem of Estimating Interface Conductance During Solidification via Conjugate Gradient Method", *J. Materials Proc. Manufact. Science*, **1**, 213-225, 1992.
86. Beck, J. V., "Criteria for Comparison of Methods of Solution of the Inverse Heat Conduction Problems", *Nucl. Eng. Design*, **53**, 11-22, 1979.
87. Ruperti Jr., N., Raynaud, M. and Sacadura, J. F., "A Method for the Solution of the Coupled Inverse Heat Conduction-Radiation Problem", *ASME J. Heat Transfer*, **118**, 10-17, 1996.
88. Matsevityi, Y.M. and Multanoviskii, A. V., "Pointwise Identification of Thermophysical Characteristics", *J. Eng. Phys*., **49**(6), 1392-1397, 1986.
89. Zabaras, N. and Ngugen, T. H., "Control of the Freezing Interface Morphology in Solidification Processes in the Presence of Natural Convection", *Int. J. Num. Meth. Eng.* , **38**, 1555-1578, 1995.
90. Zabaras, N. and Yang, G., "A Functional Optimization Formulation and Implementation of an Inverse Natural Convection Problem", *Comput. Methods Appl. Mech. Engrg*., 144, 245-274, 1997.
91. Yang, G. and Zabaras, N., "An Adjoint Method for the Inverse Design of Solidification Processes with Natural Convection", *Int. J. Num Meth. Eng*., **42**, 1121-1144, 1998.
92. Huang, C.H. and Tsai, C.C., "An Inverse Heat Conduction Problem of Estimating Boundary Fluxes in an Irregular Domain with Conjugate Gradient Method", *Heat and Mass Transfer*, **34**, 47-54, 1998.
93. Huang, C.H. and Tsai, C.C., "A Shape Identification Problem in Estimating Time-Dependent Irregular Boundary Configurations", *HTD-Vol.340 ASME National Heat Transfer Conference vol. 2*, 41-48, Dulikravich, G.S. and Woodburry, K. (eds.), 1997.
94. Huang, C.H. and Chiang, C.C., "Shape Identification Problem in Estimating Geometry of Multiple Cavities", *AIAA J. Therm. and Heat Transfer*, **12**, 270-277, 1998.
95. Lesnic, D., Elliot, L. and Ingham D. B., "Application of the Boundary Element Method to Inverse Heat Conduction Problems", *Int. J. Heat Mass Transfer*, **39**, 1503-1517, 1996.
96. Jurkowsky, T., Jarny, Y. and Delaunay, D., "Estimation of Thermal Conductivity of Thermoplastics Under Molding Conditions: an Apparatus and an Inverse Algorithm", *Int. J. Heat Mass Transfer*, **40**, 4169-4181, 1997.

97. Brizaut, J.S., Delaunay, D., Garnier, B. and Jarny, Y., "Implementation of an Inverse Method for Identification of Reticulation Kinetics from Temperature Measurements on a Thick Sample", *Int. J. Heat Mass Transfer*, **36**, 4039-4097, 1993.
98. Bailleul, J. L., Delaunay, D. and Jarny, Y., "Optimal Thermal Processing of Composite Materials: An Inverse Algorithm and Its Experimental Validation", *11th International Heat Transfer Conference*, **5**, 87-92, Kyongju, Korea, 1998.
99. Jarny, Y., Delaunay, D. and Brizaut, J.S., "Inverse Analysis of the Elastomer Cure Control of the Vulcanization Degree", *1st International Conference on Inverse Problems in Engineering – Proceedings*, 291-298, Palm Coast, Fl, 1993.
100. Prud'homme, M. and Nguen, T.H., "On the Iterative Regularization of Inverse Heat Conduction Problems by Conjugate Gradient Method", *Int. Comm. Heat and Mass Transfer*, **25**, 999-1008, 1998.
101. Martin, T.J. and Dulikravich, G.S., "Inverse Determination of Steady Convective Local Heat Transfer Coefficients", *ASME J. Heat Transfer*, **120**, 328-334, 1998.
102. Martin, T.J. and Dulikravich, G.S., "Inverse Determination of Boundary Conditions in Steady Heat Conduction with Heat Generation", *ASME J. Heat Transfer*, **118**, 546-554, 1996.
103. Martin, T.J. and Dulikravich, G.S., "Non-Iterative Inverse Determination of Temperature-Dependent Heat Conductivities", *HTD-Vol.340 ASME National Heat Transfer Conference vol. 2*, 141-150, Dulikravich, G.S. and Woodburry, K. (eds.), 1997.
104. Orlande, H.R.B. and Özisik, M.N., "Inverse Problems Computations in Heat and Mass Transfer", *International Conf. On Comput. Heat and Mass Transfer – Keynote Lecture*, North Cyprus, April, 1999.
105. Alencar Jr., J.P., Orlande, H.R.B. and Özisik, M.N., "A Generalized Coordinates Approach for the Solution of Inverse Heat Conduction Problems", *11th International Heat Transfer Conference*, **7**, 53-58, Kyongju, Korea, 1998.
106. Carvalho, R.N., Orlande, H.R.B. and Özisik, M.N., "Estimation of the Boundary Heat Flux in Grinding via the Conjugate Gradient Method", *Heat Transfer Engr.*, (to appear), 2000.
107. Colaço, M.J. and Orlande, H.R.B., "A Comparison of Different Versions of the Conjugate Gradient Method of Function Estimation", *Num. Heat Transfer – Part A*, (to appear), 1999.
108. Machado, H.A. and Orlande, H.R.B., "Estimation of the Timewise and Spacewise Variation of the Wall Heat Flux to a Non-Newtonian Fluid in a Parallel Plate Channel", *Int. Symposium on Transient Convective Heat Transfer – Proccedings*, J. Padet (ed.), 587-596, Turkey, 1996.
109. Machado, H.A. and Orlande, H.R.B., "Inverse Problem of Estimating the Heat Flux to a Non-Newtonian Fluid in a Parallel Plate Channel", *RBCM – J. Braz. Soc. Mech. Sciences*, **20**, 51-61, 1998.

110. Mejias, M. M., Orlande, H. R. B. and Özisik, M. N., "Design of Optimum Experiments for the Estimation of Thermal Conductivity Components of Orthotropic Solids", *Hyb. Meth Engr.*, **1**, 37-53, 1999.
111. Goldberg, D. E., *Genetic Algorithms in Search, Optimization and Machine Learning*, Addison Wesley, Reading, MA, 1989.
112. Dulikravich, G. S. and Martin, T. J., "Inverse Shape and Boundary Condition Problems and Optimization in Heat Conduction", *Chapter 10 in Advances in Numerical Heat Transfer*, **1**, 381-426, Minkowycz, W. J. and Sparrow, E. M. (eds.), Taylor and Francis, 1996.
113. Özisik, M. N., *Heat Conduction*, Chapter 14, Wiley, 1994.
114. Sabatier, P. C., (ed.), *Applied Inverse Problems*, Springer Verlag, Hamburg, 1978.
115. Morozov, V. A., *Methods for Solving Incorrectly Posed Problems*, Springer Verlag, New York, 1984.
116. Murio, D. A., *The Mollification Method and the Numerical Solution of Ill-Posed Problems*, Wiley Interscience, New York, 1993.
117. Trujillo, D. M. and Busby, H. R., *Practical Inverse Analysis in Engineering*, CRC Press, Boca Raton, 1997.
118. Beck, J. V. and Blackwell, B., in *Handbook of Numerical Heat Transfer*, Minkowycz, W. J., Sparrow, E. M., Schneider, G. E., and Pletcher, R. H., (eds.), Wiley Intersc., New York, 1988.
119. Hensel, E., *Inverse Theory and Applications for Engineers*, Prentice Hall, New Jersey, 1991.
120. Kurpisz, K. and Nowak, A. J., *Inverse Thermal Problems*, WIT Press, Southampton, UK, 1995.
121. Özisik, M. N., *Boundary Value Problems of Heat Conduction*, Dover, New York, 1989.
122. Farebrother, R. W., *Linear Least Squares Computations*, Marcel Dekker, New York, 1988.
123. Bard, Y. B., *Nonlinear Parameter Estimation*, Academic Press, New York, 1974.
124. Levenberg, K., "A Method for the Solution of Certain Non-linear Problems in Least-Squares", *Quart. Appl. Math*, **2**, 164-168, 1944.
125. Marquardt, D. W., "An Algorithm for Least Squares Estimation of Nonlinear Parameters", *J. Soc. Ind. Appl. Math*, **11**, 431-441, 1963.
126. Huang, C. H. and Özisik, M. N., "Optimal Regularization Method to Determine the Strength of a Plane Surface Heat Source", *Int. J. Heat and Fluid Flow*, **12**, 173-178, 1991.
127. Craven, P. and Wahba, G., "Smoothing Noisy Data with Spline Functions", *Numerische Mathematics*, **31**, 377-403, 1979.
128. Busby, H. R. and Trujillo, D. M., "Numerical Solution of a Two Dimensional Inverse Heat Conduction Problems", *Int. J. Num. Meth. Eng.*, **21**, 349-359, 1985.
129. Box, G. E., Hunter, W. and Hunter, J., *Statistics for Experimenters*, John Wiley, 1978.

130. Moore, D. S. and McCabe, G. P., *Introduction to the Practice of Statistics*, 2nd ed., W.H. Freeman and Co., New York, 1993.

NOTE 1: STATISTICAL CONCEPTS

The purpose of this note is to present some basic statistical material, needed in the analysis and solution of IHCP, that is generally not covered in regular courses in engineering. Readers should consult references [4,122,123,129,130] for a more in depth discussion of such matters.

Random Variable

A *random variable* is a variable whose value is a numerical outcome of a random phenomenon. A *phenomenon is denoted random* if its individual outcomes are unpredictable, although a regular pattern of outcomes emerges in many repetitions.

Let the capital letter X denote a random variable. It is called a *discrete random variable* if it can only assume a set of discrete numbers x_n, $n = 1,2,...,N$. On the other hand, X is called a *continuous random variable* if it can assume all values in an interval of real numbers.

Probability Distribution

The assignment of probabilities to the values of a random variable X gives the *probability distribution* of X. Depending on whether the random variable X is discrete or continuous, the probability distribution $f(x)$ is a non-negative number or function, respectively, satisfying

$$\sum_{n=1}^{N} f(x_n) = 1 \qquad \text{when } X \text{ is discrete} \qquad \text{(N1.1.1.a)}$$

$$\int_{-\infty}^{+\infty} f(x)dx = 1 \qquad \text{when } X \text{ is continuous} \qquad \text{(N1.1.1.b)}$$

Expected Value of X

Let X be a random variable, discrete or continuous, with the corresponding probability distributions $f(x_n)$ or $f(x)$, respectively. The *expected value* of X, denoted by $E(X)$, is defined as

$$E(X) = \begin{cases} \sum_{n=1}^{N} x_n f(x_n) & \text{when } X \text{ is discrete} \quad \text{(N1.1.2.a)} \\ \int_{-\infty}^{\infty} x f(x)\, dx & \text{when } X \text{ is continuous} \quad \text{(N1.1.2.b)} \end{cases}$$

The expected value of any random variable X is obtained by multiplying its value by the corresponding probability distribution and then summing up the results if X is discrete, or integrating the results if X is continuous. Clearly, the expected value of X is a *weighted mean* of all possible values with the weight factor $f(x)$. If the weights are equal, that is, $f(x) = 1$, then the expected value becomes the *arithmetic mean* of X. Usually, the expected value is simply referred to as the *mean* of the random variable X.

Expected Value of a Function $g(X)$

Consider a random variable X and the probability distribution $f(x)$ associated with it. The expected value of the function $g(X)$, denoted by $E[g(X)]$, is given by

$$E[g(X)] = \begin{cases} \sum_{n=1}^{N} g(x_n) f(x_n) & \text{when } X \text{ is discrete} \quad \text{(N1.1.3.a)} \\ \int_{-\infty}^{\infty} g(x) f(x)\, dx & \text{when } X \text{ is continuous} \quad \text{(N1.1.3.b)} \end{cases}$$

Variance of a Random Variable X

The *variance* of a random variable X, denoted by σ^2, is a measure of the spread of X around its mean μ. It is defined by

$$\sigma^2 \equiv E[(x-\mu)^2] \qquad \text{where } \mu = E(x) \qquad \text{(N1.1.4.a)}$$

or an alternative form is obtained by expanding this expression, that is,

$$\sigma^2 = E(x^2) - \mu^2 \qquad \text{(N1.1.4.b)}$$

since $E(\mu^2) = \mu^2$.

The positive square root σ of the variance is called the *standard deviation*.

Covariance of Two Random Variables X and Y

The *covariance* of two random variables X and Y is a measure of the linear dependence between them. It is defined as:

$$\text{cov}(X,Y) \equiv E\left[(x-\mu_x)(y-\mu_y)\right] \tag{N1.1.5}$$

where $\mu_x = E(x)$ and $\mu_y = E(y)$.

The covariance cov(X, Y) is zero if X and Y are independent.

Normal Distribution

The most frequently used continuous probability distribution function is the normal (Gaussian) distribution, which has a bell-shaped curve about its mean value. The normal probability distribution function with a mean μ and variance σ^2 is given by

$$f(x) = \frac{1}{\sigma\sqrt{2\pi}}\exp\left[-\frac{1}{2}\left(\frac{x-\mu}{\sigma}\right)^2\right] \tag{N1.1.6}$$

The area below this function from $-\infty$ to x represents the probability $P(-\infty < X \le x)$ that a random variable X with mean μ and variance σ^2 assumes a value between $-\infty$ and x. Therefore, $P(-\infty < X \le x)$ is defined by

$$P(-\infty < X \le x) = \frac{1}{\sigma\sqrt{2\pi}}\int_{-\infty}^{x}\exp\left[-\frac{1}{2}\left(\frac{X-\mu}{\sigma}\right)^2\right]dX \tag{N1.1.7}$$

To alleviate the difficulty in the calculation of this integral for each given set of values of σ, μ and x, a new independent variable Z was defined as

$$Z = \frac{X-\mu}{\sigma} \quad \text{or} \quad z = \frac{x-\mu}{\sigma} \tag{N1.1.8}$$

Then, the integral in equation (N1.1.7) becomes

$$P(-\infty < Z \le z) = \frac{1}{\sqrt{2\pi}}\int_{-\infty}^{z} e^{-Z^2/2}dZ \tag{N1.1.9}$$

The results of this integration were tabulated, as given in Table N1.1.1. Readers should consult references [129-130] for more comprehensive tabulation of the probability function $P(-\infty < Z \le z)$.

TABLE N1.1.1 - Probability $P(-\infty < Z \leq z)$ given by equation (N1.1.9) for a normal distribution function.

z	$P(-\infty < Z \leq z)$	z	$P(-\infty < Z \leq z)$
-2.9	0.0019	0.0	0.5000
-2.8	0.0026	0.1	0.5398
-2.7	0.0035	0.2	0.5793
-2.6	0.0047	0.3	0.6179
-2.5	0.0062	0.4	0.6554
-2.4	0.0082	0.5	0.6915
-2.3	0.0107	0.6	0.7257
-2.2	0.0139	0.7	0.7580
-2.1	0.0179	0.8	0.7881
-2.0	0.0228	0.9	0.8159
-1.9	0.0287	1.0	0.8413
-1.8	0.0359	1.1	0.8643
-1.7	0.0446	1.2	0.8849
-1.6	0.0548	1.3	0.9032
-1.5	0.0668	1.4	0.9192
-1.4	0.0808	1.5	0.9332
-1.3	0.0968	1.6	0.9452
-1.2	0.1151	1.7	0.9554
-1.1	0.1357	1.8	0.9641
-1.0	0.1587	1.9	0.9713
-0.9	0.1841	2.0	0.9772
-0.8	0.2119	2.1	0.9821
-0.7	0.2420	2.2	0.9861
-0.6	0.2743	2.3	0.9893
-0.5	0.3085	2.4	0.9918
-0.4	0.3446	2.5	0.9938
-0.3	0.3821	2.6	0.9953
-0.2	0.4207	2.7	0.9965
-0.1	0.4602	2.8	0.9974
		2.9	0.9981

Table N1.1.1 can be used as follows to determine the normal probability $P(x_1 \leq X \leq x_2)$, of a random variable X having a mean μ and variance σ^2, to assume a value between x_1 and x_2:

$$P(x_1 \leq X \leq x_2) = P\left(\frac{x_1 - \mu}{\sigma} \leq Z \leq \frac{x_2 - \mu}{\sigma}\right) =$$

$$= P(z_1 \leq Z \leq z_2) =$$

$$= P(-\infty < Z \leq z_2) - P(-\infty < Z \leq z_1) \qquad \text{(N1.1.10)}$$

where $P(-\infty < Z \le z_2)$ and $P(-\infty < Z \le z_1)$ are determined from Table N1.1.1.

The normal distribution function is useful in obtaining confidence intervals for estimated parameters, as will be discussed later in Chapter 2.

Chi-Square Distribution

Let $Z_1, Z_2, \ldots, Z_N$ be independent random variables normally distributed, with mean zero and unitary standard deviation. In this case, the summation

$$\chi_N^2 = \sum_{n=1}^{N} Z_n^2 \tag{N1.1.11}$$

has a *chi-square probability distribution with N degrees of freedom*, given by

$$f(\chi_N^2) = \frac{(\chi_N^2)^{\frac{1}{2}(N-2)}}{2^{N/2}\Gamma\left(\frac{N}{2}\right)} e^{-\chi_N^2/2} \qquad \text{for } 0 < \chi_N^2 < \infty \tag{N1.1.12}$$

where $\Gamma(\cdot)$ is the gamma function defined as

$$\Gamma(n) = \int_{x=0}^{\infty} e^{-x} x^{n-1} dx \quad , \quad n > 0 \tag{N1.1.13.a}$$

and, for n integer, we have

$$\Gamma(n+1) = n! \tag{N1.1.13.b}$$

The mean and the variance of the chi-square distribution are N and $2N$, respectively. Such distribution is skewed to the right, but it tends to the normal distribution as $N \to \infty$. This behavior is shown in figure N1.1.1.

The probability of having a value x smaller than χ_N^2 is obtained by integrating equation (N1.1.12) from zero to χ_N^2, that is,

$$F(\chi_N^2) \equiv P(0 \le x \le \chi_N^2) = \int_0^{\chi_N^2} \frac{(x)^{\frac{1}{2}(N-2)}}{2^{N/2}\Gamma(N/2)} e^{-x/2} dx \tag{N1.1.14}$$

Table N1.1.2 shows the values of χ^2_N for various probabilities $F(\chi^2_N)$, as a function of the number of degrees of freedom N. The values of χ^2_N shown in table N1.1.2 are useful in obtaining confidence regions for estimated parameters. A discussion on confidence regions and other quantities of importance to assess the accuracy of the estimated parameters will be presented in Chapter 2.

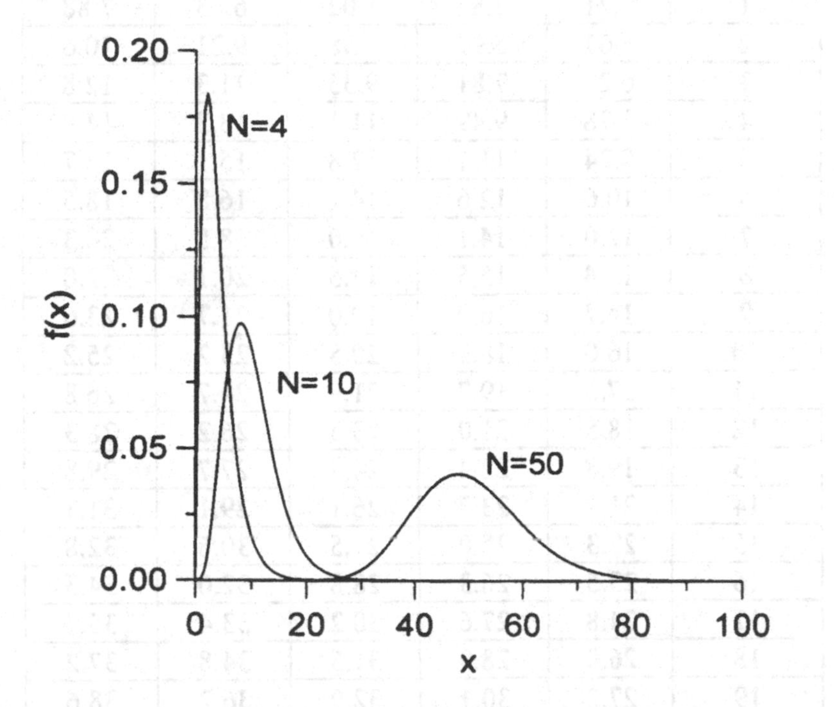

Figure N1.1.1 - Chi-Square Probability Distribution Function given by equation (N1.1.12).

TABLE N1.1.2 - Values of χ_N^2 for various degrees of freedom N and probabilities $F(\chi_N^2) \equiv P(0 \le x \le \chi_N^2)$.

$F(\chi_N^2)$ → / N ↓	0.900	0.950	0.975	0.990	0.995
1	2.71	3.84	5.02	6.63	7.88
2	4.61	5.99	7.38	9.21	10.6
3	6.25	7.81	9.35	11.3	12.8
4	7.78	9.49	11.1	13.3	14.9
5	9.24	11.1	12.8	15.1	16.7
6	10.6	12.6	14.4	16.8	18.5
7	12.0	14.1	16.0	18.5	20.3
8	13.4	15.5	17.5	20.1	22.0
9	14.7	16.9	19.0	21.7	23.6
10	16.0	18.3	20.5	23.2	25.2
11	17.3	19.7	21.9	24.7	26.8
12	18.5	21.0	23.3	26.2	28.3
13	19.8	22.4	24.7	27.7	29.8
14	21.1	23.7	26.1	29.1	31.3
15	22.3	25.0	27.5	30.6	32.8
16	23.5	26.3	28.8	32.0	34.3
17	24.8	27.6	30.2	33.4	35.7
18	26.0	28.9	31.5	34.8	37.2
19	27.2	30.1	32.9	36.2	38.6
20	28.4	31.4	34.2	37.6	40.0

Chapter 2

TECHNIQUES FOR SOLVING INVERSE HEAT TRANSFER PROBLEMS

In the previous chapter we discussed general principles related to the formulation and solution of inverse heat transfer problems. The main objective of this chapter is to provide the necessary mathematical background needed in the use of some powerful techniques for solving inverse heat transfer problems. The following four techniques are considered:

Technique I: Levenberg-Marquardt Method for Parameter Estimation
Technique II: Conjugate Gradient Method for Parameter Estimation
Technique III: Conjugate Gradient Method with Adjoint Problem for Parameter Estimation
Technique IV: Conjugate Gradient Method with Adjoint Problem for Function Estimation

Although other techniques are available, the above four are chosen for use in this book because they are sufficiently general, versatile, straightforward and powerful to overcome the difficulties associated with the solution of inverse heat transfer problems.

Technique I is an *iterative method* for solving nonlinear least squares problems of parameter estimation. The technique was first derived by Levenberg [1] in 1944, by modifying the ordinary least squares norm. Later, in 1963, Marquardt [2] derived basically the same technique by using a different approach. Marquardt's intention was to obtain a method that would tend to the Gauss method in the neighborhood of the minimum of the ordinary least squares norm, and would tend to the steepest descent method in the neighborhood of the initial guess used for the iterative procedure [2-4]. The so called *Levenberg-Marquardt Method* [1-9] has been applied to the solution of a variety of inverse problems involving the estimation of unknown parameters.

The solution of inverse parameter estimation problems by Technique I requires the computation of the *Sensitivity Matrix*, **J**, the elements of which are the *Sensitivity Coefficients*, J_{ij}, defined as

$$J_{ij} = \frac{\partial T_i}{\partial P_j}$$

where $i = 1, 2, \ldots, I$
$j = 1, 2, \ldots, N$
I = number of measurements
N = number of unknown parameters
T_i is the ith estimated temperature
P_j is the jth unknown parameter

Technique I is quite efficient for solving linear and nonlinear parameter estimation problems. However, difficulties may arise in nonlinear estimation problems involving a large number of unknown parameters, because of the time spent in the computation of the sensitivity matrix.

Technique II utilizes the *Conjugate Gradient Method of Minimization* to solve parameter estimation problems. As with Technique I, it requires the computation of the sensitivity matrix, which is a time-consuming process when the number of parameters to be estimated becomes large, specially in nonlinear problems.

Techniques III and IV utilize the *Conjugate Gradient Method of Minimization with Adjoint Problem* [9-21]. **Technique III** is specially suitable for problems involving the estimation of the coefficients of trial functions used to approximate an unknown function. The use of the adjoint problem in Technique III results in an expression for the gradient direction involving a Lagrange Multiplier, thus alleviating the need for the computation of the sensitivity matrix. **Technique IV** is a *function estimation* approach and is useful when no *a priori* information is available on the functional form of the unknown quantity. Techniques II, III and IV, together with appropriate stopping criteria for their iterative procedures, belong to the class of *iterative regularization techniques*.

In this chapter we describe the basic steps and present the solution algorithms for each of these four techniques, by using a whole time domain approach and by assuming that the eight assumptions described previously in Section 1-4 remain valid. We also present the solution of a test-problem by using Techniques I through IV to illustrate their applications. The sensitivity coefficients play an important role in the application of Techniques I, II and III to parameter estimation problems. Hence, we also discuss in this chapter the physical and mathematical significance of the sensitivity coefficients and describe three different methods for their computation.

2-1 TECHNIQUE I: THE LEVENBERG-MARQUARDT METHOD FOR PARAMETER ESTIMATION

The Levenberg-Marquardt method, originally devised for application to nonlinear parameter estimation problems, has also been successfully applied to the solution of linear problems that are too ill-conditioned to permit the application of linear algorithms.

The solution of inverse heat transfer problems with the Levenberg-Marquardt method can be suitably arranged in the following basic steps:

- The Direct Problem
- The Inverse Problem
- The Iterative Procedure
- The Stopping Criteria
- The Computational Algorithm

We present below the details of each of these steps as applied to the solution of an inverse heat conduction test-problem, involving the following physical situation:

> Consider the linear transient heat conduction in a plate of unitary dimensionless thickness. The plate is initially at zero temperature and both boundaries at x=0 and x=1 are kept insulated. For times $t > 0$, a plane heat source of strength $g_p(t)$ per unit area, placed in the mid-plane x=0.5, releases its energy as depicted in figure 2.1.1.

The mathematical formulation of this heat conduction problem is given in dimensionless form by:

$$\frac{\partial^2 T(x,t)}{\partial x^2} + g_p(t)\delta(x-0.5) = \frac{\partial T(x,t)}{\partial t} \quad \text{in } 0 < x < 1, \text{ for } t > 0 \tag{2.1.1.a}$$

$$\frac{\partial T(0,t)}{\partial x} = 0 \quad \text{at } x = 0, \text{ for } t > 0 \tag{2.1.1.b}$$

$$\frac{\partial T(1,t)}{\partial x} = 0 \quad \text{at } x = 1, \text{ for } t > 0 \tag{2.1.1.c}$$

$$T(x,0) = 0 \quad \text{for } t = 0, \text{ in } 0 < x < 1 \tag{2.1.1.d}$$

where $\delta(\cdot)$ is the Dirac delta function.

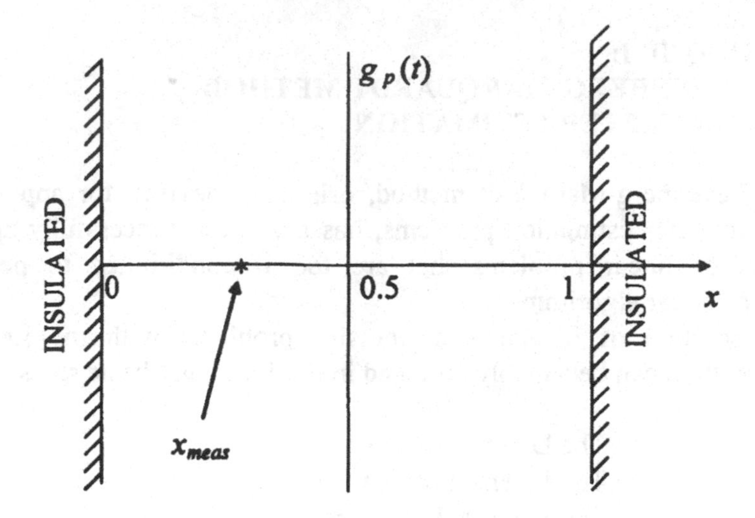

Figure 2.1.1. Geometry and coordinates for a plane heat source $g_p(t)$.

The Direct Problem

In the **Direct Problem** associated with the physical problem described above, the time-varying strength $g_p(t)$ of the plane heat source is known. The objective of the direct problem is then to determine the transient temperature field $T(x,t)$ in the plate.

The Inverse Problem

For the **Inverse Problem** considered of interest here, the time-varying strength $g_p(t)$ of the plane heat source is regarded as unknown. The additional information obtained from transient temperature measurements taken at a location $x=x_{meas}$, at times t_i, $i = 1, 2, \ldots, I$, is then used for the estimation of $g_p(t)$.

For the solution of the present inverse problem, we consider the unknown energy generation function $g_p(t)$ to be parameterized in the following general linear form:

$$g_p(t) = \sum_{j=1}^{N} P_j C_j(t) \tag{2.1.2}$$

Here, P_j are unknown parameters and $C_j(t)$ are known trial functions (e.g., polynomials, B-Splines, etc). In addition, the total number of parameters, N, is specified.

The problem given by equations (2.1.1) with $g_p(t)$ unknown, but parameterized as given by equation (2.1.2), is an inverse heat conduction problem in which the coefficients P_j are to be estimated. The solution of this inverse heat conduction problem for the estimation of the N unknown parameters P_j,

$j = 1, ..., N$, is based on the minimization of the *ordinary least squares norm* given by (see equation 1.5.2.b):

$$S(\mathbf{P}) = \sum_{i=1}^{I} [Y_i - T_i(\mathbf{P})]^2 \tag{2.1.3.a}$$

where
S = sum of squares error or objective function
$\mathbf{P}^T \equiv [P_1, P_2, ..., P_N]$ = vector of unknown parameters
$T_i(\mathbf{P}) \equiv T(\mathbf{P}, t_i)$ = estimated temperature at time t_i
$Y_i \equiv Y(t_i)$ = measured temperature at time t_i
N = total number of unknown parameters
I = total number of measurements, where $I \geq N$.

The estimated temperatures $T_i(\mathbf{P})$ are obtained from the solution of the direct problem at the measurement location, x_{meas}, by using the current estimate for the unknown parameters P_j, $j = 1, ..., N$.

Equation (2.1.3.a) can be written in matrix form as (see equation 1.5.1)

$$S(\mathbf{P}) = [\mathbf{Y} - \mathbf{T}(\mathbf{P})]^T [\mathbf{Y} - \mathbf{T}(\mathbf{P})] \tag{2.1.3.b}$$

where the superscript T denotes the transpose, and $[\mathbf{Y}-\mathbf{T}(\mathbf{P})]^T$ is defined as

$$[\mathbf{Y} - \mathbf{T}(\mathbf{P})]^T \equiv [Y_1 - T_1, Y_2 - T_2, ..., Y_I - T_I] \tag{2.1.4}$$

The Iterative Procedure for Technique I

To minimize the least squares norm given by equations (2.1.3), we need to equate to zero the derivatives of $S(\mathbf{P})$ with respect to each of the unknown parameters $[P_1, P_2, ..., P_N]$, that is,

$$\frac{\partial S(\mathbf{P})}{\partial P_1} = \frac{\partial S(\mathbf{P})}{\partial P_2} = \cdots = \frac{\partial S(\mathbf{P})}{\partial P_N} = 0 \tag{2.1.5.a}$$

Such necessary condition for the minimization of $S(\mathbf{P})$ can be represented in matrix notation by equating the *gradient of* $S(\mathbf{P})$ with respect to the vector of parameters $\mathbf{P}$ to zero, that is,

$$\nabla S(\mathbf{P}) = 2\left[-\frac{\partial \mathbf{T}^T(\mathbf{P})}{\partial \mathbf{P}}\right][\mathbf{Y} - \mathbf{T}(\mathbf{P})] = 0 \tag{2.1.5.b}$$

where

$$\frac{\partial \mathbf{T}^T(\mathbf{P})}{\partial \mathbf{P}} = \begin{bmatrix} \frac{\partial}{\partial P_1} \\ \frac{\partial}{\partial P_2} \\ \vdots \\ \frac{\partial}{\partial P_N} \end{bmatrix} [T_1 T_2 \cdots T_I] \tag{2.1.6}$$

The ***Sensitivity*** ***or Jacobian matrix***, **J(P)**, is defined as the transpose of equation (2.1.6), that is,

$$\mathbf{J}(\mathbf{P}) = \left[\frac{\partial \mathbf{T}^T(\mathbf{P})}{\partial \mathbf{P}} \right]^T \tag{2.1.7.a}$$

In explicit form, the sensitivity matrix is written as

$$\mathbf{J}(\mathbf{P}) = \left[\frac{\partial \mathbf{T}^T(\mathbf{P})}{\partial \mathbf{P}} \right]^T = \begin{bmatrix} \frac{\partial T_1}{\partial P_1} & \frac{\partial T_1}{\partial P_2} & \frac{\partial T_1}{\partial P_3} & \cdots & \frac{\partial T_1}{\partial P_N} \\ \frac{\partial T_2}{\partial P_1} & \frac{\partial T_2}{\partial P_2} & \frac{\partial T_2}{\partial P_3} & \cdots & \frac{\partial T_2}{\partial P_N} \\ \vdots & \vdots & \vdots & & \vdots \\ \frac{\partial T_I}{\partial P_1} & \frac{\partial T_I}{\partial P_2} & \frac{\partial T_I}{\partial P_3} & \cdots & \frac{\partial T_I}{\partial P_N} \end{bmatrix} \tag{2.1.7.b}$$

where N = total number of unknown parameters
I = total number of measurements

The elements of the sensitivity matrix are called the ***Sensitivity Coefficients***. The sensitivity coefficient J_{ij} is thus defined as the first derivative of the estimated temperature at time t_i with respect to the unknown parameter P_j, that is,

$$J_{ij} = \frac{\partial T_i}{\partial P_j} \tag{2.1.7.c}$$

By using the definition of the sensitivity matrix given by equation (2.1.7.a), equation (2.1.5.b) becomes

$$-2\mathbf{J}^T(\mathbf{P})[\mathbf{Y} - \mathbf{T}(\mathbf{P})] = 0 \tag{2.1.8}$$

For *linear inverse problems*, the sensitivity matrix is not a function of the unknown parameters. In such a case, equation (2.1.8) can be solved in explicit form for the vector of unknown parameters **P** as [4]:

$$\mathbf{P} = (\mathbf{J}^T\mathbf{J})^{-1}\mathbf{J}^T\mathbf{Y} \tag{2.1.9}$$

In the case of a *nonlinear inverse problem*, the sensitivity matrix has some functional dependence on the vector of unknown parameters **P**. The solution of equation (2.1.8) for nonlinear estimation problems then requires an iterative procedure, which is obtained by linearizing the vector of estimated temperatures, **T(P)**, with a Taylor series expansion around the current solution $\mathbf{P}^k$ at iteration k. Such a linearization is given by

$$\mathbf{T}(\mathbf{P}) = \mathbf{T}(\mathbf{P}^k) + \mathbf{J}^k(\mathbf{P} - \mathbf{P}^k) \tag{2.1.10}$$

where $\mathbf{T}(\mathbf{P}^k)$ and $\mathbf{J}^k$ are the estimated temperatures and the sensitivity matrix evaluated at iteration k, respectively. Equation (2.1.10) is substituted into equation (2.1.8) and the resulting expression is rearranged to yield the following iterative procedure to obtain the vector of unknown parameters **P**[4]:

$$\mathbf{P}^{k+1} = \mathbf{P}^k + [(\mathbf{J}^k)^T\mathbf{J}^k]^{-1}(\mathbf{J}^k)^T[\mathbf{Y} - \mathbf{T}(\mathbf{P}^k)] \tag{2.1.11}$$

The iterative procedure given by equation (2.1.11) is called the Gauss method. Such method is actually an approximation for the Newton (or Newton-Raphson) method [3].

We note that equation (2.1.9), as well as the implementation of the iterative procedure given by equation (2.1.11), require the matrix $\mathbf{J}^T\mathbf{J}$ to be non-singular, or

$$\left|\mathbf{J}^T\mathbf{J}\right| \neq 0 \tag{2.1.12}$$

where $|\,.\,|$ is the determinant.

Equation (2.1.12) gives the so called *Identifiability Condition,* that is, if the determinant of $\mathbf{J}^T\mathbf{J}$ is zero, or even very small, the parameters P_j, for $j = 1, \ldots, N$, cannot be determined by using the iterative procedure of equation (2.1.11).

Problems satisfying $\left|\mathbf{J}^T\mathbf{J}\right| \approx 0$ are denoted *ill-conditioned*. Inverse heat transfer problems are generally very ill-conditioned, especially near the initial guess used for the unknown parameters, creating difficulties in the application of equations (2.1.9) or (2.1.11). The *Levenberg-Marquardt Method* [1-9] alleviates such difficulties by utilizing an iterative procedure in the form:

$$\mathbf{P}^{k+1} = \mathbf{P}^k + [(\mathbf{J}^k)^T\mathbf{J}^k + \mu^k\mathbf{\Omega}^k]^{-1}(\mathbf{J}^k)^T[\mathbf{Y} - \mathbf{T}(\mathbf{P}^k)] \tag{2.1.13}$$

where μ^k is a positive scalar named *damping parameter*, and
Ω^k is a *diagonal matrix*.

The purpose of the matrix term $\mu^k \Omega^k$, included in equation (2.1.13), is to damp oscillations and instabilities due to the ill-conditioned character of the problem, by making its components large as compared to those of $\mathbf{J}^T\mathbf{J}$ if necessary. The damping parameter is made large in the beginning of the iterations, since the problem is generally ill-conditioned in the region around the initial guess used for the iterative procedure, which can be quite far from the exact parameters. With such an approach, the matrix $\mathbf{J}^T\mathbf{J}$ is not required to be non-singular in the beginning of iterations and the Levenberg-Marquardt Method tends to the *Steepest Descent Method*, that is, a very small step is taken in the negative gradient direction. The parameter μ^k is then gradually reduced as the iteration procedure advances to the solution of the parameter estimation problem, and then the Levenberg-Marquardt Method tends to the *Gauss Method* given by equation (2.1.11) [4].

The Stopping Criteria for Technique I

The following criteria were suggested by Dennis and Schnabel [7] to stop the iterative procedure of the Levenberg-Marquardt Method given by equation (2.1.13):

$$\text{(i)} \qquad S(\mathbf{P}^{k+1}) < \varepsilon_1 \tag{2.1.14.a}$$

$$\text{(ii)} \qquad \left\| (\mathbf{J}^k)^T \left[\mathbf{Y} - \mathbf{T}(\mathbf{P}^k) \right] \right\| < \varepsilon_2 \tag{2.1.14.b}$$

$$\text{(iii)} \qquad \left\| \mathbf{P}^{k+1} - \mathbf{P}^k \right\| < \varepsilon_3 \tag{2.1.14.c}$$

where ε_1, ε_2 and ε_3 are user prescribed tolerances and $\| \cdot \|$ is the vector Euclidean norm, i.e., $\|\mathbf{x}\| = (\mathbf{x}^T\mathbf{x})^{1/2}$, where the superscript T denotes transpose.

The criterion given by equation (2.1.14.a) tests if the least squares norm is sufficiently small, which is expected to be in the neighborhood of the solution for the problem. Similarly, equation (2.1.14.b) checks if the norm of the gradient of $S(\mathbf{P})$ is sufficiently small, since it is expected to vanish at the point where $S(\mathbf{P})$ is minimum. Although such a condition of vanishing gradient is also valid for maximum and saddle points of $S(\mathbf{P})$, the Levenberg-Marquardt method is very unlike to converge to such points. The last criterion given by equation (2.1.14.c) results from the fact that changes in the vector of parameters are very small when the method has converged. The use of a stopping criterion based on small changes of the least squares norm $S(\mathbf{P})$ could also be used, but with extreme caution. It may happen that the method stalls for a few iterations and then starts advancing to the point of minimum afterwards[3,4,7].

The Computational Algorithm for Technique I

Different versions of the Levenberg-Marquardt method can be found in the literature, depending on the choice of the diagonal matrix Ω^k and on the form chosen for the variation of the damping parameter μ^k [1-9]. We illustrate here a procedure with the matrix Ω^k taken as

$$\Omega^k = diag[(\mathbf{J}^k)^T \mathbf{J}^k] \tag{2.1.15}$$

The algorithm described below is available as the subroutine MRQMIN of the Numerical Recipes [6]. The reader should consult the reference for further details on the use of such subroutine.

Suppose that temperature measurements $\mathbf{Y}=(Y_1,Y_2,...,Y_I)$ are given at times t_i, $i=1,...,I$. Also, suppose an initial guess $\mathbf{P}^0$ is available for the vector of unknown parameters $\mathbf{P}$. Choose a value for μ^0, say, $\mu^0 = 0.001$ and set $k=0$ [6]. Then,

Step 1. Solve the direct heat transfer problem given by equations (2.1.1) with the available estimate $\mathbf{P}^k$ in order to obtain the temperature vector $\mathbf{T}(\mathbf{P}^k)=(T_1,T_2,...,T_I)$.
Step 2. Compute $S(\mathbf{P}^k)$ from equation (2.1.3.b).
Step 3. Compute the sensitivity matrix $\mathbf{J}^k$ defined by equation (2.1.7.a) and then the matrix Ω^k given by equation (2.1.15), by using the current values of $\mathbf{P}^k$.
Step 4. Solve the following linear system of algebraic equations, obtained from the iterative procedure of the Levenberg-Marquardt Method, equation (2.1.13):

$$[(\mathbf{J}^k)^T \mathbf{J}^k + \mu^k \Omega^k] \Delta\mathbf{P}^k = (\mathbf{J}^k)^T [\mathbf{Y} - \mathbf{T}(\mathbf{P}^k)] \tag{2.1.16}$$

in order to compute $\Delta\mathbf{P}^k = \mathbf{P}^{k+1} - \mathbf{P}^k$.
Step 5. Compute the new estimate $\mathbf{P}^{k+1}$ as

$$\mathbf{P}^{k+1} = \mathbf{P}^k + \Delta\mathbf{P}^k \tag{2.1.17}$$

Step 6. Solve the direct problem (2.1.1) with the new estimate $\mathbf{P}^{k+1}$ in order to find $\mathbf{T}(\mathbf{P}^{k+1})$. Then compute $S(\mathbf{P}^{k+1})$, as defined by equation (2.1.3.b).
Step 7. If $S(\mathbf{P}^{k+1}) \geq S(\mathbf{P}^k)$, replace μ^k by $10\mu^k$ and return to step 4.
Step 8. If $S(\mathbf{P}^{k+1}) < S(\mathbf{P}^k)$, accept the new estimate $\mathbf{P}^{k+1}$ and replace μ^k by $0.1\mu^k$.

Step 9. Check the stopping criteria given by equations (2.1.14.a-c). Stop the iterative procedure if any of them is satisfied; otherwise, replace k by $k+1$ and return to step 3.

In another version of the Levenberg-Marquardt method due to Moré [8], the matrix Ω^k is taken as the *identity matrix* and the damping parameter μ^k is chosen based on the so-called *trust region algorithm* [7,8]. The subroutines UNLSF, UNLSJ, BCLSF and BCLSJ in the IMSL [5] are based on this version of the Levenberg-Marquardt Method.

After computing $P_1, P_2, ..., P_N$ with the above computational procedure, a *Statistical Analysis* can be performed in order to obtain estimates for the standard deviations and other quantities of interest to assess the accuracy of the estimated parameters. The basic steps of such an analysis are included in Note 1 at the end of this chapter.

Sensitivity Coefficient Concept

The sensitivity matrix (2.1.7.a) plays an important role in parameter estimation problems. Therefore, we present below a discussion of the physical and mathematical significance of the sensitivity coefficients and the methods for their computation.

The sensitivity coefficient J_{ij} , as defined in equation (2.1.7.c), is a measure of the sensitivity of the estimated temperature T_i with respect to changes in the parameter P_j . A small value of the magnitude of J_{ij} indicates that large changes in P_j yield small changes in T_i . It can be easily noticed that the estimation of the parameter P_j is extremely difficult in such a case, because basically the same value for temperature would be obtained for a wide range of values of P_j . In fact, when the sensitivity coefficients are small, we have $\left| \mathbf{J}^T\mathbf{J} \right| \approx 0$ and the inverse problem is ill-conditioned. It can also be shown that $\left| \mathbf{J}^T\mathbf{J} \right|$ is null if any column of $\mathbf{J}$ can be expressed as a linear combination of the other columns [4]. Therefore, *it is desirable to have linearly-independent sensitivity coefficients J_{ij} with large magnitudes*, so that the inverse problem is not very sensitive to measurement errors and accurate estimates of the parameters can be obtained. The maximization of $\left| \mathbf{J}^T\mathbf{J} \right|$ is generally aimed in order to *design optimum experiments* for the estimation of the unknown parameters, because the confidence region of the estimates is then minimized. Some details on such an approach are presented in Note 2 at the end of this chapter.

Generally, *the timewise variations of the sensitivity coefficients and of $\left| \mathbf{J}^T\mathbf{J} \right|$ must be examined before a solution for the inverse problem is attempted.* Such examinations give an indication of the best sensor location and measurement times to be used in the inverse analysis, which correspond to linearly-independent sensitivity coefficients with large absolute values and large magnitudes of $\left| \mathbf{J}^T\mathbf{J} \right|$.

Methods of Determining the Sensitivity Coefficients

There are several different approaches for the computation of the sensitivity coefficients. We present below, with illustrative examples, three such approaches, including: (i) The direct analytic solution, (ii) The boundary value problem, and (iii) The finite-difference approximation.

1. Direct Analytic Solution for Determining Sensitivity Coefficients. If the direct heat conduction problem is linear and an analytic solution is available for the temperature field, the sensitivity coefficient with respect to an unknown parameter P_j is determined by differentiating the solution with respect to P_j. This approach is illustrated in the following examples.

Example 2-1. Consider the test-problem given by equations (2.1.1). The analytical solution of this problem at the measurement position is obtained as [9]:

$$T(x_{meas},t) = \int_{t'=0}^{t} g_p(t')dt' + 2\sum_{m=1}^{\infty} e^{-\beta_m^2 t} \cos(\beta_m x_{meas}) \cos(0.5\beta_m) \int_{t'=0}^{t} e^{\beta_m^2 t'} g_p(t')dt' \tag{2.1.18.a}$$

where $\beta_m = m\pi$ are the eigenvalues.

The first integral term on the right-hand side of equation (2.1.18.a) is due to the fact that both boundary conditions for the problem are homogeneous of the second kind. Suppose $g_p(t)$ is parameterized in the general linear form as

$$g_p(t) = \sum_{j=1}^{N} P_j C_j(t) \tag{2.1.18.b}$$

Find an analytic expression for the sensitivity coefficient $J_j \equiv \dfrac{\partial T}{\partial P_j}$, with respect to the parameter P_j.

Solution: By substituting the strength of the source term $g_p(t)$ given by equation (2.1.18.b) into equation (2.1.18.a) and differentiating the resulting expression with respect to P_j, we find the expression for the sensitivity coefficient for the parameter P_j as

$$J_j \equiv \frac{\partial T}{\partial P_j} = \int_{t'=0}^{t} C_j(t')dt' + 2\sum_{m=1}^{\infty} e^{-\beta_m^2 t} \cos(\beta_m x_{meas}) \cos(0.5\beta_m) \int_{t'=0}^{t} e^{\beta_m^2 t'} C_j(t')dt'$$

(2.1.18.c)

The above inverse problem is linear because the sensitivity coefficients do not depend on P_j.

Figure 2.1.2 presents the timewise variation of the sensitivity coefficients given by equation (2.1.18.c), for a sensor located at x_{meas}=1 and for a case involving N=5 unknown parameters, where the trial functions were taken in the form of polynomials as

$$C_j(t) = t^{(j-1)} \tag{2.1.18.d}$$

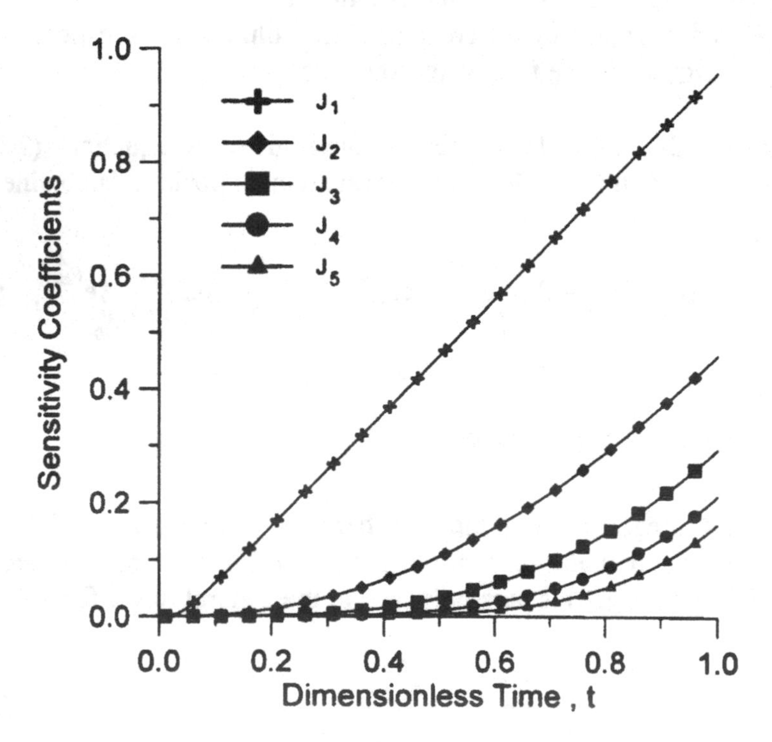

Figure 2.1.2. Sensitivity coefficients for polynomial trial functions given by equation (2.1.18.d)

Figure 2.1.2 shows that the sensitivity coefficients J_2, J_3, J_4 and J_5, with respect to the parameters P_2, P_3, P_4 and P_5, respectively, tend to be linearly dependent in the time interval $0 < t < 1$. Therefore, the estimation of the five coefficients of the polynomial used to approximate the unknown source function is difficult in such a case. This figure also shows that the sensitivity coefficient J_1 with respect to the parameter P_1 does not seem to be linearly dependent with the others in this time interval. Hence, the estimation of any pair of parameters, which necessarily includes P_1 as one of them, appears to be feasible with a sensor located a x_{meas}=1 and with measurements taken in the time interval $0 < t < 1$. In fact, for 100 transient measurements taken in this time interval, the determinant of $\mathbf{J}^T\mathbf{J}$ assumes the values 7.7 and 3.2x10^{-11}, for 2 (P_1 and P_2) and 5 (P_1 through P_5) unknown parameters, respectively, indicating that linearly dependent sensitivity coefficients yield small values of $\left| \mathbf{J}^T\mathbf{J} \right|$.

Figure 2.1.3 shows the sensitivity coefficients for a sensor located at x_{meas}=1 and for the first five coefficients of trial functions in the form

$$C_j(t) = \cos\left[(j-1)\frac{\pi}{2}t \right] \qquad \text{for } j=1,3,5,... \qquad (2.1.18.e)$$

$$C_j(t) = \sin\left[j\frac{\pi}{2}t \right] \qquad \text{for } j=2,4,6,... \qquad (2.1.18.f)$$

where the source function was approximated by a Fourier series.

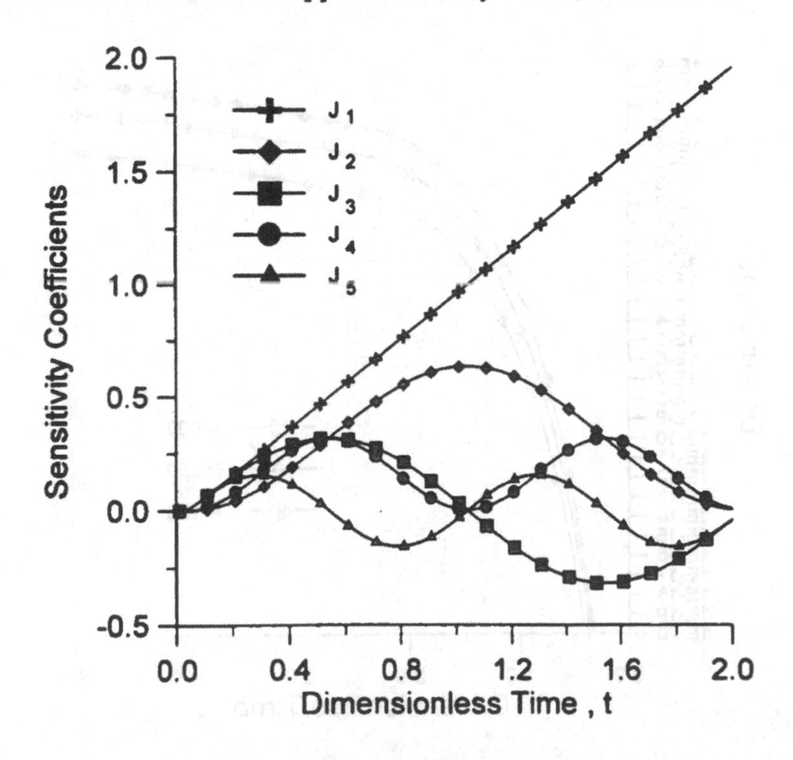

Figure 2.1.3. Sensitivity coefficients for the trial functions given by equations (2.1.18.e,f)

We notice in figure 2.1.3 that the sensitivity coefficients are not linearly dependent in the time interval $0.3 < t < 2$. Some linear-dependence is noticed among the sensitivity coefficients J_1, J_3 and J_5 for $t < 0.3$. Therefore, the conditions for the estimation of the five unknown parameters are not adequate if measurements taken only in the interval $0 < t < 0.3$ are used in the analysis; but it appears that the parameters can be estimated if the measurements are taken up to $t_f = 2$.

Figure 2.1.4 illustrates the time variation of the determinant of $\mathbf{J}^T\mathbf{J}$ up to a final experimental time $t_f = 5$, by considering $I = 100$, 250 and 500 measurements available from a sensor located at $x_{meas} = 1$. The trial functions are given by equations (2.1.18.e,f). For the three number of measurements considered, we notice a large increase in the magnitude of $\left|\mathbf{J}^T\mathbf{J}\right|$ up to about $t_f = 2$. The magnitude of $\left|\mathbf{J}^T\mathbf{J}\right|$ continues to grow for larger times, but at a much smaller rate. As expected, $\left|\mathbf{J}^T\mathbf{J}\right|$ increases with the number of measurements, since more information is available for the estimation of the unknown parameters. However, such increase is not as significant as increasing the experimental time from $t_f = 1$ to $t_f = 2$. In the example, $t_f = 2$ is a suitable duration for the experiment, since the value of $\left|\mathbf{J}^T\mathbf{J}\right|$ has already approached a reasonably large magnitude and the experiment duration is not too long.

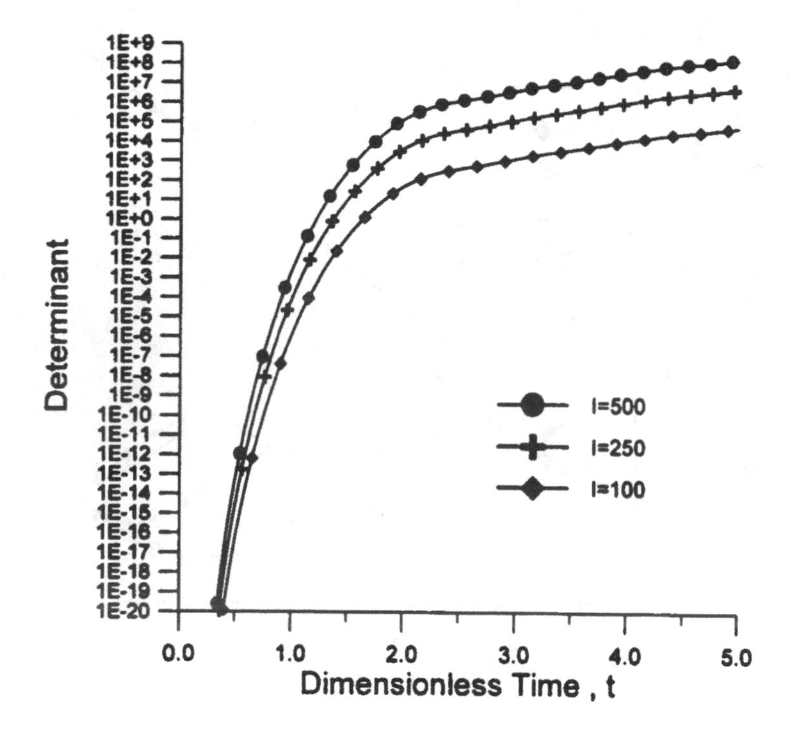

Figure 2.1.4. Determinant of $\mathbf{J}^T\mathbf{J}$ for the trial functions given by equations (2.1.18.e,f)

Example 2-2. Consider a semi-infinite medium initially at zero temperature, and for times $t > 0$ the boundary surface at $x = 0$ is subjected to a constant heat flux q_0, W/m^2. Develop an analytic expression for the sensitivity coefficient $Jq_0(0, t)$ with respect to the applied heat flux q_0 and for the sensitivity coefficient $J_\alpha(0,t)$ with respect to the thermal diffusivity α, based on the temperature at the boundary surface $x = 0$.

Solution. The temperature of the boundary surface at $x = 0$ is given by [9]:

$$T(0,t) = \frac{2q_0}{k}\left(\frac{\alpha t}{\pi}\right)^{1/2} \tag{2.1.19.a}$$

Then the sensitivity coefficient with respect to q_0 is determined from its definition as

$$J_{q_0}(0,t) \equiv \frac{\partial T(0,t)}{\partial q_0} = \frac{2}{k}\left(\frac{\alpha t}{\pi}\right)^{1/2} \tag{2.1.19.b}$$

which is independent of the applied heat flux q_0. Then, the inverse problem of estimating q_0 is linear.

The sensitivity coefficient with respect to α is determined as

$$J_\alpha(0,t) \equiv \frac{\partial T(0,t)}{\partial \alpha} = \frac{q_0}{k}\left(\frac{t}{\pi \alpha}\right)^{1/2} \tag{2.1.19.c}$$

which depends on α, and, hence the inverse problem of estimating α is nonlinear.

Example 2-3. Consider the transient heat conduction problem in a plate of thickness L, initially at a uniform temperature T_0. For times $t > 0$, the boundary $x=L$ is maintained at a temperature T_0. A constant heat flux q_0 is applied on the boundary $x = 0$, during the period $0 < t \le t_h$. For $t > t_h$ this boundary is kept insulated. The mathematical formulation of this problem in dimensionless form is given by:

$$\frac{\partial \theta}{\partial \tau} = \frac{\partial^2 \theta}{\partial \xi^2} \quad \text{in} \quad 0 < \xi < 1, \quad \text{for } \tau > 0 \tag{2.1.20.a}$$

$$\frac{\partial \theta}{\partial \xi} = \begin{cases} -1 & \text{for } 0 < \tau \le \tau_h \\ 0 & \text{for } \tau > \tau_h \end{cases} \quad \text{at } \xi = 0 \tag{2.1.20.b}$$

$$\theta = 0 \quad \text{at} \quad \xi = 1, \quad \text{for } \tau > 0 \tag{2.1.20.c}$$

$$\theta = 0 \quad \text{for} \quad \tau = 0, \quad 0 < \xi < 1 \tag{2.1.20.d}$$

where various dimensionless groups are defined as

$$\theta = \frac{T - T_0}{(q_0 L / k)}, \quad \tau = \frac{\alpha t}{L^2}, \quad \tau_h = \frac{\alpha t_h}{L^2}, \quad \xi = \frac{x}{L}, \quad \tau_f = \frac{\alpha t_f}{L^2} \tag{2.1.21.a-e}$$

where
α = thermal diffusivity
k = thermal conductivity
τ_h = dimensionless heating time
τ_f = dimensionless final time

The dimensionless sensitivity coefficients with respect to the thermal conductivity k and heat capacity $\rho c_p \equiv C$ are defined respectively as

$$J_k \equiv \frac{k}{(q_0 L / k)} \frac{\partial T}{\partial k} \tag{2.1.22.a}$$

and

$$J_C \equiv \frac{C}{(q_0 L / k)} \frac{\partial T}{\partial C} \tag{2.1.22.b}$$

Develop analytic expressions for the dimensionless sensitivity coefficients J_k and J_C, with respect to thermal conductivity and heat capacity, respectively.

Solution. The transient heat conduction problem given by equations (2.1.20) has been solved in reference [22] and the resulting expressions for the dimensionless temperature field $\theta(\xi,\tau)$ are given in the form

$$\theta(\xi,\tau) = (1 - \xi) - 2 \sum_{n=0}^{\infty} \frac{(-1)^n}{\lambda_n^2} \sin[\lambda_n (1 - \xi)]\, e^{-\lambda_n^2 \tau} \qquad \text{for } 0 < \tau < \tau_h$$

(2.1.23.a)

$$\theta(\xi,\tau) = -2 \sum_{n=0}^{\infty} \frac{(-1)^n}{\lambda_n^2} \sin[\lambda_n (1 - \xi)][e^{-\lambda_n^2 \tau} - e^{-\lambda_n^2 (\tau - \tau_h)}] \qquad \text{for } \tau_h < \tau < \tau_f$$

(2.1.23.b)

where $\lambda_n = (2n+1)\frac{\pi}{2}$ are the eigenvalues.

Since the temperature field $\theta(\xi,\tau)$ is known explicitly, analytic expressions can be developed for the dimensionless sensitivity coefficients according to equations (2.1.22). We obtain the sensitivity coefficients for the thermal conductivity as

$$J_k = -(1-\xi) + 2\sum_{n=0}^{\infty} \frac{(-1)^n}{\lambda_n^2} \sin[\lambda_n(1-\xi)] e^{-\lambda_n^2 \tau} (1+\lambda_n^2 \tau) \quad \text{for} \quad 0 < \tau < \tau_h$$

(2.1.24.a)

and

$$J_k = 2\sum_{n=0}^{\infty} \frac{(-1)^n}{\lambda_n^2} \sin[\lambda_n(1-\xi)][e^{-\lambda_n^2 \tau}(1+\lambda_n^2 \tau)] - e^{-\lambda_n^2(\tau-\tau_h)}[1+\lambda_n^2(\tau-\tau_h)]]$$

$$\text{for } \tau_h < \tau < \tau_f \qquad (2.1.24.b)$$

Similar expressions can be developed for the dimensionless sensitivity coefficients with respect to the heat capacity as

$$J_C = -2\sum_{n=0}^{\infty} (-1)^n \sin[\lambda_n(1-\xi)] \tau e^{-\lambda_n^2 \tau} \quad \text{for } 0 < \tau < \tau_h \qquad (2.1.25.a)$$

and

$$J_C = 2\sum_{n=0}^{\infty} (-1)^n \sin[\lambda_n(1-\xi)] [-\tau e^{-\lambda_n^2 \tau} + (\tau-\tau_h) e^{-\lambda_n^2(\tau-\tau_h)}] \quad \text{for } \tau_h < \tau < \tau_f$$

(2.1.25.b)

Figures 2.1.5.a-c show the plots of, respectively, the dimensionless temperature θ, the dimensionless sensitivity coefficients J_k and J_C , as a function of the dimensionless time τ, for $\tau_h = \tau_f = 7$ and at several different dimensionless locations $\xi = \frac{x}{L} = 0, 0.25, 0.5$ and 0.75. These figures show that for this particular case the magnitude of sensitivity coefficients attain relatively large values, i.e., of the order of dimensionless temperature θ. The magnitude of the sensitivity coefficients for the volumetric heat capacity are smaller than those for the thermal conductivity and they approach zero for $\tau > 2$. Thus, basically no information can be obtained from measurements taken for $\tau > 2$ for the estimation of C. Also, note that the magnitude of the sensitivity coefficients for k and C decrease as the sensor is placed farther from the boundary $\xi=0$. For each sensor location, the shapes of the sensitivity coefficient curves for J_k and J_C are different except at the very early times. In fact, J_k tends to a finite value after the steady-state is reached, while J_C becomes zero. Therefore, they are not linearly-dependent on each other, and, as a result, the conditions are good for the estimation of the unknown parameters with a single sensor. Such sensor should be located as close to the boundary $\xi=0$ as possible, because the magnitudes of the sensitivity coefficients are larger in this region.

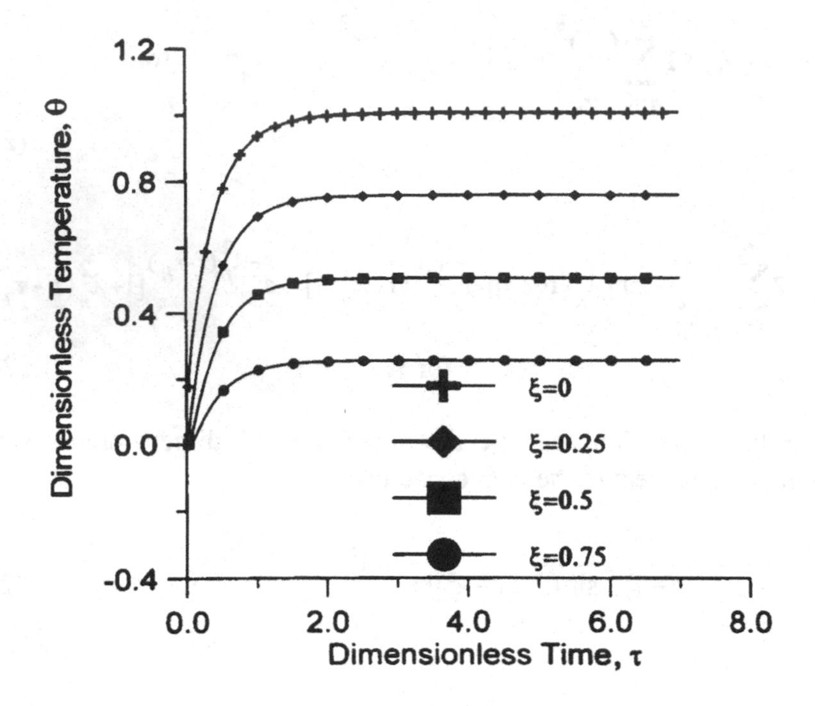

Figure 2.1.5.a Dimensionless temperature

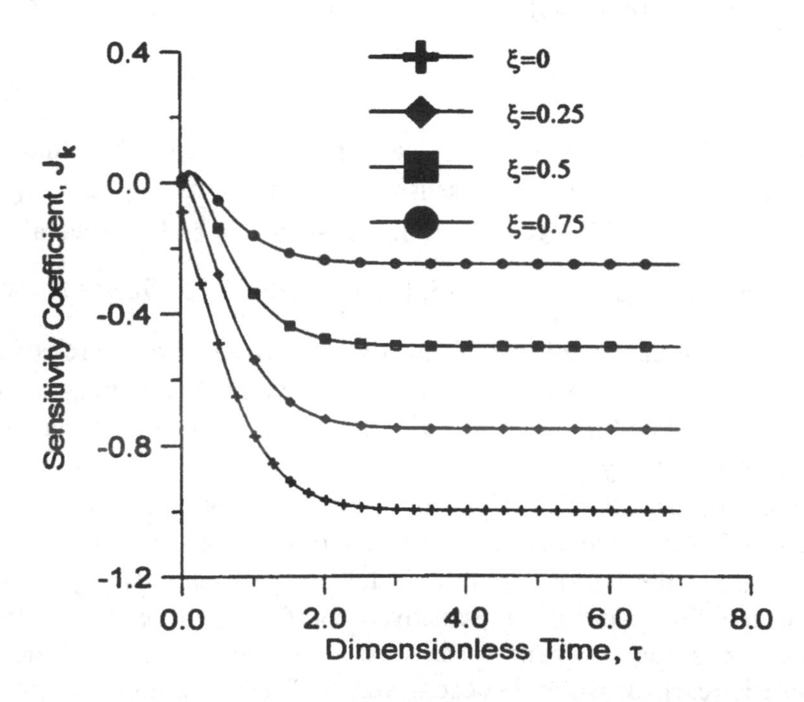

Figure 2.1.5.b Dimensionless sensitivity coefficient for thermal conductivity

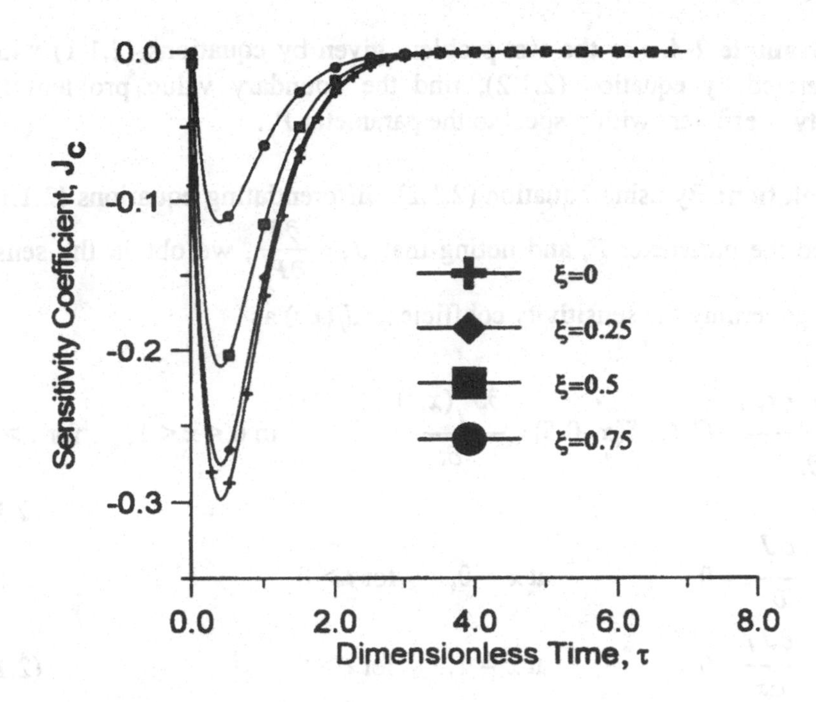

Figure 2.1.5.c Dimensionless sensitivity coefficient for volumetric heat capacity

In problems involving parameters with different orders of magnitude, the sensitivity coefficients with respect to the various parameters may also differ by several orders of magnitude, creating difficulties in their comparison and identification of linear dependence. These difficulties can be alleviated through the analysis of either dimensionless sensitivity coefficients (given in the example above by equations 2.1.22.a,b), or *relative sensitivity coefficients* defined as

$$J_{P_j} \equiv P_j \frac{\partial T}{\partial P_j} \tag{2.1.26}$$

where P_j , j = 1, ..., N, are the unknown parameters. Note that the relative sensitivity coefficients have the units of temperature; hence, they are compared as having the magnitude of the measured temperature as a basis.

2. The Boundary Value Problem Approach For Determining The Sensitivity Coefficients. A boundary value problem can be developed for the determination of the sensitivity coefficients by differentiating the original direct problem with respect to the unknown coefficients. If the direct heat conduction problem is linear, the construction of the corresponding sensitivity problem is a relatively simple and straightforward matter. To illustrate this approach, we use the following examples.

Example 2-4. For the test-problem given by equations (2.1.1) with $g_p(t)$ parameterized by equation (2.1.2), find the boundary value problem for the sensitivity coefficient with respect to the parameter P_j.

Solution: By using equation (2.1.2), differentiating equations (2.1.1) with respect to the parameter P_j and noting that $J_j = \frac{\partial T}{\partial P_j}$, we obtain the sensitivity problem governing the sensitivity coefficients $J_j(x,t)$ as

$$\frac{\partial^2 J_j(x,t)}{\partial x^2} + C_j(t)\,\delta(x-0.5) = \frac{\partial J_j(x,t)}{\partial t} \qquad \text{in } 0 < x < 1, \quad \text{for } t > 0 \tag{2.1.27.a}$$

$$\frac{\partial J_j}{\partial x} = 0 \qquad \text{at } x = 0, \quad \text{for } t > 0 \tag{2.1.27.b}$$

$$\frac{\partial J_j}{\partial x} = 0 \qquad \text{at } x = 1, \quad \text{for } t > 0 \tag{2.1.27.c}$$

$$J_j(x, 0) = 0 \qquad \text{for } t = 0, \quad \text{in } 0 < x < 1 \tag{2.1.27.d}$$

Note that problem (2.1.27) is similar to problem (2.1.1). The problem (2.1.27) needs to be solved N times, in order to compute the sensitivity coefficients with respect to each parameter P_j, $j = 1, \ldots, N$. For this particular case, the analytical solution of problem (2.1.27) can be easily obtained with equation (2.1.18.c). For more involved cases, the solution of the boundary value problem for determining the sensitivity coefficients may require numerical techniques, such as finite-differences. Thus, the computation of the sensitivity coefficients may become very time-consuming.

Example 2-5: Consider the following heat conduction problem

$$k\frac{\partial^2 T}{\partial x^2} = C\frac{\partial T}{\partial t} \qquad \text{in } 0 < x < L, \qquad \text{for } t > 0 \tag{2.1.28.a}$$

$$-k\frac{\partial T}{\partial x} = q_0 \qquad \text{at} \quad x = 0, \qquad \text{for } t > 0 \tag{2.1.28.b}$$

$$\frac{\partial T}{\partial x} = 0 \qquad \text{at} \quad x = L, \qquad \text{for } t > 0 \tag{2.1.28.c}$$

$$T = T_i \qquad \text{for} \quad t = 0, \qquad \text{in } 0 < x < L \tag{2.1.28.d}$$

where $\rho c_p \equiv C$, heat capacity
q_0 = applied heat flux

k = thermal conductivity

Construct the *sensitivity problem* for determining the sensitivity coefficients with respect to thermal conductivity, i.e.,

$$J_k \equiv \frac{\partial T}{\partial k} \tag{2.1.29}$$

Solution: By differentiating problem (2.1.28) with respect to k and utilizing the definition of J_k, we obtain the following boundary value problem for determination of the sensitivity coefficients

$$k\frac{\partial^2 J_k}{\partial x^2} + \frac{\partial^2 T}{\partial x^2} = C\frac{\partial J_k}{\partial t} \quad \text{in} \quad 0 < x < L, \quad \text{for } t > 0 \tag{2.1.30.a}$$

$$k\frac{\partial J_k}{\partial x} + \frac{\partial T}{\partial x} = 0 \quad \text{at} \quad x = 0, \quad \text{for } t > 0 \tag{2.1.30.b}$$

$$\frac{\partial J_k}{\partial x} = 0 \quad \text{at} \quad x = L, \quad \text{for } t > 0 \tag{2.1.30.c}$$

$$J_k = 0 \quad \text{for} \quad t = 0, \quad \text{in } 0 < x < L \tag{2.1.30.d}$$

We note that problem (2.1.30) contains the non-homogeneous terms $\partial^2 T/\partial x^2$ and $\partial T/\partial x$ in equations (2.1.30.a) and (2.1.30.b), respectively. Also, the unknown parameter k appears in these two equations; thus, the problem of estimating k is nonlinear. The solution of problem (2.1.30) yields the sensitivity coefficients J_k, with respect to thermal conductivity k. By following a similar procedure, the sensitivity problem for determining the sensitivity coefficient J_C, with respect to heat capacity C, can be developed.

3. Finite Difference Approximation For Determining Sensitivity Coefficients. The first derivative appearing in the definition of the sensitivity coefficient, equation (2.1.7.c), can be computed by finite differences. If a *forward difference* is used, the sensitivity coefficient with respect to the parameter P_j is approximated by

$$J_{ij} \cong \frac{T_i(P_1, P_2, \ldots, P_j + \varepsilon P_j, \ldots, P_N)' - T_i(P_1, P_2, \ldots, P_j, \ldots, P_N)}{\varepsilon P_j} \tag{2.1.31.a}$$

where $\varepsilon \approx 10^{-5}$ or 10^{-6}. If the first-order approximation given by equation (2.1.31.a) is not sufficiently accurate, the sensitivity coefficients can be approximated by using *central differences* in the form

$$J_{ij} \cong \frac{T_i(P_1,P_2,...,P_j+\varepsilon P_j,...,P_N)-T_i(P_1,P_2,...,P_j-\varepsilon P_j,...,P_N)}{2\varepsilon P_j} \tag{2.1.31.b}$$

We note that the approximation of the sensitivity coefficients given by equation (2.1.31.a) requires the computation of N additional solutions of the direct problem, while equation (2.1.31.b) requires $2N$ additional solutions of the problem. Therefore, the computation of the sensitivity coefficients by using finite-differences can be very time-consuming.

The Use of Multiple Sensors

The computational algorithm of the Levenberg-Marquardt Method, as given above, can also be used with few modifications in cases involving the measurements of multiple sensors. The quantities requiring modifications include the explicit forms of the vector [**Y**-**T**(**P**)], of the objective function S(**P**) and of the sensitivity matrix **J**.

In cases where the measurements of M sensors are available for the analysis, the vector containing the differences between measured and estimated temperatures is written as (see equation 1.5.3.a):

$$[\mathbf{Y}-\mathbf{T}(\mathbf{P})]^T = [\vec{Y}_1-\vec{T}_1(\mathbf{P}),\vec{Y}_2-\vec{T}_2(\mathbf{P}),\cdots,\vec{Y}_I-\vec{T}_I(\mathbf{P})] \tag{2.1.32.a}$$

where $[\vec{Y}_i-\vec{T}_i(\mathbf{P})]$ is a row vector which contains the difference between measured and estimated temperatures for each of the M sensors at time t_i, $i = 1, ..., I$. It is given in the form(see equation 1.5.3.b):

$$[\vec{Y}_i-\vec{T}_i(\mathbf{P})] = [Y_{i1}-T_{i1}(\mathbf{P}),Y_{i2}-T_{i2}(\mathbf{P}),\cdots,Y_{iM}-T_{iM}(\mathbf{P})] \quad \text{for } i=1,...,I \tag{2.1.32.b}$$

In the vector element $[Y_{im}-T_{im}(\mathbf{P})]$, the subscript i refers to time t_i, while the subscript m refers to the sensor number, where $i = 1, ..., I$ and $m = 1, ..., M$.

By substituting equations (2.1.32.a,b) into equation (2.1.3.b), the least-squares norm can be expressed in the following explicit form (see equation 1.5.3.c):

$$S(\mathbf{P})=\sum_{m=1}^{M}\sum_{i=1}^{I}[Y_{im}-T_{im}(\mathbf{P})]^2 \tag{2.1.32.c}$$

The sensitivity matrix defined by equation (2.1.7.a) needs also to be modified in order to accommodate the measurements of M sensors. The transpose of the sensitivity matrix, defined by equation (2.1.6), is then written as

$$\frac{\partial \mathbf{T}^T(\mathbf{P})}{\partial \mathbf{P}} = \begin{bmatrix} \frac{\partial}{\partial P_1} \\ \frac{\partial}{\partial P_2} \\ \vdots \\ \frac{\partial}{\partial P_N} \end{bmatrix} \begin{bmatrix} \vec{T}_1 & \vec{T}_2 & \cdot & \cdot & \vec{T}_I \end{bmatrix} \tag{2.1.33.a}$$

where $\vec{T}_i = [T_{i1}, T_{i2}, \cdots, T_{iM}]$ for $i = 1,\ldots,I$ (2.1.33.b)

Therefore, we can write the sensitivity matrix in the form

$$\mathbf{J}(\mathbf{P}) \equiv \left[\frac{\partial \mathbf{T}^T(\mathbf{P})}{\partial \mathbf{P}}\right]^T = \begin{bmatrix} \frac{\partial \vec{T}_1^T}{\partial P_1} & \frac{\partial \vec{T}_1^T}{\partial P_2} & \frac{\partial \vec{T}_1^T}{\partial P_3} & \cdots & \frac{\partial \vec{T}_1^T}{\partial P_N} \\ \frac{\partial \vec{T}_2^T}{\partial P_1} & \frac{\partial \vec{T}_2^T}{\partial P_2} & \frac{\partial \vec{T}_2^T}{\partial P_3} & \cdots & \frac{\partial \vec{T}_2^T}{\partial P_N} \\ \vdots & \vdots & \vdots & & \vdots \\ \frac{\partial \vec{T}_I^T}{\partial P_1} & \frac{\partial \vec{T}_I^T}{\partial P_2} & \frac{\partial \vec{T}_I^T}{\partial P_3} & \cdots & \frac{\partial \vec{T}_I^T}{\partial P_N} \end{bmatrix} \tag{2.1.34.a}$$

where

$$\frac{\partial \vec{T}_i^T}{\partial P_j} = \begin{bmatrix} \frac{\partial T_{i1}}{\partial P_j} \\ \frac{\partial T_{i2}}{\partial P_j} \\ \vdots \\ \frac{\partial T_{iM}}{\partial P_j} \end{bmatrix} \quad \text{for } i = 1, \ldots, I \text{ and } j = 1, \ldots, N \tag{2.1.34.b}$$

I = number of transient measurements per sensor
M = number of sensors
N = number of unknown parameters

The elements of the sensitivity matrix, as given by equation (2.1.34.a), can be suitably written in the form

$$J_{kj} = \frac{\partial T_k}{\partial P_j} \tag{2.1.35.a}$$

where the subscripts k and j refer to the row number and to the column number of the sensitivity matrix, respectively. The row number k is then related to the measurement time t_i and to the sensor number m by the expression

$$k = (i-1)M + m \tag{2.1.35.b}$$

With the modifications above, the computational algorithm for Technique I can be applied to cases involving the measurements of multiple sensors.

2-2 TECHNIQUE II: THE CONJUGATE GRADIENT METHOD FOR PARAMETER ESTIMATION

We present in this section an alternative technique for the estimation of unknown parameters. **Technique II**, the *Conjugate Gradient Method,* is a straightforward and powerful iterative technique for solving linear and nonlinear inverse problems of parameter estimation. In the iterative procedure of the Conjugate Gradient Method, at each iteration a suitable step size is taken along a direction of descent in order to minimize the objective function. The direction of descent is obtained as a linear combination of the negative gradient direction at the current iteration with the direction of descent of the previous iteration. The linear combination is such that the resulting angle between the direction of descent and the negative gradient direction is less than 90° and the minimization of the objective function is assured. Theorems regarding the convergence of the Conjugate Gradient Method can be found in references [14,15,17,19]. The Conjugate Gradient Method with an appropriate stopping criterion belongs to the class of iterative regularization techniques, in which the number of iterations is chosen so that stable solutions are obtained for the inverse problem.

Similarly to Technique I, the application of Technique II to inverse heat transfer problems of parameter estimation can be conveniently organized in the following steps:

- The Direct Problem
- The Inverse Problem
- The Iterative Procedure
- The Stopping Criterion
- The Computational Algorithm

We present below the details of each of such steps, as applied to the heat conduction test-problem described in section 2-1, involving the estimation of the unknown source term function $g_p(t)$.

The Direct Problem

In the **Direct Problem** related to the physical problem described above in section 2-1, which is mathematically formulated by equations (2.1.1), the time-varying strength $g_p(t)$ of the plane heat source is known. The objective of the direct problem is then to determine the transient temperature field $T(x,t)$ in the region.

The Inverse Problem

In the **Inverse Problem** considered here, the time-varying strength $g_p(t)$ of the plane heat source is regarded as unknown and transient temperature measurements taken at a location $x = x_{meas}$, at times t_i, $i = 1, 2, ..., I$, are considered available for the analysis.

For the solution of such inverse problem, we consider the unknown energy generation function $g_p(t)$ to be parameterized in the general linear form given by equation (2.1.2). The estimation of the unknown function $g_p(t)$ then reduces to the estimation of the N unknown parameters P_j, $j = 1, ..., N$. Such parameter estimation problem is solved by the minimization of the ordinary least squares norm:

$$S(\mathbf{P}) = [\mathbf{Y} - \mathbf{T}(\mathbf{P})]^T [\mathbf{Y} - \mathbf{T}(\mathbf{P})] \tag{2.2.1}$$

The Iterative Procedure for Technique II

The *iterative procedure* of the Conjugate Gradient Method for the minimization of the above norm $S(\mathbf{P})$ is given by [9-19]

$$\mathbf{P}^{k+1} = \mathbf{P}^k - \beta^k \mathbf{d}^k \tag{2.2.2}$$

where β^k *is the search step size,* $\mathbf{d}^k$ *is the direction of descent* and the superscript k is the number of iterations. The direction of descent is a conjugation of the *gradient direction,* $\nabla S(\mathbf{P}^k)$, and the direction of descent of the previous iteration, $\mathbf{d}^{k-1}$. It is given as

$$\mathbf{d}^k = \nabla S(\mathbf{P}^k) + \gamma^k \mathbf{d}^{k-1} \tag{2.2.3}$$

Different expressions are available for the conjugation coefficient γ^k. The Polak-Ribiere [12,14] expression is given in the form:

$$\gamma^k = \frac{\sum_{j=1}^{N}\left\{\left[\nabla S(\mathbf{P}^k)\right]_j\left[\nabla S(\mathbf{P}^k)-\nabla S(\mathbf{P}^{k-1})\right]_j\right\}}{\sum_{j=1}^{N}\left[\nabla S(\mathbf{P}^{k-1})\right]_j^2} \quad \text{for } k=1,2,... \tag{2.2.4.a}$$

$$\text{with} \qquad \gamma^0 = 0 \quad \text{for } k=0$$

while the Fletcher-Reeves [12,14,15] expression is given as

$$\gamma^k = \frac{\sum_{j=1}^{N}\left[\nabla S(\mathbf{P}^k)\right]_j^2}{\sum_{j=1}^{N}\left[\nabla S(\mathbf{P}^{k-1})\right]_j^2} \quad \text{for } k = 1,2, ... \tag{2.2.4.b}$$

$$\text{with} \qquad \gamma^0 = 0 \quad \text{for } k=0$$

Here, $[\nabla S(\mathbf{P}^k)]_j$ is the jth *component of the gradient direction* evaluated at iteration k. The expression for the gradient direction is obtained by differentiating equation (2.2.1) with respect to the unknown parameters **P**, i.e.,

$$\nabla S(\mathbf{P}^k) = -2(\mathbf{J}^k)^T[\mathbf{Y} - \mathbf{T}(\mathbf{P}^k)] \tag{2.2.5.a}$$

where $\mathbf{J}^k$ is the sensitivity matrix defined by equation (2.1.7.a). The jth component of the gradient direction can be obtained in explicit form as

$$\left[\nabla S(\mathbf{P}^k)\right]_j = -2\sum_{i=1}^{I}\frac{\partial T_i^k}{\partial P_j}[Y_i - T_i(\mathbf{P}^k)] \quad \text{for } j = 1,...,N \tag{2.2.5.b}$$

Either expressions (2.2.4.a,b) for the computation of the conjugation coefficient γ^k assure that the angle between the direction of descent and the negative gradient direction is less than 90°, so that the function $S(\mathbf{P})$ is minimized [14]. They are equivalent on linear estimation problems; but there is some evidence that the Polak-Ribiere expression provides improved convergence in nonlinear estimation problems [6,14].

We note that if γ^k=0 for all iterations k, the direction of descent becomes the gradient direction in equation (2.2.3) and the *steepest-descent method* is obtained. Although simpler, the steepest descent method does not converge as fast as the conjugate gradient method [10-21].

The *search step size* β^k appearing in equation (2.2.2) is obtained by minimizing the function $S(\mathbf{P}^{k+1})$ with respect to β^k, that is,

$$\min_{\beta^k} S(\mathbf{P}^{k+1}) = \min_{\beta^k} [\mathbf{Y} - \mathbf{T}(\mathbf{P}^{k+1})]^T [\mathbf{Y} - \mathbf{T}(\mathbf{P}^{k+1})] \tag{2.2.6.a}$$

By substituting $\mathbf{P}^{k+1}$ as given by equation (2.2.2) into equation (2.2.6.a), we obtain

$$\min_{\beta^k} S(\mathbf{P}^{k+1}) = \min_{\beta^k} [\mathbf{Y} - \mathbf{T}(\mathbf{P}^k - \beta^k \mathbf{d}^k)]^T [\mathbf{Y} - \mathbf{T}(\mathbf{P}^k - \beta^k \mathbf{d}^k)] \tag{2.2.6.b}$$

The temperature vector $\mathbf{T}(\mathbf{P}^k - \beta^k \mathbf{d}^k)$ can be linearized with a Taylor series expansion and then the minimization with respect to β^k is performed to yield the following expression for the search step size

$$\beta^k = \frac{\sum_{i=1}^{I} \left[\left(\frac{\partial T_i}{\partial \mathbf{P}^k} \right)^T \mathbf{d}^k \right] \left[T_i(\mathbf{P}^k) - Y_i \right]}{\sum_{i=1}^{I} \left[\left(\frac{\partial T_i}{\partial \mathbf{P}^k} \right)^T \mathbf{d}^k \right]^2} \tag{2.2.7.a}$$

where

$$\left(\frac{\partial T_i}{\partial \mathbf{P}^k} \right)^T = \left[\frac{\partial T_i}{\partial P_1^k}, \frac{\partial T_i}{\partial P_2^k}, \dots, \frac{\partial T_i}{\partial P_N^k} \right] \tag{2.2.7.b}$$

We note that the vector in equation (2.2.7.b) is the ith row of the sensitivity matrix (see equation 2.1.7.b). Hence, we can write equation (2.2.7.a) in matrix form as

$$\beta^k = \frac{[\mathbf{J}^k \mathbf{d}^k]^T [\mathbf{T}(\mathbf{P}^k) - \mathbf{Y}]}{[\mathbf{J}^k \mathbf{d}^k]^T [\mathbf{J}^k \mathbf{d}^k]} \tag{2.2.7.c}$$

For further details on the derivation of equations (2.2.7.a,c), the reader should refer to Note 3 at the end of this chapter.

After computing the sensitivity matrix $\mathbf{J}^k$, the gradient direction $\nabla S(\mathbf{P}^k)$, the conjugation coefficient γ^k and the search step size β^k, the iterative procedure given by equation (2.2.2) is implemented until a stopping criterion based on the *Discrepancy Principle* described below is satisfied. The sensitivity matrix may be computed by using one of the appropriate methods described in section 2-1.

The Stopping Criterion for Technique II

The iterative procedure given by equations (2.2.2-4), with the search step size β^k given by equation (2.2.7.c), does not provide the conjugate gradient method with the stabilization necessary for the minimization of the objective function (2.2.1) to be classified as well-posed. Such is the case because of the random errors inherent to the measured temperatures. As the estimated temperatures approach the measured temperatures containing errors, during the minimization of the function (2.2.1), large oscillations may appear in the inverse problem solution, resulting in an ill-posed character for the inverse problem. However, the conjugate gradient method may become well-posed if the *Discrepancy Principle* [12] is used to stop the iterative procedure.

In the discrepancy principle, the iterative procedure is stopped when the following criterion is satisfied

$$S(\mathbf{P}^{k+1}) < \varepsilon \tag{2.2.8}$$

where the value of the tolerance ε is chosen so that sufficiently stable solutions are obtained. In this case, we stop the iterative procedure when the residuals between measured and estimated temperatures are of the same order of magnitude of the measurement errors, that is,

$$\left| Y(t_i) - T(x_{meas}, t_i) \right| \approx \sigma_i \tag{2.2.9}$$

where σ_i is the standard deviation of the measurement error at time t_i. For constant standard deviations, i.e., $\sigma_i = \sigma = \text{constant}$, we can then obtain the following value for ε by substituting equation (2.2.9) into equation (2.2.1)

$$\varepsilon = \sum_{i=1}^{I} \sigma_i^2 = I\sigma^2 \tag{2.2.10}$$

The above assumption for the temperature residuals in the discrepancy principle was also used by Tikhonov [12], in order to find the optimal regularization parameter. Such a procedure gives the conjugate gradient method an *iterative regularization character*. If the measurements are regarded as errorless, the tolerance ε can be chosen as a sufficiently small number, since the expected minimum value for the objective function (2.2.1) is zero.

At this point it is important for the reader to notice that the use of the discrepancy principle is not required to provide Technique I, the Levenberg-Marquardt method, with the regularization necessary to obtain stable solutions for those cases involving measurements with random errors. Computational experiments revealed that the Levenberg-Marquardt method, through its automatic control of the damping parameter μ^k, reduces drastically the increment in the vector of estimated parameters, at the iteration where the measurement errors start to cause instabilities in the inverse problem solution. The iterative procedure of the Levenberg-Marquardt method is then stopped by the criterion given by equation (2.1.14.c).

We note that the above stopping criterion approach, based on the discrepancy principle, requires the *a priori* knowledge of the standard deviation of the measurement errors. For those cases involving measurements with unknown standard deviations, an alternative approach based on an additional measurement can be used for the stopping criterion, as described in Note 4 at the end of this chapter. The stopping criterion approach based on the additional measurement also provides the conjugate gradient method with an iterative regularization character.

The Computational Algorithm for Technique II

Suppose that temperature measurements $\mathbf{Y}=(Y_1,Y_2,...,Y_I)$ are given at times t_i, $i = 1, ..., I$, and an initial guess $\mathbf{P}^0$ is available for the vector of unknown parameters $\mathbf{P}$. Set $k = 0$ and then

Step 1. Solve the direct heat transfer problem (2.1.1) by using the available estimate $\mathbf{P}^k$ and obtain the vector of estimated temperatures $\mathbf{T}(\mathbf{P}^k)=(T_1,T_2,...,T_I)$.

Step 2. Check the stopping criterion given by equation (2.2.8). Continue if not satisfied.

Step 3. Compute the sensitivity matrix $\mathbf{J}^k$ defined by equation (2.1.7.a), by using one of the appropriate methods described in section 2-1.

Step 4. Knowing $\mathbf{J}^k$, $\mathbf{Y}$ and $\mathbf{T}(\mathbf{P}^k)$, compute the gradient direction $\nabla S(\mathbf{P}^k)$ from equation (2.2.5.a) and then the conjugation coefficient γ^k from either equation (2.2.4.a) or (2.2.4.b).

Step 5. Compute the direction of descent $\mathbf{d}^k$ by using equation (2.2.3).

Step 6. Knowing $\mathbf{J}^k$, $\mathbf{Y}$, $\mathbf{T}(\mathbf{P}^k)$ and $\mathbf{d}^k$, compute the search step size β^k from equation (2.2.7.c).

Step 7. Knowing $\mathbf{P}^k$, β^k and $\mathbf{d}^k$, compute the new estimate $\mathbf{P}^{k+1}$ using equation (2.2.2).

Step 8. Replace k by $k+1$ and return to step 1.

The Use of Multiple Sensors

As for the case of Technique I, the above computational algorithm for Technique II can be applied to cases involving the measurements of multiple sensors, with modifications in the explicit forms of few quantities as described next.

For those cases involving transient measurements of M sensors, the vector $[\mathbf{Y}-\mathbf{T}(\mathbf{P})]$ is given by equations (2.1.32.a,b). Hence, the objective function or least-squares norm $S(\mathbf{P})$ is obtained from equation (2.1.32.c), while the sensitivity matrix $\mathbf{J}$ is given by equations (2.1.34.a,b).

Since the objective function was modified, the computation of the tolerance ε for the stopping criterion based on the discrepancy principle needs also to be modified; it is now obtained from equation (2.1.32.c) as

$$\varepsilon=\sum_{m=1}^{M}\sum_{i=1}^{I}\sigma_{im}^{2}=IM\sigma^{2} \tag{2.2.11}$$

where I = number of transient measurements taken per sensor
M = number of sensors
$\sigma_{im}=\sigma$ = constant standard deviation of the measurements

In cases involving the readings of M sensors, the search step size β^k can still be obtained from the matrix form given by equation (2.2.7.c). However, the explicit form of β^k, equation (2.2.7.a), needs to be modified since the sensitivity matrix is now given by equations (2.1.34.a,b) instead of equation (2.1.7.b). Then, the search step size takes the form

$$\beta^{k}=\frac{\displaystyle\sum_{m=1}^{M}\sum_{i=1}^{I}\left[\left(\frac{\partial T_{im}}{\partial \mathbf{P}^{k}}\right)^{T}\mathbf{d}^{k}\right]\left[T_{im}\left(\mathbf{P}^{k}\right)-Y_{im}\right]}{\displaystyle\sum_{m=1}^{M}\sum_{i=1}^{I}\left[\left(\frac{\partial T_{im}}{\partial \mathbf{P}^{k}}\right)^{T}\mathbf{d}^{k}\right]^{2}} \tag{2.2.12}$$

Similarly, the expression in matrix form for the gradient of the objective function, equation (2.2.5.a), remains valid for the case involving multiple sensors; but the explicit form for the components of the gradient, equation (2.2.5.b), becomes

$$\left[\nabla S(\mathbf{P}^{k})\right]_{j}=-2\sum_{m=1}^{M}\sum_{i=1}^{I}\frac{\partial T_{im}^{k}}{\partial P_{j}}\left[Y_{im}-T_{im}(\mathbf{P}^{k})\right] \quad \text{for } j=1,\ldots,N \tag{2.2.13}$$

Continuous Measurements

So far we have considered the measured data in the time domain to be *discrete*. In cases where the measured data are so many that it can be approximated as *continuous*, some modifications are needed in the expressions of the objective function (2.2.1), the gradient vector (2.2.5.a), the search step size

(2.2.7) and the tolerance for the discrepancy principle (2.2.10). This matter is discussed next.

We now consider a situation in which the measured temperature data of a single sensor located at x_{meas} can be approximated as being continuous. For such a case, the objective function involves an integration over the time domain $0 \le t \le t_f$, where t_f is the duration of the experiment. It can be written as

$$S(\mathbf{P}) = \int_{t=0}^{t_f} [Y(t) - T(x_{meas}, t; \mathbf{P})]^2 dt \tag{2.2.14}$$

Thus, the gradient of equation (2.2.14) becomes

$$\nabla S(\mathbf{P}) = -2 \int_{t=0}^{t_f} [Y(t) - T(x_{meas}, t; \mathbf{P})] \frac{\partial T}{\partial \mathbf{P}} dt \tag{2.2.15.a}$$

or, more explicitly, each component of the gradient vector is given by

$$[\nabla S(\mathbf{P})]_j = -2 \int_{t=0}^{t_f} [Y(t) - T(x_{meas}, t; \mathbf{P})] \frac{\partial T}{\partial P_j} dt \quad \text{for } j=1,...,N \tag{2.2.15.b}$$

In addition, equation (2.2.7.a) for the search step size β^k should also be expressed in the continuous form in the time domain. This is accomplished by minimizing the objective function (2.2.14) with respect to β^k to obtain

$$\beta^k = \frac{\displaystyle\int_{t=0}^{t_f} [T(x_{meas}, t; \mathbf{P}^k) - Y(t)] \left[\frac{\partial T}{\partial \mathbf{P}^k}\right]^T \mathbf{d}^k \, dt}{\displaystyle\int_{t=0}^{t_f} \left\{\left[\frac{\partial T}{\partial \mathbf{P}^k}\right]^T \mathbf{d}^k\right\}^2 dt} \tag{2.2.16}$$

Note that the expressions for β^k given by equations (2.2.7.a) and (2.2.16) are very similar. The summation appearing in equation (2.2.7.a) changes into an integral for the continuous measurements case of equation (2.2.16).

The tolerance ε for the stopping criterion based on the Discrepancy Principle, for cases involving continuous measurements, is obtained from equation (2.2.14) as

$$\varepsilon = \sigma^2 t_f \tag{2.2.17}$$

The remaining quantities appearing in the computational algorithm for Technique II stay unaltered for cases involving continuous measurements.

We note that in cases involving continuous measurements of multiple sensors, the integral on the right-hand side of equation (2.2.14) is summed-up over the number of sensors, that is,

$$S(\mathbf{P}) = \sum_{m=1}^{M} \int_{t=0}^{t_f} [Y_m(t) - T(x_m, t; \mathbf{P})]^2 dt \tag{2.2.18}$$

Thus, the other quantities appearing in equations (2.2.15-17) also involve summations over the number of sensors. The derivations of the expressions for the gradient equation, the search step size and the tolerance for the stopping criterion, for cases involving continuous measurements of multiple sensors, are straightforward. They are left as an exercise to the reader.

In order to implement the iterative algorithm of the conjugate gradient method as presented above, the sensitivity matrix needs to be computed for each iteration. For linear problems, it is quite easy to compute such matrix with an analytical solution. Indeed, the sensitivity matrix being constant for linear problems, it has to be computed only once. On the other hand, for general nonlinear inverse problems, the sensitivity matrix needs to be computed most likely by finite-differences. This might be very time-consuming when the problem involves a large number of parameters and/or measurements. For cases involving the estimation of the coefficients of unknown functions parameterized, we present below an alternative implementation of the conjugate gradient method, which does not require the computation of the sensitivity matrix in order to obtain the gradient direction and the search step size.

2-3 TECHNIQUE III: THE CONJUGATE GRADIENT METHOD WITH ADJOINT PROBLEM FOR PARAMETER ESTIMATION

In this section we present an alternative implementation of the conjugate gradient method where two auxiliary problems, known as the *sensitivity problem* and the *adjoint problem*, are solved in order to compute the search step size β^k and the gradient equation $\nabla S(\mathbf{P}^k)$. The technique is specially suitable for problems involving the estimation of the coefficients of trial functions used to approximate an unknown functional form.

For convenience in the subsequent analysis, we consider the measured data to be continuous, rather than discrete. Thus, the ordinary least squares norm, equation (2.1.3.a), is rewritten as

$$S(\mathbf{P}) = \int_{t=0}^{t_f} [Y(t) - T(x_{meas}, t; \mathbf{P})]^2 \, dt \tag{2.3.1}$$

where $Y(t)$ is the measured temperature, $T(x_{meas}, t;\mathbf{P})$ is the estimated temperature at the single measurement location x_{meas} and t_f is the duration of the experiment.

The basic steps for the solution of parameter estimation problems, by using the conjugate gradient method with adjoint problem, include:

- The Direct Problem
- The Inverse Problem
- The Sensitivity Problem
- The Adjoint Problem
- The Gradient Equation
- The Iterative Procedure
- The Stopping Criterion
- The Computational Algorithm

We present below the details of each of these basic steps as applied to our test-problem.

The Direct Problem

For the test-problem considered here, involving the estimation of the strength $g_p(t)$ of a plane heat source, the direct problem is given by equations (2.1.1). It is rewritten below in order to facilitate the analysis.

$$\frac{\partial^2 T(x,t)}{\partial x^2} + g_p(t)\delta(x - 0.5) = \frac{\partial T(x,t)}{\partial t} \quad \text{in } 0 < x < 1, \text{ for } t > 0 \tag{2.3.2.a}$$

$$\frac{\partial T(0,t)}{\partial x} = 0 \quad \text{at } x = 0, \text{ for } t > 0 \tag{2.3.2.b}$$

$$\frac{\partial T(1,t)}{\partial x} = 0 \quad \text{at } x = 1, \text{ for } t > 0 \tag{2.3.2.c}$$

$$T(x, 0) = 0 \quad \text{for } t = 0, \text{ in } 0 < x < 1 \tag{2.3.2.d}$$

The direct problem is concerned with the determination of the temperature field $T(x,t)$ in the region $0 < x < 1$, when the strength of the source term $g_p(t)$ is known.

The Inverse Problem

The inverse problem, on the other hand, is concerned with the estimation of the unknown strength of the source term by using the readings taken by a sensor located at $x=x_{meas}$. We consider the unknown function $g_p(t)$ to be parameterized in a general linear form given by

$$g_p(t)=\sum_{j=1}^{N} P_j C_j(t) \tag{2.3.3.a}$$

where $C_j(t)$, j = 1, ..., N are known trial functions. Thus, the objective of the inverse problem is to estimate the N unknown parameters P_j, j = 1, ..., N.

The Sensitivity Problem

The *sensitivity function* $\Delta T(x,t)$, solution of the sensitivity problem, is defined as the directional derivative of the temperature $T(x,t)$ in the direction of the perturbation of the unknown function [12,21]. The sensitivity function is needed for the computation of the search step size β^k, as will be apparent later in this section.

The *sensitivity problem* can be obtained by assuming that the temperature $T(x,t)$ is perturbed by an amount $\Delta T(x,t)$, when the unknown strength $g_p(t)$ of the source term is perturbed by $\Delta g_p(t)$. Since the strength was parameterized in the form given by equation (2.3.3.a), the function $\Delta g_p(t)$ is obtained by perturbing each of the unknown parameters P_j by an amount ΔP_j, that is,

$$\Delta g_p(t)=\sum_{j=1}^{N} \Delta P_j C_j(t) \tag{2.3.3.b}$$

By replacing $T(x,t)$ by $[T(x,t)+\Delta T(x,t)]$ and $g_p(t)$ by $[g_p(t)+\Delta g_p(t)]$ in the direct problem given by equations (2.3.2), and then subtracting the original direct problem from the resulting expressions, we obtain the following *sensitivity problem:*

$$\frac{\partial^2 \Delta T(x,t)}{\partial x^2}+\Delta g_p(t)\delta(x-0.5)=\frac{\partial \Delta T(x,t)}{\partial t} \qquad \text{in } 0<x<1, \text{ for } t>0 \tag{2.3.4.a}$$

$$\frac{\partial \Delta T(0,t)}{\partial x}=0 \qquad \text{at } x=0, \text{ for } t>0 \tag{2.3.4.b}$$

$$\frac{\partial \Delta T(1,t)}{\partial x}=0 \qquad \text{at } x=1, \text{ for } t>0 \tag{2.3.4.c}$$

$$\Delta T(x,0)=0 \qquad \text{for } t=0\,,\text{ in } 0<x<1 \qquad (2.3.4.d)$$

In the sensitivity problem (2.3.4), $\Delta g_p(t)$ as given by equation (2.3.3.b) is the only forcing function needed for the solution. The computation of $\Delta g_p(t)$ will be addressed later in this section.

The Adjoint Problem

A Lagrange multiplier $\lambda(x,t)$ comes into picture in the minimization of the function (2.3.1) because the temperature $T(x_{meas},t;\mathbf{P})$ appearing in such function needs to satisfy a constraint, which is the solution of the direct problem. Such Lagrange multiplier, needed for the computation of the gradient equation (as will be apparent below), is obtained through the solution of a problem *adjoint* to the sensitivity problem given by equations (2.3.4). For the definition and properties of adjoint problems, the reader should consult references [12,21].

In order to derive the *adjoint problem*, we write the following extended function:

$$S(\mathbf{P})=\int_{t=0}^{t_f}[Y(t)-T(x_{meas},t;\mathbf{P})]^2\,dt+ \\ +\int_{x=0}^{1}\int_{t=0}^{t_f}\lambda(x,t)\left[\frac{\partial^2 T}{\partial x^2}+g_p(t)\delta(x-0.5)-\frac{\partial T}{\partial t}\right]dt\,dx \qquad (2.3.5)$$

which is obtained by multiplying the partial differential equation of the direct problem, equation (2.3.2.a), by the Lagrange multiplier, $\lambda(x,t)$, integrating over the time and space domains and adding the resulting expression to the function $S(\mathbf{P})$ given by equation (2.3.1).

An expression for the variation $\Delta S(\mathbf{P})$ of the function $S(\mathbf{P})$ can be developed by perturbing $T(x,t)$ by $\Delta T(x,t)$ and $g_p(t)$ by $\Delta g_p(t)$ in equation (2.3.5). We note that $\Delta S(\mathbf{P})$ is the directional derivative of $S(\mathbf{P})$ in the direction of the perturbation $\Delta\mathbf{P}=[\Delta P_1, \Delta P_2,\ldots,\Delta P_N]$ [12,21]. Then, by replacing $T(x,t)$ by $[T(x,t)+\Delta T(x,t)]$, $g_p(t)$ by $[g_p(t)+\Delta g_p(t)]$ and $S(\mathbf{P})$ by $[S(\mathbf{P})+\Delta S(\mathbf{P})]$ in equation (2.3.5), subtracting from the resulting expression the original equation (2.3.5), and neglecting second order terms, we find

$$\Delta S(\mathbf{P}) = \int_{t=0}^{t_f}\int_{x=0}^{1} 2\,[T(x,t;\mathbf{P}) - Y(t)]\,\Delta T(x,t)\delta(x - x_{meas})\,dx\,dt$$
$$+ \int_{t=0}^{t_f}\int_{x=0}^{1} \lambda(x,t)\left[\frac{\partial^2 \Delta T}{\partial x^2} + \Delta g_p(t)\delta(x-0.5) - \frac{\partial \Delta T}{\partial t}\right]dx\,dt \tag{2.3.6}$$

where δ(.) is the Dirac delta function.

The second integral term on the right-hand side of this equation is simplified by integration by parts and by utilizing the boundary and initial conditions of the sensitivity problem. The integration by parts of the term involving the second derivative in the spatial variable yields

$$\int_{x=0}^{1} \lambda(x,t)\frac{\partial^2 \Delta T}{\partial x^2}\,dx = \left[\lambda\frac{\partial \Delta T}{\partial x} - \frac{\partial \lambda}{\partial x}\Delta T\right]_{x=0}^{1} + \int_{x=0}^{1} \Delta T(x,t)\frac{\partial^2 \lambda}{\partial x^2}\,dx \tag{2.3.7.a}$$

By substituting the boundary conditions (2.3.4.b,c) of the sensitivity problem into equation (2.3.7.a), we obtain

$$\int_{x=0}^{1} \lambda(x,t)\frac{\partial^2 \Delta T}{\partial x^2}\,dx = \frac{\partial \lambda(0,t)}{\partial x}\Delta T(0,t) - \frac{\partial \lambda(1,t)}{\partial x}\Delta T(1,t) + \int_{x=0}^{1} \Delta T(x,t)\frac{\partial^2 \lambda}{\partial x^2}\,dx \tag{2.3.7.b}$$

Similarly, the integration by parts of the term involving the time derivative in equation (2.3.6) gives

$$\int_{t=0}^{t_f} \lambda(x,t)\frac{\partial \Delta T}{\partial t}\,dt = \left[\lambda(x,t)\Delta T(x,t)\right]_{t=0}^{t_f} - \int_{t=0}^{t_f} \Delta T(x,t)\frac{\partial \lambda}{\partial t}\,dt \tag{2.3.8.a}$$

After substituting the initial condition (2.3.4.d) of the sensitivity problem, the equation above becomes

$$\int_{t=0}^{t_f} \lambda(x,t)\frac{\partial \Delta T}{\partial t}\,dt = \lambda(x,t_f)\Delta T(x,t_f) - \int_{t=0}^{t_f} \Delta T(x,t)\frac{\partial \lambda}{\partial t}\,dt \tag{2.3.8.b}$$

Equations (2.3.7.b) and (2.3.8.b) are then substituted into equation (2.3.6) to obtain

$$\Delta S(\mathbf{P})=\int_{t=0}^{t_f}\int_{x=0}^{1}\left\{\frac{\partial^2\lambda(x,t)}{\partial x^2}+\frac{\partial\lambda(x,t)}{\partial t}+2[T(x,t;\mathbf{P})-Y(t)]\,\delta(x-x_{meas})\right\}\Delta T(x,t)\,dx\,dt$$
$$+\int_{t=0}^{t_f}\frac{\partial\lambda(0,t)}{\partial x}\Delta T(0,t)\,dt-\int_{t=0}^{t_f}\frac{\partial\lambda(1,t)}{\partial x}\Delta T(1,t)\,dt \tag{2.3.9}$$
$$-\int_{x=0}^{1}\lambda(x,t_f)\,\Delta T(x,t_f)\,dx+\int_{t=0}^{t_f}\lambda(0.5,t)\,\Delta g_p(t)\,dt$$

The boundary value problem for the Lagrange multiplier $\lambda(x,t)$ is obtained by allowing the first four integral terms containing $\Delta T(x,t)$ on the right-hand side of equation (2.3.9) to vanish. This leads to the following *adjoint problem*:

$$\frac{\partial\lambda(x,t)}{\partial t}+\frac{\partial^2\lambda(x,t)}{\partial x^2}+2\,[T(x,t;\mathbf{P})-Y(t)]\,\delta(x-x_{meas})=0$$
$$\text{in } 0<x<1,\ \text{for } 0<t<t_f \tag{2.3.10.a}$$

$$\frac{\partial\lambda(0,t)}{\partial x}=0 \qquad \text{at } x=0,\ \text{for } 0<t<t_f \tag{2.3.10.b}$$

$$\frac{\partial\lambda(1,t)}{\partial x}=0 \qquad \text{at } x=1,\ \text{for } 0<t<t_f \tag{2.3.10.c}$$

$$\lambda(x,t_f)=0 \qquad \text{for } t=t_f\ ,\ \text{in } 0<x<1 \tag{2.3.10.d}$$

We note that in the adjoint problem, the condition (2.3.10.d) is the value of the function $\lambda(x,t)$ at the final time $t = t_f$. In the conventional *initial value problem*, the value of the function is specified at time $t = 0$. However, the *final value problem* (2.3.10) can be transformed into an *initial value problem* by defining a new time variable given by $\tau = t_f - t$.

The Gradient Equation

After letting the terms containing $\Delta T(x,t)$ vanish, the following integral term is left on the right-hand side of equation (2.3.9):

$$\Delta S(\mathbf{P}) = \int_{t=0}^{t_f} \lambda(0.5,t)\Delta g_p(t)dt \tag{2.3.11}$$

By substituting $\Delta g_p(t)$ in the parametric form given by equation (2.3.3.b) into equation (2.3.11), we obtain

$$\Delta S(\mathbf{P}) = \sum_{j=1}^{N} \int_{t=0}^{t_f} \lambda(0.5,t)C_j(t)dt \, \Delta P_j \tag{2.3.12}$$

By definition, the directional derivative of $S(\mathbf{P})$ in the direction of a vector $\Delta\mathbf{P}$, is given by

$$\Delta S(\mathbf{P}) = \sum_{j=1}^{N} [\nabla S(\mathbf{P})]_j \Delta P_j \tag{2.3.13}$$

where

$$\Delta\mathbf{P} = [\Delta P_1, \Delta P_2, \ldots, \Delta P_N] \tag{2.3.14}$$

We note that the magnitude of the vector $\Delta\mathbf{P}$ was omitted in equation (2.3.13), since it is not relevant for the present analysis. Therefore, by comparing equations (2.3.12) and (2.3.13), we obtain the jth component of the gradient vector $\nabla S(\mathbf{P})$ for the function $S(\mathbf{P})$ as

$$[\nabla S(\mathbf{P})]_j = \int_{t=0}^{t_f} \lambda(0.5,t)C_j(t)dt \qquad \text{for } j=1,\ldots,N \tag{2.3.15}$$

The use of an adjoint problem for the computation of the gradient vector is most useful for problems involving unknown functions that can be parameterized in a form similar to equation (2.3.3.a), especially those problems which do not have analytical expressions for the sensitivity coefficients and finite-difference approximations need to be used. With the present approach, *the gradient vector is computed with the solution of a single adjoint problem.* On the other hand, the calculation of the gradient vector in Technique II, as given by equation (2.2.5.b), may require N additional solutions of the direct problem in order to compute the sensitivity coefficients by forward finite-differences (see equation 2.1.31.a).

The Iterative Procedure for Technique III

The iterative procedure of the conjugate gradient method, for the computation of the vector of unknown parameters **P**, is given by equations (2.2.2), (2.2.3) and (2.2.4.a,b). However, the gradient vector components are now computed by using equation (2.3.15), rather than equation (2.2.5.b).

The search step size β^k is chosen as the one that minimizes the function $S(\mathbf{P})$ at each iteration k, that is,

$$\min_{\beta^k} S(\mathbf{P}^{k+1}) = \min_{\beta^k} \int_{t=0}^{t_f} [Y(t) - T(x_{meas}, t; \mathbf{P}^k - \beta^k \mathbf{d}^k)]^2 \, dt \tag{2.3.16}$$

By linearizing the estimated temperature $T(x_{meas}, t; \mathbf{P}^k - \beta^k \mathbf{d}^k)$ with a Taylor series expansion and performing the above minimization, we find

$$\beta^k = \frac{\displaystyle\int_{t=0}^{t_f} [T(x_{meas}, t; \mathbf{P}^k) - Y(t)] \, \Delta T(x_{meas}, t; \mathbf{d}^k) \, dt}{\displaystyle\int_{t=0}^{t_f} [\Delta T(x_{meas}, t; \mathbf{d}^k)]^2 \, dt} \tag{2.3.17}$$

where $\Delta T(x_{meas}, t; \mathbf{d}^k)$ is the solution of the sensitivity problem given by equations (2.3.4), obtained by setting $\Delta P_j = d_j^k$, $j=1,\ldots,N$, in the computation of the function $\Delta g_p(t)$ given by equation (2.3.3.b). Further details on the derivation of equation (2.3.17) can be found in Note 5 at the end of this chapter. The reader should note that a single sensitivity problem is solved for the computation of β^k at each iteration, because the unknown function was parameterized in the form given by equation (2.3.3.a). Therefore, in the present approach the computation of β^k does not require the computation of the sensitivity matrix as in equation (2.2.7.c). For problems not involving the estimation of coefficients of trial functions, as in equation (2.3.3.a), the use of Techniques I or II may be more appropriate.

The Stopping Criterion for Technique III

As for Technique II, the stopping criterion for Technique III is based on the *Discrepancy Principle,* when the standard deviation σ of the measurements is *a priori* known. It is given by

$$S(\mathbf{P}) < \varepsilon \tag{2.3.18}$$

where $S(\mathbf{P})$ is now computed with equation (2.3.1). The tolerance ε is then obtained from equation (2.3.1) by assuming

$$\left|Y(t) - T(x_{meas}, t; \mathbf{P})\right| \approx \sigma \tag{2.3.19}$$

where σ is the standard-deviation of the measurement errors, which is assumed to be constant. Thus, the tolerance ε is determined as

$$\varepsilon = \sigma^2 t_f \tag{2.3.20}$$

For those cases involving measurements with unknown standard deviation, the alternative approach based on a additional measurement can be used for the stopping criterion, as illustrated in Note 4 at the end of this chapter.

The Computational Algorithm for Technique III

The computational algorithm for the conjugate gradient method with adjoint problem for parameter estimation can be summarized as follows. Suppose that temperature measurements $\mathbf{Y}=(Y_1,Y_2,...,Y_I)$ are given at times t_i , $i = 1, ..., I$, and an initial guess $\mathbf{P}^0$ is available for the vector of unknown parameters $\mathbf{P}$. Set k=0 and then

Step 1. Compute $g_p(t)$ according to equation (2.3.3.a) and then solve the direct problem given by equations (2.3.2) in order to compute $T(x, t)$.

Step 2. Check the stopping criterion given by equation (2.3.18). Continue if not satisfied.

Step 3. Knowing $T(x_{meas}, t)$ and the measured temperature $Y(t)$, solve the adjoint problem (2.3.10) to compute $\lambda(0.5, t)$.

Step 4. Knowing $\lambda(0.5, t)$, compute each component of the gradient vector $\nabla S(\mathbf{P})$ from equation (2.3.15).

Step 5. Knowing the gradient $\nabla S(\mathbf{P})$, compute γ^k from either equation (2.2.4.a) or (2.2.4.b), and then the direction of descent $\mathbf{d}^k$ from equation (2.2.3).

Step 6. By setting $\Delta\mathbf{P}^k = \mathbf{d}^k$, compute $\Delta g_p(t)$ from equation (2.3.3.b) and then solve the sensitivity problem given by equations (2.3.4) to obtain $\Delta T(x_{meas},t;\mathbf{d}^k)$.

Step 7. Knowing $\Delta T(x_{meas},t;\mathbf{d}^k)$, compute the search step size β^k from equation (2.3.17).

Step 8. Knowing β^k and $\mathbf{d}^k$, compute the new estimate $\mathbf{P}^{k+1}$ from equation (2.2.2). Replace k by k+1 and return to step 1.

The Use of Multiple Sensors

The above computational algorithm can also be applied, with few modifications, to cases where the readings of M sensors are available for the inverse analysis. In such cases, the objective function (2.3.1) is modified to

$$S(\mathbf{P}) = \sum_{m=1}^{M} \int_{t=0}^{t_f} [Y_m(t) - T(x_m,t;\mathbf{P})]^2 \, dt \tag{2.3.21}$$

where $Y_m(t)$ are the continuous measurements of the sensor located at x_m, for $m=1,\ldots,M$.

Since the objective function appears in the development of the adjoint problem, such a problem needs also to be modified in order to accommodate the readings of multiple sensors. It can be easily shown that the differential equation for the adjoint problem, equation (2.3.10.a), then becomes

$$\frac{\partial \lambda(x,t)}{\partial t} + \frac{\partial^2 \lambda(x,t)}{\partial x^2} + 2\sum_{m=1}^{M} [T(x_m,t;\mathbf{P}) - Y(t)]\delta(x - x_m) = 0$$

$$\text{in } 0 < x < 1, \text{ for } 0 < t < t_f \tag{2.3.22}$$

while the final and boundary conditions, equations (2.3.10.b-d), remain unaltered for multiple sensors.

The objective function also appears in the development of the search step size, equation (2.3.17), and of the tolerance for the stopping criterion, equation (2.3.20). For the readings of M sensors, such quantities are respectively obtained from the following expressions:

$$\beta^k = \frac{\sum_{m=1}^{M} \int_{t=0}^{t_f} [T(x_m,t;\mathbf{P}^k) - Y_m(t)] \Delta T(x_m,t;\mathbf{d}^k) \, dt}{\sum_{m=1}^{M} \int_{t=0}^{t_f} [\Delta T(x_m,t;\mathbf{d}^k)]^2 \, dt} \tag{2.3.23}$$

and

$$\varepsilon = M\sigma^2 t_f \tag{2.3.24}$$

There are many practical situations in which no information is available on the functional form of the unknown quantity. Therefore, it should not be parameterized as in equation (2.3.3.a), since wrong trial functions can be used in such process. Although general trial functions, such as B-Splines, could be used in the parameterization, there would still remain the question of how many trial functions need to be used for a correct approximation of the unknown quantity. For cases with no prior information on the functional form of the unknown quantity, the minimization of equation (2.3.1) should be preferably performed in a space of functions. In other words, in this section the function (2.3.1) was minimized in the space of all possible N parameters P_j. On the other hand, for a function estimation approach, equation (2.3.1) will be minimized in an infinite space of functions. In the next section we present the Conjugate Gradient Method as applied to the function estimation approach.

2-4 TECHNIQUE IV: THE CONJUGATE GRADIENT METHOD WITH ADJOINT PROBLEM FOR FUNCTION ESTIMATION

In this section we present a powerful iterative minimization scheme called the *Conjugate Gradient Method of Minimization with Adjoint Problem*, for solving inverse heat transfer problems of *function estimation*. In this approach, no *a priori* information on the functional form of the unknown function is available [9-21], except for the functional space which it belongs to.

To illustrate Technique IV, we consider the test-problem given by equations (2.3.2) for the estimation of the unknown time-varying strength $g_p(t)$ of a plane energy source, by using the transient readings of a single sensor located at x_{meas}. We assume that the unknown function belongs to the Hilbert space of square-integrable functions in the time domain [12,14,21], denoted as $L_2(0,t_f)$, where t_f is the duration of the experiment. Functions in such space satisfy the following property:

$$\int_{t=0}^{t_f} [g_p(t)]^2 dt < \infty \tag{2.4.1}$$

For some definitions and properties regarding Hilbert spaces, the reader is referred to Note 6 at the end of the chapter.

In order to solve the present function estimation problem, the functional $S[g_p(t)]$ defined as

$$S[g_p(t)] = \int_{t=0}^{t_f} \{Y(t) - T[x_{meas}, t; g_p(t)]\}^2 dt \tag{2.4.2}$$

is minimized under the constraint specified by the corresponding direct heat conduction problem.

The basic steps of Technique IV for the solution of function estimation problems, obtained through the minimization of functional (2.4.2), are very similar to the basic steps of Technique III for parameter estimation problems. They include:

- The Direct Problem
- The Inverse Problem
- The Sensitivity Problem
- The Adjoint Problem
- The Gradient Equation
- The Iterative Procedure
- The Stopping Criterion
- The Computational Algorithm

We now present some details for each of these distinct steps.

The Direct Problem

The direct problem involves the determination of the temperature field in the medium when the source term is known. The formulation of the direct problem is given by equations (2.3.2).

The Inverse Problem

In the inverse problem considered here, the source term $g_p(t)$ is an unknown function of time, while measured transient temperature data $Y(t)$, taken at the location x_{meas}, are available over the time domain $0 \leq t \leq t_f$, where t_f is the final time. However, differently from Technique III where $g_p(t)$ was parameterized by equation (2.3.3.a), no functional form is now *a priori* assumed for the source-term. The only assumption is that $g_p(t)$ belongs to the space $L_2(0,t_f)$. The *sensitivity function* $\Delta T(x,t)$ and the *Lagrange multiplier* $\lambda(x,t)$ are needed to implement the iterative procedure of Technique IV. Therefore, we need to develop two auxiliary problems, i.e., the *sensitivity problem* and the *adjoint problem*, in order to determine these two functions, as described below.

The Sensitivity Problem

The derivation of the sensitivity problem for Technique IV is very similar to that for Technique III. It is assumed that when $g_p(t)$ undergoes an increment $\Delta g_p(t)$, the temperature $T(x,t)$ changes by an amount $\Delta T(x,t)$. Therefore, we replace $T(x,t)$ by $[T(x,t)+\Delta T(x,t)]$ and $g_p(t)$ by $[g_p(t) + \Delta g_p(t)]$ in the direct problem (2.3.2) and subtract from it the original problem (2.3.2), in order to obtain the *sensitivity problem* given by equations (2.3.4). However, in Technique

IV the variation of the unknown function, $\Delta g_p(t)$, is not given in the parameterized form of equation (2.3.3.b). Such variation of $g_p(t)$ is now a general function in the space $L_2(0,t_f)$, as the unknown function itself. The choice of $\Delta g_p(t)$ will be described later in the analysis.

The Adjoint Problem

As for the sensitivity problem, the derivation procedure of the adjoint problem for Technique IV is very similar to the one for Technique III. To develop the *adjoint problem*, we introduce a *Lagrange multiplier* $\lambda(x,t)$. We multiply equation (2.3.2.a) by $\lambda(x,t)$, integrate the resulting expression over the spatial domain from $x = 0$ to $x = 1$, and then over the time domain from $t = 0$ to $t = t_f$. The expression obtained in this manner is added to the functional $S[g_p(t)]$ given by equation (2.4.2) in order to obtain the following extended functional

$$S[g_p(t)] = \int_{t=0}^{t_f} \{Y(t) - T[x_{meas}, t; g_p(t)]\}^2 dt + \int_{x=0}^{1} \int_{t=0}^{t_f} \lambda(x,t) \left[\frac{\partial^2 T}{\partial x^2} + g_p(t)\delta(x-0.5) - \frac{\partial T}{\partial t} \right] dt\, dx \tag{2.4.3}$$

which is the equivalent form of equation (2.3.5) for parameter estimation.

An expression for the variation $\Delta S[g_p(t)]$ of the functional $S[g_p(t)]$ can be developed by assuming that $T(x,t)$ is perturbed by $\Delta T(x,t)$ when $g_p(t)$ is perturbed by $\Delta g_p(t)$. The variation $\Delta S[g_p(t)]$ gives the directional derivative of $S[g_p(t)]$ in the direction of the perturbation $\Delta g_p(t)$ [12,21]. By replacing $T(x,t)$ by $[T(x,t)+\Delta T(x,t)]$, $g_p(t)$ by $[g_p(t)+\Delta g_p(t)]$ and $S[g_p(t)]$ by $\{S[g_p(t)]+\Delta S[g_p(t)]\}$ in equation (2.4.3), subtracting from the resulting expression the original equation (2.4.3), and neglecting second order terms, we obtain

$$\Delta S[g_p(t)] = \int_{t=0}^{t_f} \int_{x=0}^{1} 2\left\{ T\left[x,t;g_p(t)\right] - Y(t) \right\} \Delta T(x,t)\delta(x - x_{meas})\, dx\, dt + \int_{t=0}^{t_f} \int_{x=0}^{1} \lambda(x,t) \left[\frac{\partial^2 \Delta T}{\partial x^2} + \Delta g_p(t)\delta(x-0.5) - \frac{\partial \Delta T}{\partial t} \right] dx\, dt \tag{2.4.4}$$

where $\delta(.)$ is the Dirac delta function. Equation (2.4.4) is analogous to equation (2.3.6) for parameter estimation.

The second integral term on the right-hand side of equation (2.4.4) is simplified with integration by parts, and by utilizing the boundary and initial conditions of the sensitivity problem. The integral terms containing $\Delta T(x,t)$ in the resulting expression are then allowed to go to zero, in order to obtain the *adjoint problem* given by equations (2.3.10) for the determination of the Lagrange Multiplier $\lambda(x,t)$.

The Gradient Equation

In the limiting process used above to obtain the adjoint problem, the following term is left:

$$\Delta S[g_p(t)] = \int_{t=0}^{t_f} \lambda(0.5,t)\,\Delta g_p(t)\,dt \tag{2.4.5.a}$$

The reader should recall that in Technique III the parameterized form of $\Delta g_p(t)$, equation (2.3.3.b), was substituted into the equation above in order to obtain the components of the gradient vector given by equation (2.3.15). Such an approach cannot be used here, since we are now dealing with function estimation, rather than with parameter estimation as in Technique III. However, by invoking the hypothesis that the unknown function $g_p(t)$ belongs to the space of square-integrable functions in the domain $0 < t < t_f$, we can write [12,14,21]:

$$\Delta S[g_p(t)] = \int_{t=0}^{t_f} \nabla S[g_p(t)]\,\Delta g_p(t)\,dt \tag{2.4.5.b}$$

where $\nabla S[g_p(t)]$ is the gradient of the functional $S[g_p(t)]$.

From the comparison of equations (2.4.5.a) and (2.4.5.b), we conclude that

$$\nabla S[g_p(t)] = \lambda(0.5,t) \tag{2.4.6}$$

which is the *gradient equation* for the functional.

The Iterative Procedure for Technique IV

The mathematical development given above provides three distinct problems defined by equations (2.3.2), (2.3.4) and (2.3.10), called the *direct, sensitivity and adjoint problems*, for the computation of the functions $T(x,t)$, $\Delta T(x,t)$ and $\lambda(x,t)$, respectively. The measured data $Y(t)$ are considered available from a sensor located at x_{meas} and the gradient $\nabla S[g_p(t)]$ is given by equation (2.4.6).

The unknown function $g_p(t)$ is estimated through the minimization of the functional $S[g_p(t)]$ given by equation (2.4.2). This is achieved with an iterative procedure by proper selection of the direction of descent and of the step size in going from iteration k to $k+1$. The iterative procedure of the conjugate gradient method [9-21] for the estimation of the function $g_p(t)$ is given by:

$$g_p^{k+1}(t)=g_p^k(t)-\beta^k d^k(t) \tag{2.4.7}$$

where β^k is the *search step size* and $d^k(t)$ is the *direction of descent*, defined as

$$d^k(t)=\nabla S[g_p^k(t)]+\gamma^k d^{k-1}(t) \tag{2.4.8}$$

The *conjugation coefficient* γ^k can be computed either from the Polak-Ribiere [12,14] expression:

$$\gamma^k=\frac{\int_{t=0}^{t_f}\nabla S[g_p^k(t)]\{\nabla S[g_p^k(t)]-\nabla S[g_p^{k-1}(t)]\}\,dt}{\int_{t=0}^{t_f}\{\nabla S[g_p^{k-1}(t)]\}^2\,dt} \quad \text{for } k=1,2,\ldots \tag{2.4.9.a}$$

$$\text{with } \gamma^0=0 \quad \text{for } k=0$$

or from the Fletcher-Reeves [12,14,15] expression:

$$\gamma^k=\frac{\int_{t=0}^{t_f}\{\nabla S[g_p^k(t)]\}^2\,dt}{\int_{t=0}^{t_f}\{\nabla S[g_p^{k-1}(t)]\}^2\,dt} \quad \text{for } k=1,2,\ldots \tag{2.4.9.b}$$

$$\text{with } \gamma^0=0 \quad \text{for } k=0$$

The step size β^k is determined by minimizing the functional $S[g_p^{k+1}(t)]$ given by equation (2.4.2) with respect to β^k, that is,

$$\min_{\beta^k} S[g_p^{k+1}(t)] = \min_{\beta^k} \int_{t=0}^{t_f} \left\{ Y(t) - T\left[x_{meas}, t; g_p^k(t) - \beta^k d^k(t)\right] \right\}^2 dt$$

(2.4.10.a)

Then, by a Taylor series expansion equation (2.4.10.a) takes the form

$$\min_{\beta^k} S[g_p^{k+1}(t)] = \min_{\beta^k} \int_{t=0}^{t_f} \left\{ Y(t) - T\left[x_{meas}, t; g_p^k(t)\right] + \beta^k \Delta T\left[x_{meas}, t; d^k(t)\right] \right\}^2 dt$$

(2.4.10.b)

where $\Delta T\,[x_{meas}, t; d^k(t)]$ is the solution of the sensitivity problem given by equations (2.3.4), obtained by setting $\Delta g_p^k(t) = d^k(t)$. To minimize equation (2.4.10.b), we differentiate it with respect to β^k and set the resulting expression equal to zero. After some manipulations, the following expression is obtained for the step size β^k

$$\beta^k = \frac{\int_{t=0}^{t_f} \left\{ T\left[x_{meas}, t; g_p^k(t)\right] - Y(t) \right\} \Delta T\left[x_{meas}, t; d^k(t)\right] dt}{\int_{t=0}^{t_f} \left\{ \Delta T\left[x_{meas}, t; d^k(t)\right] \right\}^2 dt} \tag{2.4.11}$$

The reader should refer to Note 7 at the end of this chapter for more details on the derivation of the above expression for β^k. We note that equations (2.4.7-9) and (2.4.11) for function estimation are analogous to equations (2.2.2-4) and (2.3.17) for parameter estimation, respectively.

By examining equations (2.3.10.d) and (2.4.6), it can be noticed that the gradient equation is null at the final time t_f. Therefore, the initial guess used for $g_p(t)$ at $t = t_f$ is never changed by the iterative procedure of the conjugate gradient method for function estimation, given by equations (2.4.7-9,11). The estimated function can deviate from the exact solution in a neighborhood of t_f, if the initial guess used is too different from the exact $g_p(t_f)$. This apparent drawback of the method can be easily overcome by using a final time larger than that of interest, so that the effects of the initial guess are not noticeable in the time interval that the solution is sought. Another approach to overcome this difficulty is to repeat the solution of the inverse problem, by using as initial guess a previously estimated value for $g_p(t)$ in the neighborhood of t_f. Both approaches will be illustrated with examples later in the book.

The iterative procedure given by equations (2.4.7-9,11) is applied until a stopping criterion based on the *Discrepancy Principle* is satisfied, as described below.

The Stopping Criterion for Technique IV

Similarly to Techniques II and III, the stopping criterion based on the *Discrepancy Principle* gives the Conjugate Gradient Method of Function Estimation an iterative regularization character. The stopping criterion is given by

$$S[g_p(t)] < \varepsilon \tag{2.4.12}$$

where $S[g_p(t)]$ is computed with equation (2.4.2). The tolerance ε is chosen so that smooth solutions are obtained with measurements containing random errors. It is assumed that the solution is sufficiently accurate when

$$\left| Y(t) - T[x_{meas}, t; g_p(t)] \right| \approx \sigma \tag{2.4.13}$$

where σ is the standard deviation of measurement errors.

Thus, ε is obtained from equation (2.4.2) as

$$\varepsilon = \sigma^2 t_f \tag{2.4.14}$$

For cases involving errorless measurements, ε can be specified *a priori* as a sufficiently small number. For those cases involving measurements with unknown standard deviation, an alternative approach based on an additional measurement can be used, as described in Note 4 at the end of this chapter.

The Computational Algorithm for Technique IV

Suppose an initial guess $g_p^0(t)$ is available for the function $g_p(t)$. Set $k = 0$ and then:

Step 1. Solve the direct problem (2.3.2) and compute $T(x,t)$, based on $g_p^k(t)$.

Step 2. Check the stopping criterion (2.4.12). Continue if not satisfied.

Step 3. Knowing $T(x_{meas},t)$ and measured temperature $Y(t)$, solve the adjoint problem (2.3.10) and compute $\lambda(0.5, t)$.

Step 4. Knowing $\lambda(0.5, t)$, compute $\nabla S[g_p^k(t)]$ from equation (2.4.6).

Step 5. Knowing the gradient $\nabla S[g_p^k(t)]$, compute γ^k from either equations (2.4.9.a) or (2.4.9.b) and the direction of descent $d^k(t)$ from equation (2.4.8).

Step 6. Set $\Delta g_p^k(t) = d^k(t)$ and solve the sensitivity problem (2.3.4) to obtain $\Delta T[x_{meas},t;d^k(t)]$.

Step 7. Knowing $\Delta T[x_{meas},t;d^k(t)]$, compute the search step size β^k from equation (2.4.11).

Step 8. Knowing the search step size β^k and the direction of descent $d^k(t)$, compute the new estimate $g_p^{k+1}(t)$ from equation (2.4.7), and return to step 1.

The extension of the above algorithm to the use of multiple sensors is analogous to that described in the previous section for Technique III. It is a straightforward matter and will not be repeated here.

2-5 SOLUTION OF A TEST-PROBLEM

In the previous sections of this Chapter, we developed the relevant equations and introduced the computational algorithms for Techniques I, II, III and IV. In this section, we present the results obtained with such techniques as applied to the solution of our test-problem, involving the estimation of the strength of a plane heat source term.

As apparent from the analysis of figure 2.1.2, the problem of estimating the coefficients of polynomial trial functions used to approximate the unknown source term is quite difficult, due to the linear dependence of the sensitivity coefficients. Therefore, we consider here the source term to be approximated by Fourier series, where the trial functions are given in the form of equations (2.1.18.e,f). The duration of the experiment is taken as $t_f = 2$, since the rate of increase in $|\mathbf{J}^T\mathbf{J}|$ is strongly reduced for $t > 2$, as shown in figure 2.1.4 for a case involving 5 unknown parameters. During the time interval $0 < t \le 2$, we consider available for the inverse analysis 100 transient measurements of a single sensor located at $x_{meas}=1$. Techniques I, II and III are applied to the estimation of the coefficients of the trial functions (2.1.18.e,f), while Technique IV is applied to the estimation of the source term function itself, by assuming that no information regarding its functional form is available.

We use simulated measurements in the forthcoming analysis, as described next.

Simulated Measurements

Simulated measurements are obtained from the solution of the Direct Problem at the sensor location, by using *a priori* prescribed values for the unknown parameters or functions.

Consider, as an example, that 5 trial functions are to be used in the analysis. Hence, the number of parameters to be estimated is N=5. Also, consider for generating the simulated measurements that the five parameters are equal to 1, that is, $P_1=P_2=P_3=P_4=P_5=1$. By using the trial functions (2.1.18.e,f), the source term function is then given by

$$g_p(t) = 1 + \sin \pi t + \cos \pi t + \sin 2\pi t + \cos 2\pi t \tag{2.5.1}$$

The solution of the direct problem (2.1.1) at the measurement location $x_{meas}=1$, by using the source term given by equation (2.5.1), provides the exact (errorless) measurements $Y_{ex}(t_i)$, $i = 1, \ldots, I$. Measurements containing random errors are simulated by adding an error term to $Y_{ex}(t_i)$ in the form:

$$Y(t_i) = Y_{ex}(t_i) + \omega\sigma \tag{2.5.2}$$

where $Y(t_i)$ = simulated measurements containing random errors
$Y_{ex}(t_i)$ = exact (errorless) simulated measurements
σ = standard deviation of the measurement errors
ω = random variable with normal distribution, zero mean and unitary standard deviation. For the 99% confidence level we have $-2.576 < \omega < 2.576$. This variable can be generated with the subroutine DRRNOR of the IMSL [5].

With the use of such simulated measurements as the input data for the inverse analysis, we expect the solution of the estimation problem to be $P_1=P_2=P_3=P_4=P_5=1$, if Techniques I, II or III of parameter estimation are utilized; or the function given by equation (2.5.1) itself, if Technique IV of function estimation is utilized. We note that the stability of the inverse problem solution can be examined for various levels of measurement errors, by generating measurements with different standard deviations σ and by comparing the estimated quantities with those used to generate the simulated measurements.

Solution

We now consider the inverse problem of estimating the parameters $P_1=P_2=P_3=P_4=P_5=1$ of the function (2.5.1) by Techniques I, II and III, and the estimation of the function itself by Technique IV. The IMSL [5] version of the Levenberg-Marquardt method in the form of subroutine DBCLSJ was used for Technique I. The other techniques were programmed in FORTRAN, in accordance with the computational algorithms described above. The direct, sensitivity and adjoint problems were solved with finite-volumes by using an implicit discretization in time. The spatial domain $0 \le x \le 1$ was discretized with 100 volumes, while 100 time-steps were used to advance the solutions from $t = 0$ to $t_f = 2$. The sensitivity coefficients, needed for the solutions with Techniques I and II, were evaluated with finite-differences by utilizing the forward approximation of equation (2.1.31.a) with $\varepsilon=10^{-6}$.

We note that the computational algorithms of Techniques I and II as presented above could be simplified, since the present parameter estimation problem is linear, as shown by equation (2.1.18.c). In this case, the sensitivity matrix could be computed only once because it is constant, instead of being recomputed every iteration as suggested in the algorithms. However, for the sake of generality we preferred to use the computational algorithms as presented above in sections 2.1 and 2.2, instead of their simplified versions for linear cases. For the same reason, we preferred to use a numerical method of solution for the direct, sensitivity and adjoint problems, as well as for the computation of the sensitivity coefficients.

The initial guesses for the unknown parameters and for the unknown function were taken as zero, that is,

$$P_1^0 = P_2^0 = P_3^0 = P_4^0 = P_5^0 = 0 \qquad \text{for Techniques I, II and III}$$

and

$$g_p^0(t) = 0 \qquad \text{in } 0 \le t \le t_f \text{ , for Technique IV}$$

Since the gradient equation is null at the final time for Technique IV, the initial guess used for $g_p(t_f)$ is never changed by the iterative procedure, generating instabilities on the solution in the neighborhood of t_f. One approach to overcome such difficulties is to consider a final time larger than that of interest. We illustrate such an approach by considering for Technique IV t_f= 2.0, 2.2 and 2.4. The number of measurements and number of time-steps were increased accordingly in such cases.

Table 2.5.1 presents the results obtained for the estimated parameters, RMS errors, CPU time and number of iterations for Techniques I, II, III and IV. The computer used was a Pentium 166 Mhz with 32 Mbytes of RAM memory. Two different levels of measurement errors considered for the analysis included σ = 0 (errorless) and $\sigma = 0.01T_{max}$, where T_{max} is the maximum measured temperature. The RMS error is defined here as

$$e_{RMS} = \sqrt{\frac{1}{I}\sum_{i=1}^{I}[g_{est}(t_i) - g_{ex}(t_i)]^2} \tag{2.5.3}$$

where $g_{est}(t_i)$ is the estimated source term function at time t_i,
$g_{ex}(t_i)$ is the exact source term function (used to generate the simulated measurements) at time t_i and
I is the number of measurements.

Table 2.5.1. Results obtained with parameter and function estimation. Source term function given by equation (2.5.1).

Technique	σ	Estimated Parameters P_1, P_2 P_3, P_4 P_5	RMS error	CPU Time (s)	Iterations
I	0.0	1.000, 1.000, 1.000, 1.000, 1.000	0.0	0.11	2
	$0.01T_{max}$	0.999, 1.003, 0.997, 1.004, 1.009	0.008	0.17	5
II	0.0	1.000, 1.000, 1.000, 1.000, 1.000	0.0	0.33	10
	$0.01T_{max}$	0.968, 1.020, 0.918, 1.130, 0.894	0.139	0.11	5
III	0.0	1.000, 1.000, 1.000, 1.000, 1.000	0.0	0.93	26
	$0.01T_{max}$	0.981, 1.016, 0.949, 1.065, 0.916	0.087	0.28	8
IV, $t_f = 2$	0.0	-	0.476	1.37	101
	$0.01T_{max}$	-	0.553	0.17	12
IV, $t_f = 2.2$	0.0	-	0.054	1.48	101
	$0.01T_{max}$	-	0.157	0.27	17
IV, $t_f = 2.4$	0.0	-	0.042	2.59	101
	$0.01T_{max}$	-	0.157	0.17	11

Let us consider first in the analysis of table 2.5.1, Techniques I, II and III of parameter estimation. Table 2.5.1 shows that the exact values $P_1=P_2=P_3=P_4=P_5=1$ were recovered with these 3 techniques, when errorless measurements ($\sigma = 0$) were used. In such cases, we had the smallest number of iterations and the smallest computational time for Technique I. For cases involving measurement errors ($\sigma = 0.01T_{max}$), the smallest RMS error was also obtained with Technique I; but the smallest CPU time was obtained with Technique II, which had the largest RMS error.

The foregoing analysis reveals that Technique I, among those examined for parameter estimation, provided the best results in the estimation of the five coefficients of the Fourier series utilized to approximate the source term function. Technique I had the smallest CPU time for errorless measurements ($\sigma = 0$) and the smallest RMS error for measurements with random errors ($\sigma = 0.01T_{max}$). The reader must be aware that such conclusion is not general and should not be extended directly to other problems of parameter estimation. The results may depend on the physical character of the problem, number of parameters to be estimated, initial guess, etc [28]. In fact, the computational times and the RMS errors shown in table 2.5.1 were quite small for all cases considered, as a result of the simplicity of the present test-problem.

Table 2.5.1 shows that the number of iterations and the CPU time decreased for Techniques II and III, when measurements containing random errors were used instead of errorless measurements. This is because of the discrepancy principle used to obtain the tolerance for the stopping criterion, when measurements with errors were used in the analysis. The value obtained with equation (2.2.10) for the tolerance in the stopping criterion of Technique II was 0.032. For Technique III, the tolerance obtained with equation (2.3.20) was 0.00064. On the other hand, a much smaller value could be prescribed for the tolerance when errorless measurements were used in the analysis, since the solution was not affected by the measurement errors. For the results shown in table 2.5.1, we prescribed the tolerances of 10^{-9} and $2\text{x}10^{-11}$ for techniques II and III, respectively. Such values were not set identical because of the different definitions of the objective function for techniques II and III (see equations 2.2.1 and 2.3.1, respectively). We note that the tolerances ε_1, ε_2 and ε_3 appearing in equations (2.1.14) for Technique I are set internally by the subroutine DBCLSJ of the IMSL [5].

We note in table 2.5.1 the larger number of iterations for Technique III, as compared to Technique II, for both cases of errorless measurements and measurements with random errors. This is probably due to the different calculations that are performed with both techniques, in order to compute the gradient equation $\nabla S(\mathbf{P}^k)$ and the search step size β^k. While the computation of these two quantities with equations (2.2.5.a) and (2.2.7) in Technique II involves the sensitivity matrix, the expressions for $\nabla S(\mathbf{P}^k)$ and β^k for Technique III, equations (2.3.15) and (2.3.17), respectively, involve the solutions of the adjoint and sensitivity problems.

Figures 2.5.1-3 present the results for the source term function, obtained with Techniques I, II and III, respectively. These figures clearly show the better results obtained with Technique I when measurements with random errors were used in the analysis, although the results obtained with techniques II and III were also quite good.

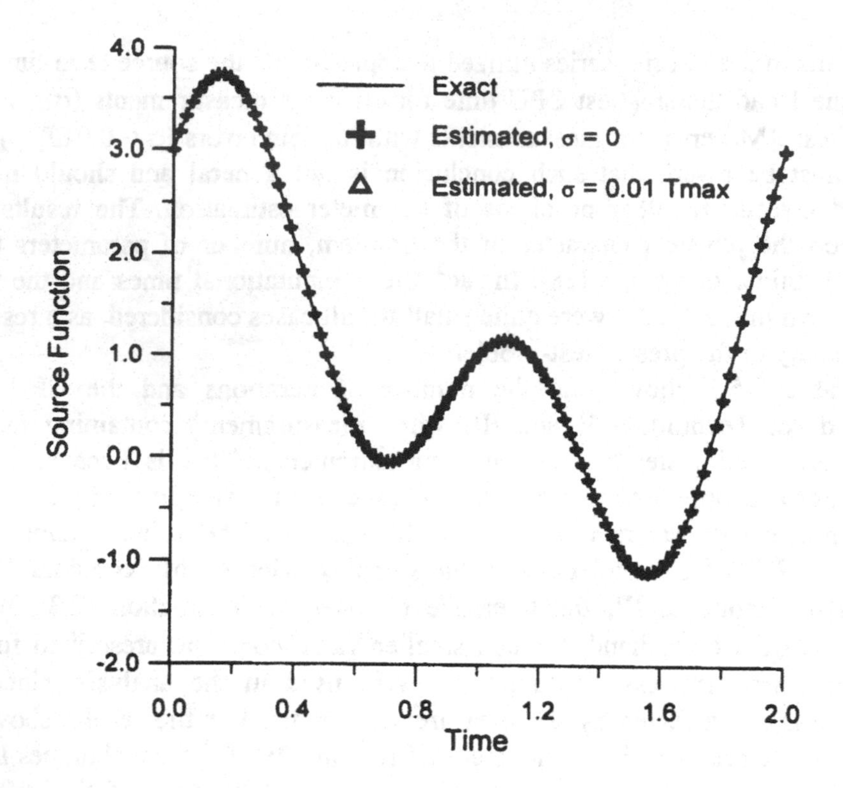

Figure 2.5.1. Estimation of the source term given by equation (2.5.1) with Technique I

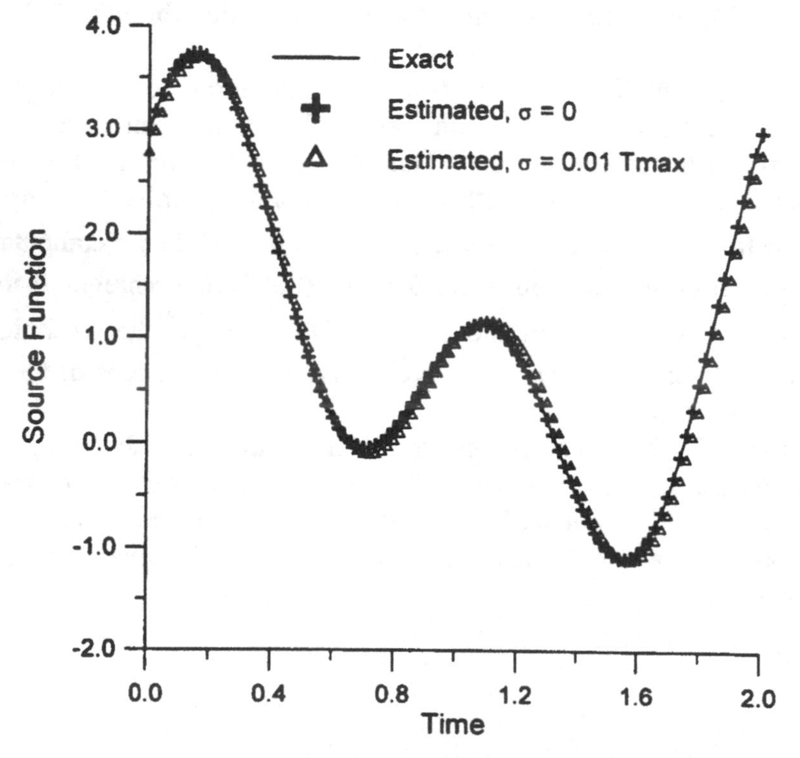

Figure 2.5.2. Estimation of the source term given by equation (2.5.1) with Technique II

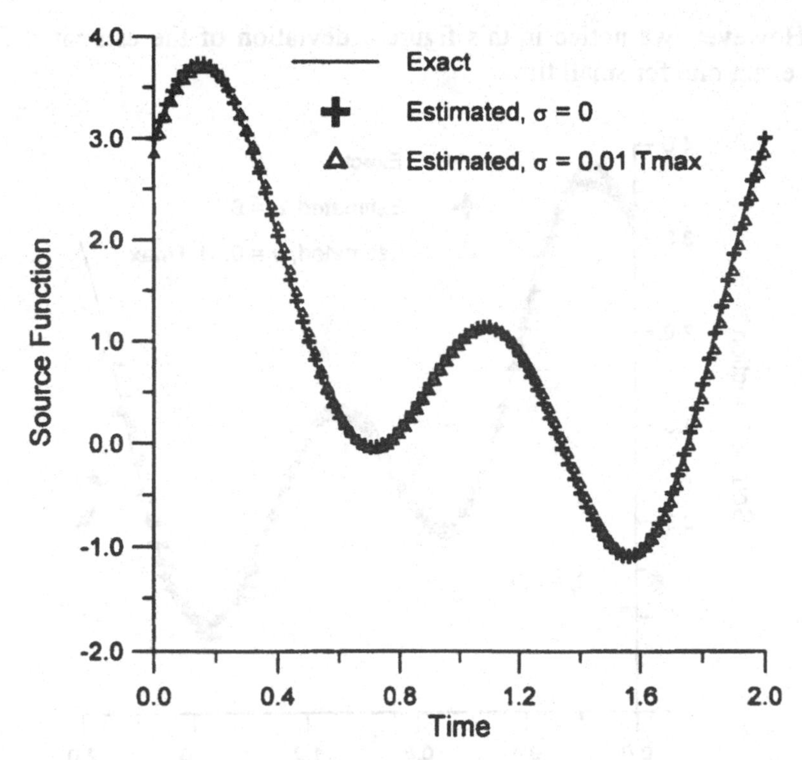

Figure 2.5.3. Estimation of the source term given by equation (2.5.1) with Technique III

After discussing the solution of the inverse problem of estimating the source term function given by equation (2.5.1) as a parameter estimation approach, by using techniques I, II and III, let us consider now the results obtained with the function estimation approach of Technique IV. Table 2.5.1 shows a large reduction on the RMS errors for both errorless measurements and measurements with random errors, when the final time was increased from $t_f = 2$ to $t_f = 2.2$. Such a reduction on the RMS errors is due to the effects of the initial guess on the solution, because of the null gradient at the final time. The RMS errors were computed for $0 \le t \le 2$, for both cases involving $t_f = 2$ and $t_f = 2.2$, since this is our time domain of interest. We also note in table 2.5.1 that the RMS errors were very little affected when t_f was increased from 2.2 to 2.4.

Figures 2.5.4.a-c show the results obtained with Technique IV for final times of 2.0, 2.2 and 2.4, respectively. The deviation of the estimated function from the exact one in the neighborhood of t_f, caused by the null gradient at t_f, is apparent in figure 2.5.4.a. Note in this figure that the estimated function is zero for $t = t_f$, which is exactly the initial guess used for the iterative procedure of Technique IV. As t_f was increased to 2.2, the effects caused by the null gradient at the final time are practically not noticeable in the time domain of interest, $0 \le t \le 2$, as can be seen in figure 2.5.4.b. In fact, quite accurate estimates were obtained in this case with errorless measurements, as well as with measurements containing random errors. The solution in the neighborhood of $t_f = 2$ can be further improved by increasing the final time from 2.2 to 2.4, as apparent in figure

2.5.4.c. However, we notice in this figure a deviation of the estimated function from the exact one for small times.

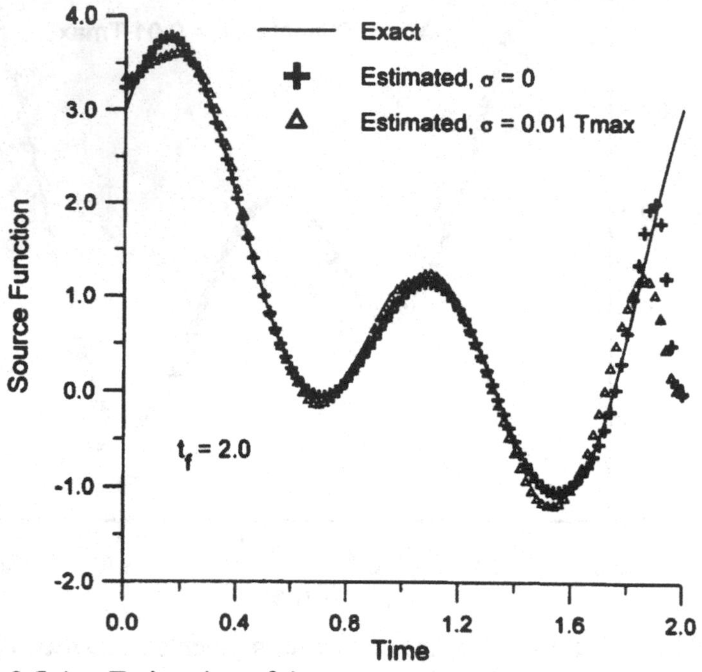

Figure 2.5.4.a. Estimation of the source term given by equation (2.5.1) with Technique IV for $t_f = 2.0$

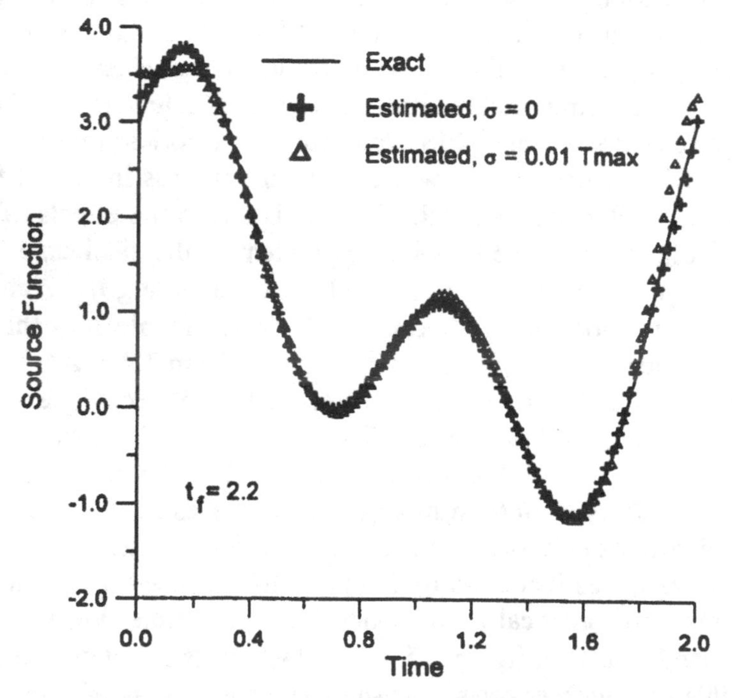

Figure 2.5.4.b. Estimation of the source term given by equation (2.5.1) with Technique IV for $t_f = 2.2$

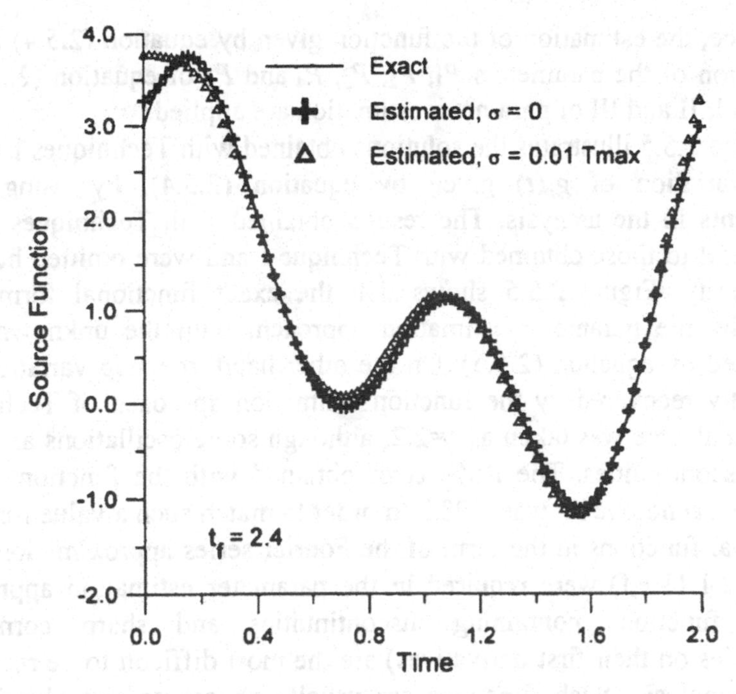

Figure 2.5.4.c. Estimation of the source term given by equation (2.5.1) with Technique IV for t_f=2.4

It is interesting to note in table 2.5.1 that generally more accurate results were obtained with parameter estimation (Techniques I, II and III) rather than with function estimation (Technique IV). However, such parameter estimation results were based on the *a priori* available information that the function could be exactly approximated by 5 trial functions in the form given by equations (2.1.18.e,f). Unfortunately, this is not generally the case. In several applications there is no prior information regarding the functional form of the unknown. In such cases, the use of parameter estimation approach can yield completely wrong solutions, because either wrong trial functions or an insufficient number of them can be chosen to approximate the unknown function. As an example, consider a step variation for the source term in the form

$$g_p(t) = \begin{cases} 1 \ , & \text{for} \quad t < 2/3 \quad \text{and} \quad t > 4/3 \\ 2 \ , & \text{for} \quad 2/3 \le t \le 4/3 \end{cases} \tag{2.5.4}$$

Also, consider for the parameter estimation approach that the unknown function is approximated by 5 trial functions in the form of equations (2.1.18.e,f), that is,

$$g_p(t) = P_1 + P_2 \sin \pi t + P_3 \cos \pi t + P_4 \sin 2\pi t + P_5 \cos 2\pi t \tag{2.5.5}$$

Hence, the estimation of the function given by equation (2.5.4) reduces to the estimation of the parameters P_1, P_2, P_3, P_4 and P_5 of equation (2.5.5), when Techniques I, II and III of parameter estimation are applied.

Figure 2.5.5 illustrates the solutions obtained with Techniques I and IV for the step variation of $g_p(t)$ given by equation (2.5.4), by using errorless measurements in the analysis. The results obtained with Techniques II and III were identical to those obtained with Technique I and were omitted here for the sake of clarity. Figure 2.5.5 shows that the exact functional form was not recovered by the parameter estimation approach, with the unknown function approximated by equation (2.5.5). On the other hand, the step variation of $g_p(t)$ was correctly recovered by the function estimation approach of Technique IV, when the final time was taken as t_f=2.2, although some oscillations are observed near the discontinuities. The RMS error obtained with the function estimation approach of Technique IV was 0.085. In order to match such a value for the RMS error, 30 trial functions in the form of the Fourier series approximation given by equations (2.1.18.e,f) were required in the parameter estimation approach. We note that functions containing discontinuities and sharp corners (i.e., discontinuities on their first derivatives) are the most difficult to be recovered by an inverse analysis. Such functions are usually chosen to test algorithms and methods of solution for inverse problems.

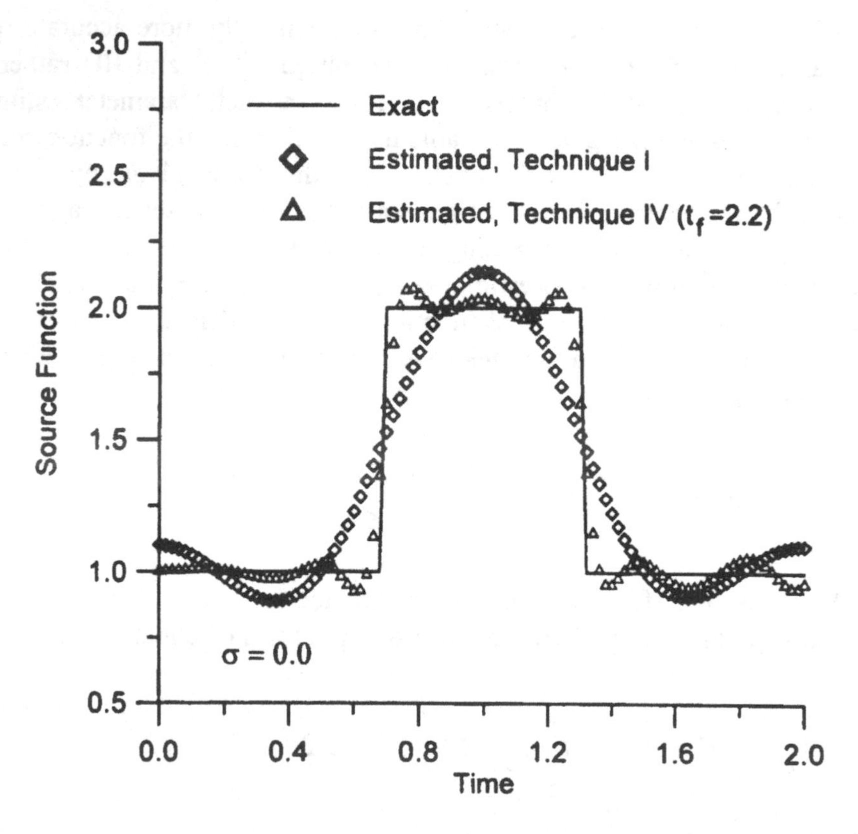

Figure 2.5.5. Inverse problem solutions for a step variation of the source function

The foregoing analysis reveals that the parameter estimation approach should only be used when there is available sufficient information regarding the functional form of the unknown. If such is not the case, a function estimation approach should be applied to the solution of the inverse problem.

In this chapter we developed the basic steps and algorithms of four powerful techniques of solution of inverse problems. The remaining chapters of this book are devoted to the application of such techniques to the solution of inverse problems involving different heat transfer modes.

PROBLEMS

2-1 Prove that, for linear parameter estimation problems, the vector of estimated temperatures can be written as $\mathbf{T} = \mathbf{JP}$, where $\mathbf{J}$ is the sensitivity matrix and $\mathbf{P}$ is the vector of parameters.

2-2 Use the relation $\mathbf{T} = \mathbf{JP}$ to derive equation (2.1.9) for linear parameter estimation problems.

2-3 Show that the linearization of the estimated temperatures $\mathbf{T(P)}$ around the vector of parameters $\mathbf{P}^k$ at iteration k, can be written in the form given by equation (2.1.10).

2-4 Derive equation (2.1.11).

2-5 Calculate the sensitivity coefficients presented in figures 2.1.2 and 2.1.3 by using forward and central finite-difference approximations, instead of using the analytical expression given by equation (2.1.18.c). How do the sensitivity coefficients calculated numerically by finite-differences compare to those calculated analytically in terms of accuracy and computational time? What is the effect of the factor ε, appearing in equations (2.1.31.a,b), on the accuracy of the finite-difference approximations?

2-6 Derive equation (2.1.34.a).

2-7 Derive the sensitivity problem given by equations (2.3.4).

2-8 Derive equation (2.3.6).

2-9 A semi-infinite medium initially at the zero temperature, has the temperature at the surface $x = 0$ suddenly changed to a constant value T_0. The formulation of such heat conduction problem is given by:

$$C\frac{\partial T}{\partial t} = k\frac{\partial^2 T}{\partial x^2} \qquad \text{for } x > 0 \quad \text{and} \quad t > 0$$

$$T = T_0 \qquad \text{at } x = 0 \quad \text{for} \quad t > 0$$

$$T = 0 \qquad \text{for } t = 0 \quad \text{and} \quad x > 0$$

Examine the transient variation of the sensitivity coefficients with respect to the volumetric heat capacity $C = \rho c_p$ and to the thermal conductivity k, for sensors located at different depths below the surface. Is the simultaneous estimation of C and k possible? What is the behavior of $|\mathbf{J}^T\mathbf{J}|$?

2-10 Repeat problem 2-9 for the surface at $x = 0$ subjected to a constant heat flux q_0, instead of being maintained at the constant temperature T_0. In this case the formulation of the heat conduction problem is given by:

$$C\frac{\partial T}{\partial t} = k\frac{\partial^2 T}{\partial x^2} \qquad \text{for } x > 0 \quad \text{and} \quad t > 0$$

$$-k\frac{\partial T}{\partial x} = q_0 \qquad \text{at } x = 0 \quad \text{for} \quad t > 0$$

$$T = 0 \qquad \text{for } t = 0 \quad \text{and} \quad x > 0$$

2-11 By using the formulation of either problem 2-9 or problem 2-10 (whichever is more appropriate) estimate simultaneously k and C with Techniques I and II. Use $C = k = 1$ and $T_0 = 1$ (or $q_0 = 1$), in order to generate the simulated measurements of a single sensor for the analysis. Examine the effects of random measurement errors, initial guess and sensor location on the estimated parameters. Is such parameter estimation problem linear or nonlinear?

2-12 For the physical situation of problem 2-10, consider k and C known and q_0 unknown. Examine the transient variation of the sensitivity coefficients with respect to q_0 for sensors located at different depths below the surface. Use $C = k = q_0 = 1$ in order to generate the simulated measurements of a single sensor for the analysis. Thus, use such measurements to estimate q_0 by using Techniques I and II. Examine the effects of random measurement errors, initial guess and sensor location on the estimated heat flux. Is it possible to estimate simultaneously k and/or C together with q_0?

2-13 Consider the following heat conduction problem in dimensionless form:

$$\frac{\partial T}{\partial t} = \frac{\partial^2 T}{\partial x^2} \qquad \text{in } 0 < x < 1 \quad \text{for } t > 0$$

$$\frac{\partial T}{\partial x} = 0 \qquad \text{at } x = 0 \quad \text{for } t > 0$$

$$\frac{\partial T}{\partial x} = q(t) \qquad \text{at } x = 1 \quad \text{for } t > 0$$

$$T = 0 \qquad \text{for } t = 0 \quad \text{in } 0 < x < 1$$

Formulate all the steps for the solution of the inverse problem of estimating the unknown heat flux $q(t)$, by using Techniques I, II and III. Consider available the transient readings Y_i, $i = 1, \ldots, I$ of a single sensor located at x_{meas}. Also, assume that $q(t)$ is given in the following general linear parametric form

$$q(t) = \sum_{j=1}^{N} P_j C_j(t)$$

where P_j are the unknown parameters and $C_j(t)$ are known trial functions.

2-14 Is the inverse problem involving the estimation of P_j in problem 2-13 linear or nonlinear?

2-15 Consider $q(t)$ in problem 2-13 to be approximated by 3 trial functions in the form of a polynomial, as given by equation (2.1.18.d). Examine the transient variation of the sensitivity coefficients with respect to the parameters P_j, $j = 1, 2, 3$, for a sensor located at $x_{meas} = 0$. Is the estimation of such parameters possible? What is the behavior of $|\mathbf{J}^T\mathbf{J}|$?

2-16 Consider $q(t)$ in problem 2-13 to be approximated by 3 trial functions in the form of a Fourier series, as given by equations (2.1.18.e,f). Examine the transient variation of the sensitivity coefficients with respect to the parameters P_j, $j = 1, 2, 3$, for a sensor located at $x_{meas} = 0$. Is the estimation of such parameters possible? What is the behavior of $|\mathbf{J}^T\mathbf{J}|$?

2-17 Repeat problem 2-13 by now assuming available the transient readings of M sensors located at $x = x_m$, $m = 1, \ldots, M$.

2-18 Repeat problem 2-13 by using the function estimation approach of Technique IV, where no information regarding the functional form of $q(t)$ is assumed available for the analysis.

REFERENCES

1. Levenberg, K., "A Method for the Solution of Certain Non-linear Problems in Least Squares", *Quart. Appl. Math.*, 2, 164-168, 1944.
2. Marquardt, D. W., "An Algorithm for Least Squares Estimation of Nonlinear Parameters", *J. Soc. Ind. Appl. Math*, 11, 431-441, 1963.
3. Bard, Y. B., *Nonlinear Parameter Estimation*, Acad. Press, New York, 1974.
4. Beck, J. V. and Arnold, K. J., *Parameter Estimation in Engineering and Science*, Wiley, New York, 1977.
5. IMSL Library, Edition 10, 1987, NBC Bldg., 7500 Ballaire Blvd., Houston, Texas.
6. Press, W. H., Flannery, B. F., Teukolsky S. A. and Wetterling W. T., *Numerical Recipes*, Cambridge University Press, New York, 1989.
7. Dennis, J. and Schnabel, R., *Numerical Methods for Unconstrained Optimization and Nonlinear Equations*, Prentice Hall, 1983.
8. Moré, J. J., "The Levenberg-Marquardt Algorithm: Implementation and Theory", *in Numerical Analysis, Lecture Notes in Mathematics, Vol. 630*, G. A. Watson (ed.), Springer-Verlag, Berlin, 105-116, 1977.
9. Özisik, M. N., *Heat Conduction*, 2nd ed., Wiley, New York, 1993.
10. Alifanov, O. M., "Determination of Heat Loads from a Solution of the Nonlinear Inverse Problem," *High Temperature*, 15, 498-504, 1977.

11. Alifanov, O. M. and Artyukhin, E. A., "Regularized Numerical Solutions of Nonlinear Inverse Heat Conduction Problem", *J. Eng. Phys.*, **29**, 934-938, 1975.
12. Alifanov, O. M., *Inverse Heat Transfer Problems*, Springer-Verlag, New York, 1994.
13. Alifanov, O. M., "Solution of an Inverse Problem of Heat-Conduction by Iterative Methods", *J. Eng. Phys.*, **26**, 471-476, 1974.
14. Daniel, J. W., *The Approximate Minimization of Functionals*, Prentice-Hall Inc., Englewood Cliffs, 1971.
15. Fletcher, R. and Reeves C. M., "Function Minimization by Conjugate Gradients," *Computer J.*, **7**, 149-154, 1964.
16. Alifanov, O. M., "Inverse Boundary Value Problems of Heat Conduction", *J. Eng. Phys.*, **25**, 1975.
17. Hestenes, M. R. and Stiefel, E., "Methods of Conjugate Gradients for Solving Linear Systems", *J. Res. NBS*, **49**, 409-436, 1952.
18. Beckman, F. S., "The Solution of Linear Equation by Conjugate Gradient Method", *Mathematical Methods for Digital Computer*, A. Ralston and H. S. Wilf (eds.) chap. 4, Wiley, New York, 1960.
19. Kammerer, W. I. and Nashed, M. Z., "On the Convergence of the Conjugate Gradient Method for Singular Linear Operator Equations", *J. Num. Anal.*, **9**, 165-181, 1972.
20. Huang, C. H. and Özisik, M. N., "Inverse Problem of Determining the Unknown Strength of an Internal Plane Heat Source", *J. Franklin Institute*, 751-764, 1992.
21. Jarny, Y., Özisik, M. N. and Bardon J. P., "A General Optimization Method Using Adjoint Equation for Solving Multidimensional Inverse Heat Conduction," *Int. J. Heat and Mass Transfer*, **34**, 2911-2919, 1991.
22. Taktak, R., *Design and Validation of Optimal Experiments for Estimating Thermal Properties of Composite Materials*, Ph.D. Thesis, Dept. of Mech. Eng., Michigan State University, 1992.
23. Box, G. E. P., Hunter, W. G. and Hunter, J. S., *Statistics for Experimenters*, Wiley, New York, 1978.
24. Artyukhin, E. A., "Optimum Planning of Experiments in the Identification of Heat Transfer Processes", *J. Eng. Phys.*, 256-259, 1989.
25. Fadale, T. D., Nenarokomov, A. V. and Emery, A. F., "Two Approaches to Optimal Sensor Locations", *J. Heat Transfer*, **117**, 373-379, 1995.
26. Mikhailov, V. V., "Arrangement of the Temperature Measurement Points and Conditionally of Inverse Thermal Conductivity Problems", *J. Eng. Phys.*, 1369-1373, 1990.
27. Kardestunner, H. and Horrie, D. (eds.), *Finite Element Handbook*, McGraw Hill, New York, 1987.
28. Bokar, J. and Özisik, M. N., "An Inverse Problem for the Estimation of Radiation Temperature Source Term in a Sphere", *Inv. Problems in Engineering*, **1**, 191-205, 1995.

NOTE 1. STATISTICAL ANALYSIS FOR PARAMETER ESTIMATION

By performing a statistical analysis it is possible to assess the accuracy of $\hat{P}_j, j = 1, \ldots, N$, which are the values estimated for the unknown parameters P_j, $j = 1, \ldots, N$. Assuming that the eight statistical assumptions discussed in section 1-4 are valid, and using the minimization of the ordinary least-squares norm for solving the parameter estimation problem, the *covariance matrix*, **V**, of the estimated parameters $\hat{P}_j, j = 1, \ldots, N$, is given by [4]

$$\mathbf{V} \equiv \begin{bmatrix} \text{cov}(\hat{P}_1,\hat{P}_1) & \text{cov}(\hat{P}_1,\hat{P}_2) & \ldots & \text{cov}(\hat{P}_1,\hat{P}_N) \\ \text{cov}(\hat{P}_2,\hat{P}_1) & \text{cov}(\hat{P}_2,\hat{P}_2) & \ldots & \text{cov}(\hat{P}_2,\hat{P}_N) \\ . & . & . & . \\ . & . & . & . \\ \text{cov}(\hat{P}_N,\hat{P}_1) & \text{cov}(\hat{P}_N,\hat{P}_2) & \ldots & \text{cov}(\hat{P}_N,\hat{P}_N) \end{bmatrix} = (\mathbf{J}^T\mathbf{J})^{-1}\sigma^2 \tag{N1.2.1}$$

where **J** is the sensitivity matrix and σ is the standard deviation of the measurement errors, which is assumed to be constant.

The *standard deviations* for the estimated parameters can thus be obtained from the diagonal elements of **V** as

$$\sigma_{\hat{P}_j} \equiv \sqrt{\text{cov}(\hat{P}_j,\hat{P}_j)} = \sqrt{V_{jj}} \qquad \text{for } j = 1, \ldots, N \tag{N1.2.2.a}$$

where V_{jj} is the jth element in the diagonal of **V**. More explicitly, we can write:

$$\sigma_{\hat{P}_j} = \sigma\sqrt{\left[\mathbf{J}^T\mathbf{J}\right]^{-1}_{jj}} \qquad \text{for } j = 1, \ldots, N \tag{N1.2.2.b}$$

Confidence intervals at the 99% confidence level for the estimated parameters are obtained as

$$\hat{P}_j - 2.576\sigma_{\hat{P}_j} \le P_j \le \hat{P}_j + 2.576\sigma_{\hat{P}_j} \qquad \text{for } j = 1, \ldots, N \tag{N1.2.3}$$

The factor 2.576 appearing in the expression above comes from table N1.1.1 in Chapter 1 for the normal distribution, so that the probability of the actual parameter P_j be in the interval $\hat{P}_j \pm 2.576\sigma_{\hat{P}_j}$ is 99%. For other confidence levels, this factor is changed accordingly. For example, for the 95% confidence level, 2.576 should be replaced by 1.96.

Confidence intervals do not provide a good approximation for a *joint confidence region* for the estimated parameters. In fact, the confidence interval is obtained for each parameter, regardless the estimation of the other parameters. Confidence regions built from confidence intervals may include areas outside the actual confidence region and not include areas that belong to the actual confidence region [4,23].

The *joint confidence region* for the estimated parameters is given by [4,23]:

$$(\hat{\mathbf{P}}-\mathbf{P})^T \mathbf{V}^{-1}(\hat{\mathbf{P}}-\mathbf{P}) \le \chi_N^2 \tag{N1.2.4}$$

where $\mathbf{V}$ = covariance matrix of the estimated parameters, given by equation (N1.2.1)

$\hat{\mathbf{P}} = [\hat{P}_1, \hat{P}_2, ..., \hat{P}_N]$ is the vector with the values estimated for the parameters

$\mathbf{P} = [P_1, P_2, ..., P_N]$ is the vector of unknown parameters

N = the number of parameters

χ_N^2 = value of the chi-square distribution with N degrees of freedom for a given probability, obtained from table N1.1.2.

The confidence region given by equation (N1.2.4) is thus the interior of a hyperellipsoid centered at the origin and with coordinates $(\hat{P}_1 - P_1), (\hat{P}_2 - P_2), \cdots, (\hat{P}_N - P_N)$. The surface of the hyperellipsoid is a constant probability density surface, obtained from the chi-square distribution for a chosen confidence level (probability). For a case involving the estimation of two parameters, the values of χ_2^2 obtained from table N1.1.2 in Chapter 1 for the 95% and 99% confidence levels are 5.99 and 9.21, respectively.

NOTE 2. DESIGN OF OPTIMUM EXPERIMENTS

a. Parameter Estimation

Optimum experiments are usually designed by minimizing the hypervolume of the confidence region of the estimated parameters, in order to ensure minimum variance for the estimates. The minimization of the confidence region given by equation (N1.2.4) can be obtained by maximizing the determinant of $\mathbf{V}^{-1}$, in the so called *D-optimum design* [4,12,22,24,25]. Since the covariance matrix $\mathbf{V}$ is given by equation (N1.2.1), we can then *design optimum experiments by maximizing the determinant of the matrix* $\mathbf{J}^T\mathbf{J}$. Therefore, experimental variables such as the duration of the experiment, location and number of sensors, are chosen based on the criterion

$$\max\left|\mathbf{J}^T\mathbf{J}\right| \tag{N2.2.1}$$

By using the definition of the sensitivity matrix for the case involving a single sensor, equation (2.1.7.b), each element $\mathbf{F}_{m,n}$, $m,n = 1, \ldots, N$, of the matrix $\mathbf{F} \equiv \mathbf{J}^T\mathbf{J}$ is given by:

$$\mathbf{F}_{m,n} \equiv \left[\mathbf{J}^T\mathbf{J}\right]_{m,n} = \sum_{i=1}^{I}\left(\frac{\partial T_i}{\partial P_m}\right)\left(\frac{\partial T_i}{\partial P_n}\right) \quad \text{for } m,n = 1, \ldots, N \tag{N2.2.2}$$

where I is the number of measurements and N is the number of unknown parameters.

Different particular cases of the general criterion (N2.2.1) can now be examined.

Case 1. A large but fixed number of equally spaced measurements is available. Then, each element $\mathbf{F}_{m,n}$ can be written as

$$\mathbf{F}_{m,n} = \frac{1}{\Delta t}\sum_{i=1}^{I}\left(\frac{\partial T_i}{\partial P_m}\right)\left(\frac{\partial T_i}{\partial P_n}\right)\Delta t \approx \frac{I}{t_f}\int_{t=0}^{t_f}\left(\frac{\partial T}{\partial P_m}\right)\left(\frac{\partial T}{\partial P_n}\right)dt \quad \text{for } m,n = 1, \ldots, N \tag{N2.2.3.a}$$

where t_f is the duration of the experiment and Δt is the constant time interval between two consecutive measurements. Since the number of measurements, I, is fixed, we can choose to maximize the determinant of $\mathbf{F}_I$ instead of maximizing the determinant of $\mathbf{F}$, where the elements of $\mathbf{F}_I$ are given by

$$[\mathbf{F}_I]_{m,n} = \frac{1}{t_f}\int_{t=0}^{t_f}\left(\frac{\partial T}{\partial P_m}\right)\left(\frac{\partial T}{\partial P_n}\right)dt \quad \text{for } m,n = 1, \ldots, N \tag{N2.2.3.b}$$

Case 2. In addition to a large and fixed number of equally spaced measurements, the maximum value for the temperature in the region, T_{max}, is known. Thus, equation (N2.2.3.b) can be written as

$$[\mathbf{F}_I]_{m,n} = \frac{T_{max}^2}{t_f P_m P_n}\int_{t=0}^{t_f}\left(\frac{P_m}{T_{max}}\frac{\partial T}{\partial P_m}\right)\left(\frac{P_n}{T_{max}}\frac{\partial T}{\partial P_n}\right)dt \quad \text{for } m,n = 1,\ldots,N \tag{N2.2.4.a}$$

Note that the quantities inside parentheses in equation (N2.2.4.a) are dimensionless. However, it is possible that T^*, and not T_{max}, is the variable suitable for the non-dimensionalization of the temperature T. In such a case, equation (N2.2.4.a) can be written as

$$[\mathbf{F}_I]_{m,n} = \frac{T_{max}^2}{t_f P_m P_n} \int_{t=0}^{t_f} \left(\frac{P_m}{T^*} \frac{\partial T}{\partial P_m} \right) \left(\frac{P_n}{T^*} \frac{\partial T}{\partial P_n} \right) \left(\frac{T^*}{T_{max}} \right)^2 dt \quad \text{for } m,n = 1, \ldots, N \tag{N2.2.4.b}$$

and the design of optimum experiments is then based on the maximization of the determinant of the dimensionless form of $\mathbf{F}_I$, $\mathbf{F}_I^*$, the elements of which are given by

$$[\mathbf{F}_I^*]_{m,n} = \frac{1}{t_f} \int_{t=0}^{t_f} \left(\frac{P_m}{T^*} \frac{\partial T}{\partial P_m} \right) \left(\frac{P_n}{T^*} \frac{\partial T}{\partial P_n} \right) \left(\frac{T^*}{T_{max}} \right)^2 dt \quad \text{for } m,n = 1, \ldots, N \tag{N2.2.4.c}$$

Case 3. Measurements of *M* sensors are available. Thus, the elements $[\mathbf{F}_I^*]_{m,n}$ become

$$[\mathbf{F}_I^*]_{m,n} = \frac{1}{M t_f} \sum_{s=1}^{M} \int_{t=0}^{t_f} \left(\frac{P_m}{T^*} \frac{\partial T_s}{\partial P_m} \right) \left(\frac{P_n}{T^*} \frac{\partial T_s}{\partial P_n} \right) \left(\frac{T^*}{T_{max}} \right)^2 dt \quad \text{for } m,n = 1, \ldots, N \tag{N2.2.5}$$

We note that for non-linear parameter estimation problems, the sensitivity matrix, and thus $|\mathbf{J}^T\mathbf{J}|$, depend on the unknown parameters P_j, $j = 1, \ldots, N$. In such cases, only a local optimum experimental design is possible by using some *a priori* information regarding the expected values for the unknown parameters.

In order to illustrate this approach for the design of optimum experiments, we consider the analysis developed in reference [22] for the physical problem described in example 2-3, involving the estimation of thermal conductivity, k, and volumetric heat capacity, C. It is assumed that a *large but fixed number of measurements is available from a single sensor*. By taking into account the maximum temperature in the medium and using the dimensionless temperature θ, as defined in equation (2.1.21.a), we can write equation (N2.2.4.c) as

$$[\mathbf{F}_I^*]_{m,n} = \frac{1}{\tau_f} \int_{\tau=0}^{\tau_f} \left(P_m \frac{\partial \theta}{\partial P_m} \right) \left(P_n \frac{\partial \theta}{\partial P_n} \right) \left(\frac{1}{\theta_{max}} \right)^2 d\tau \quad \text{for } m,n = 1, 2 \tag{N2.2.6}$$

where the unknown parameters are $P_1=k$ and $P_2=C$. The dimensionless sensitivity coefficients appearing inside parentheses in equation (N2.2.6) are computed with equations (2.1.24) and (2.1.25). Note that in the present problem, the suitable quantity for the non-dimensionalization of temperature is

$$T^* = \frac{q_0 L}{k} \tag{N2.2.7}$$

The maximum dimensionless temperature θ_{max} is obtained from equation (2.1.23.a) for $\xi = 0$ and $\tau = \tau_f$ when $\tau_f \leq \tau_h$, and for $\xi = 0$ and $\tau = \tau_h$, when $\tau_f > \tau_h$.

We can choose different experimental variables, such as the heating and final times, τ_h and τ_f, respectively, as well as the sensor position, by plotting the time variation of the determinant of the matrix $\mathbf{F}_I^*$, that is,

$$\left|\mathbf{F}_I^*\right| = [\mathbf{F}_I^*]_{11}[\mathbf{F}_I^*]_{22} - [\mathbf{F}_I^*]_{12}^2 \tag{N2.2.8}$$

Figure N2.2.1 presents the variation of $|\mathbf{F}_I^*|$ for different heating times and for a single sensor located at $\xi = 0$. This figure shows that the maximum value of $|\mathbf{F}_I^*|$ is reached with heating and final times of approximately 2.5 and 3.3, respectively. Note that a curve joining the peaks for $\tau_h = 2$ and 2.5 is rather flat, showing that any value in this range will be very close to the optimum heating time. On the other hand, the behavior of $|\mathbf{F}_I^*|$ is very sensitive to the choice of the final time τ_f. Note that $|\mathbf{F}_I^*|$ decreases very fast after its maximum value is reached. Therefore, an analysis of figure N2.2.1 reveals that the heating time for an optimum experiment should be chosen in the interval $2 \leq \tau_h \leq 2.5$, with final time given approximately by $\tau_h + 0.8$.

The reader should note that such conclusions are based on the hypothesis that a large but fixed number of measurements is available. Thus, the number of measurements remain constant when the final time is increased. Different conclusions could be obtained if the number of measurements increases with increasing final time. As a matter of fact, more accurate estimates for the parameters are generally obtained if more measurements are available for the inverse analysis. Therefore, the ***design of optimum experiments requires detailed knowledge of the experimental setup and data acquisition system*** utilized in the experiment, in order to choose the correct form of $\left|\mathbf{J}^T\mathbf{J}\right|$ to be maximized.

Figure N2.2.2 presents the transient behavior of $|\mathbf{F}_I^*|$ for a single sensor located at different positions and for $\tau_h = \tau_f$. This figure shows that the sensor should be located as close to the boundary $\xi = 0$ as possible, since $|\mathbf{F}_I^*|$ attains the largest values in this region. Such conclusion was also obtained from the analysis of the sensitivity coefficients presented in figures 2.1.5.

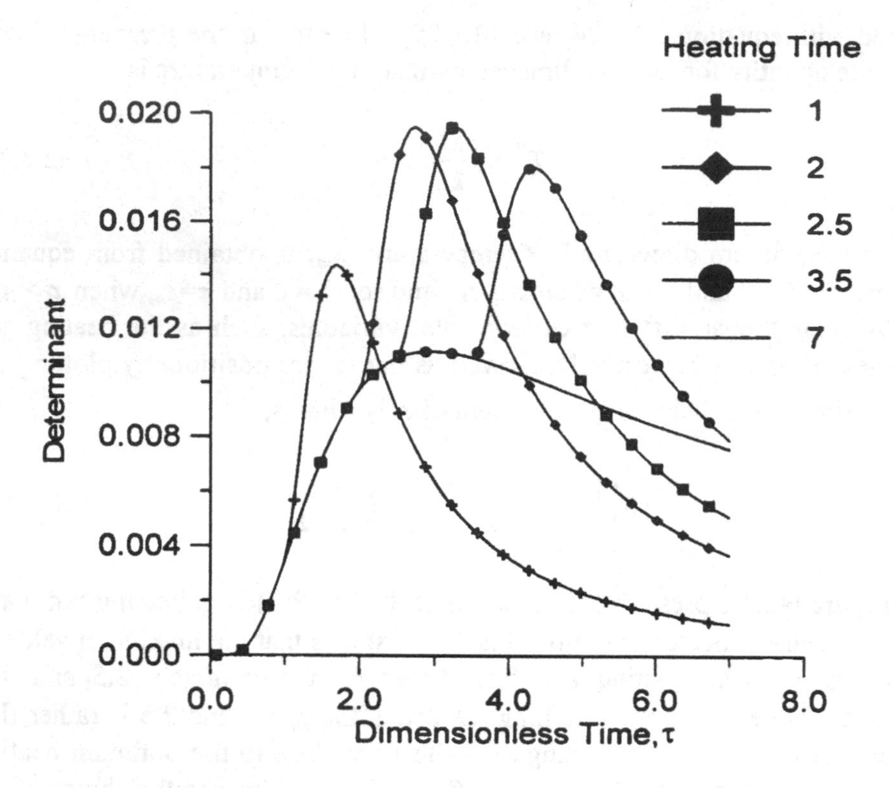

Figure N2.2.1. Effect of the heating time on the determinant $|\mathbf{F}_l^*|$.

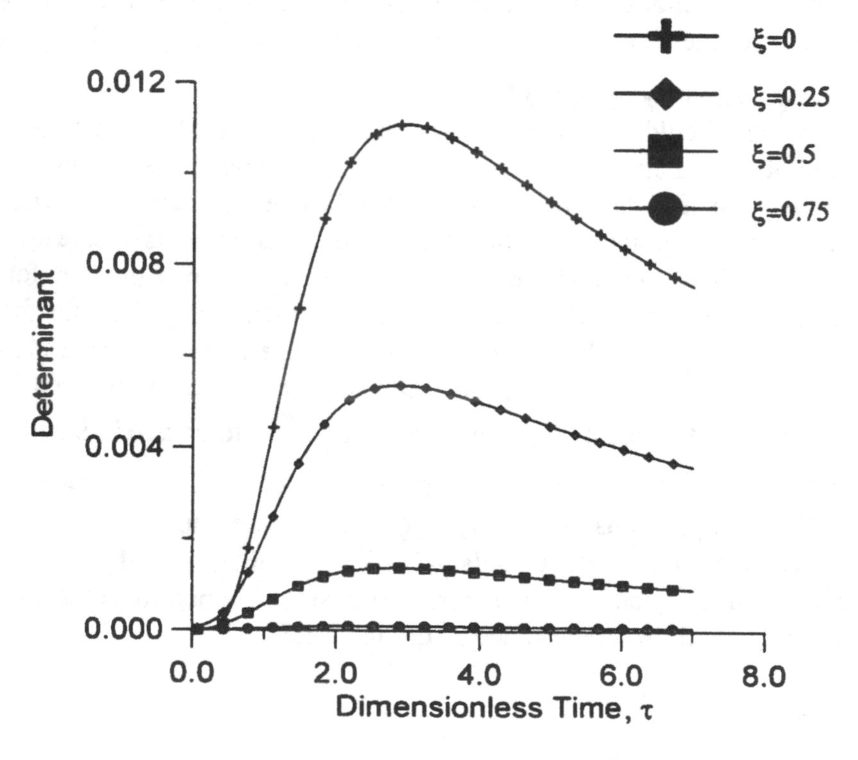

Figure N2.2.2. Effect of the sensor position on the determinant $|\mathbf{F}_l^*|$.

b. Function Estimation

Consider now the case involving the estimation of an unknown function $g_p(t)$, by using the measurements of M sensors and by minimizing the following functional

$$S[g_p(t)] = \sum_{s=1}^{M} \int_{t=0}^{t_f} \{Y_s(t) - T(x_s, t)\}^2 dt \tag{N2.2.9}$$

where $Y_s(t)$ and $T(x_s,t)$ are the measured and estimated temperatures, respectively, at the sensor positions x_s, $s = 1, \ldots, M$.

Consider now that the temperature $T(x_s,t)$ is perturbed by $\Delta T(x_s,t)$ and the functional $S[g_p(t)]$ is perturbed by $\Delta S[g_p(t)]$, when the unknown function $g_p(t)$ is perturbed by $\Delta g_p(t)$. Then we can write the perturbed form of equation (N2.2.9) as

$$S[g_p(t)] + \Delta S[g_p(t)] = \sum_{s=1}^{M} \int_{t=0}^{t_f} [Y_s(t) - T(x_s,t) - \Delta T(x_s,t)]^2 dt \tag{N2.2.10}$$

By subtracting equation (N2.2.9) from equation (N2.2.10), we find

$$\Delta S[g_p(t)] = \sum_{s=1}^{M} \int_{t=0}^{t_f} \{2\Delta T(x_s,t)[T(x_s,t) - Y_s(t)] + [\Delta T(x_s,t)]^2\} dt \tag{N2.2.11.a}$$

In the neighborhood of the minimum of the functional, we have $Y_s(t) \approx T(x_s, t)$. In this case, equation (N2.2.11.a) can be approximated by

$$\Delta S[g_p(t)] \approx \sum_{s=1}^{M} \int_{t=0}^{t_f} [\Delta T(x_s,t)]^2 dt \tag{N2.2.11.b}$$

As for the problems involving parameter estimation, the optimum design of experiments for function estimation involves locating the sensors and choosing other experimental variables, so that the measured temperatures are most affected by changes in the sought function. This is accomplished by maximizing the functional given by equation (N2.2.11.b) involving the sensitivity function $\Delta T(x_s,t)$ for, say, a unitary perturbed function $\Delta g_p(t)$. This criterion might not work properly for nonlinear estimation problems, where the sensitivity function depends on the unknown function [26].

For the linear test problem of this chapter, involving the estimation of the strength $g_p(t)$ of the source function, an analysis of the sensitivity problem given by equations (2.3.4) clearly reveals that the sensor should be located as close to the unknown source function as possible, in order to maximize equation (N2.2.11.b). Also, note that the use of several sensors ($M > 1$) may increase the value of the functional given by equation (N2.2.11.b), since more information can be available for the inverse analysis.

NOTE 3. SEARCH STEP-SIZE FOR TECHNIQUE II

The search step size, β^k, for the conjugate gradient method of parameter estimation, is obtained as the one that minimizes the least-squares norm given by equation (2.2.1) at each iteration, that is,

$$\min_{\beta^k} S(\mathbf{P}^{k+1}) = \min_{\beta^k} [\mathbf{Y} - \mathbf{T}(\mathbf{P}^{k+1})]^T [\mathbf{Y} - \mathbf{T}(\mathbf{P}^{k+1})] \tag{N3.2.1}$$

From the iterative procedure of the conjugate gradient method, we have

$$\mathbf{P}^{k+1} = \mathbf{P}^k - \beta^k \mathbf{d}^k \tag{N3.2.2}$$

Thus, by substituting $\mathbf{P}^{k+1}$ into equation (N3.2.1), we obtain

$$\min_{\beta^k} S(\mathbf{P}^{k+1}) = \min_{\beta^k} [\mathbf{Y} - \mathbf{T}(\mathbf{P}^k - \beta^k \mathbf{d}^k)]^T [\mathbf{Y} - \mathbf{T}(\mathbf{P}^k - \beta^k \mathbf{d}^k)] \tag{N3.2.3.a}$$

which can be written, for cases involving a single sensor, as

$$\min_{\beta^k} S(\mathbf{P}^{k+1}) = \min_{\beta^k} \sum_{i=1}^{I} [Y_i - T_i(\mathbf{P}^k - \beta^k \mathbf{d}^k)]^2 \tag{N3.2.3.b}$$

where I is the number of measurements.

We now linearize $T_i(\mathbf{P}^k - \beta^k \mathbf{d}^k)$ by using a Taylor series expansion in the form

$$T_i(\mathbf{P}^k - \beta^k \mathbf{d}^k) = T_i[(P_1^k - \beta^k d_1^k), (P_2^k - \beta^k d_2^k), \cdots, (P_N^k - \beta^k d_N^k)] \approx$$
$$\approx T_i(P_1^k, P_2^k, \cdots, P_N^k) - \beta^k \frac{\partial T_i}{\partial P_1^k} d_1^k - \beta^k \frac{\partial T_i}{\partial P_2^k} d_2^k - \cdots - \beta^k \frac{\partial T_i}{\partial P_N^k} d_N^k \tag{N3.2.4.a}$$

or, in vector form,

$$T_i(\mathbf{P}^k - \beta^k \mathbf{d}^k) \approx T_i(\mathbf{P}^k) - \beta^k \left[\frac{\partial T_i}{\partial \mathbf{P}^k}\right]^T \mathbf{d}^k \tag{N3.2.4.b}$$

where

$$\left[\frac{\partial T_i}{\partial \mathbf{P}^k}\right]^T = \left[\frac{\partial T_i}{\partial P_1^k}, \frac{\partial T_i}{\partial P_2^k}, \ldots, \frac{\partial T_i}{\partial P_N^k}\right] \tag{N3.2.5}$$

By substituting equation (N3.2.4.b) into equation (N3.2.3.b) and performing the minimization with respect to β^k, we obtain the following expression for the search step size for Technique II:

$$\beta^k = \frac{\sum_{i=1}^{I} \left[\left(\frac{\partial T_i}{\partial \mathbf{P}^k}\right)^T \mathbf{d}^k\right] [T_i(\mathbf{P}^k) - Y_i]}{\sum_{i=1}^{I} \left[\left(\frac{\partial T_i}{\partial \mathbf{P}^k}\right)^T \mathbf{d}^k\right]^2} \tag{N3.2.6.a}$$

or, by using the definition of the sensitivity matrix **J** given by equation (2.1.7.b), the expression above for β^k can be written in matrix form as

$$\beta^k = \frac{[\mathbf{J}^k \mathbf{d}^k]^T [\mathbf{T}(\mathbf{P}^k) - \mathbf{Y}]}{[\mathbf{J}^k \mathbf{d}^k]^T [\mathbf{J}^k \mathbf{d}^k]} \tag{N3.2.6.b}$$

NOTE 4. ADDITIONAL MEASUREMENT APPROACH FOR THE STOPPING CRITERION OF THE CONJUGATE GRADIENT METHOD

The stopping criterion approach based on the discrepancy principle, described above for Techniques II, III and IV involving the conjugate gradient method, requires the *a priori* knowledge of the standard deviation of the measurement errors. However, there are several practical situations in which scarce information is available regarding this quantity. For such cases, an alternative stopping criterion approach based on an additional measurement [12] can be used, which also provides the conjugate gradient method with an iterative regularization character.

In order to illustrate the additional measurement approach for the stopping criterion, we take as an example the estimation of the boundary heat flux $q(t)$ in a slab of unitary thickness, by using Technique IV. The formulation of the dimensionless heat conduction problem considered here is given by

$$\frac{\partial T}{\partial t}=\frac{\partial^2 T}{\partial x^2} \quad \text{in } 0<x<1, \quad \text{for } t>0 \qquad \text{(N4.2.1.a)}$$

$$\frac{\partial T}{\partial x}=0 \quad \text{at } x=0, \quad \text{for } t>0 \qquad \text{(N4.2.1.b)}$$

$$\frac{\partial T}{\partial x}=q(t) \quad \text{at } x=1, \quad \text{for } t>0 \qquad \text{(N4.2.1.c)}$$

$$T=0 \quad \text{for } t=0, \quad \text{in } 0<x<1 \qquad \text{(N4.2.1.d)}$$

The unknown function $q(t)$ is estimated with Technique IV by minimizing the following functional

$$S[q(t)]=\int_{t=0}^{t_f}\{Y(t)-T[x_{meas},t;q(t)]\}^2 dt \qquad \text{(N4.2.2)}$$

where $Y(t)$ are the measured temperatures at the location x_{meas}, while $T[x_{meas},t;q(t)]$ are the estimated temperatures at the same location.

Consider now that the additional measured data $Y_c(t)$ of a sensor located at x_c are also available for the analysis. The functional $S_c[q(t)]$ based on such data is given by

$$S_c[q(t)]=\int_{t=0}^{t_f}\{Y_c(t)-T[x_c,t;q(t)]\}^2 dt \qquad \text{(N4.2.3)}$$

The examination of the behavior of the functional $S_c[q(t)]$, as the minimization of $S[q(t)]$ is performed, can be used to detect the point where the errors in the measured data $Y(t)$ start to cause instabilities on the estimated function $q(t)$. Generally, the value of $S_c[q(t)]$ passes through a minimum and then increases, as a result of such instabilities. The iterative procedure is then stopped at the iteration number corresponding to the minimum value of $S_c[q(t)]$, so that sufficiently stable solutions can be obtained for the inverse problem.

Results for the estimation of a step variation of the boundary heat flux $q(t)$, obtained by using the stopping criterion approaches based on the discrepancy principle and on the additional measurement, are illustrated in figure N4.2.1. The simulated measured data $Y(t)$ and $Y_c(t)$ were generated with a constant standard deviation σ=0.01Y_{max}, where Y_{max} is the maximum value of $Y(t)$. The measurements $Y(t)$ used in the minimization of the functional $S[q(t)]$ were considered taken at the position x_{meas}=0.986. For the case involving the additional measurements for the stopping criterion, the additional sensor was supposed to be located at x_c=0.982. Figure N4.2.1 shows that the two approaches for the stopping criterion are equivalent. Both provide quite accurate and stable estimates for the step variation of the heat flux, which represents a very strict test function.

We note that the additional measurement approach for the stopping criterion, illustrated above as applied to Technique IV, can be readily modified to be applied to Techniques II and III.

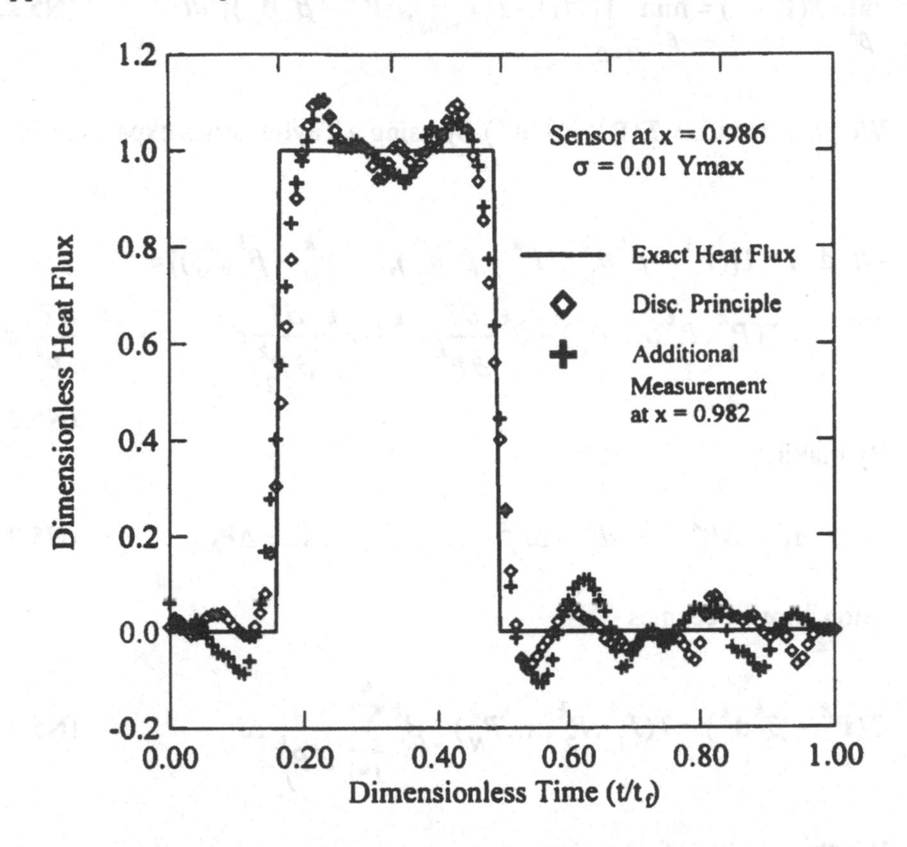

Figure N4.2.1. A comparison of the discrepancy principle and additional measurement approaches for the stopping criterion

NOTE 5. SEARCH STEP-SIZE FOR TECHNIQUE III

Similarly to Technique II, the search step size for Technique III is obtained as the one that minimizes the objective function given by equation (2.3.1) at each iteration, that is,

$$\min_{\beta^k} S(\mathbf{P}^{k+1}) = \min_{\beta^k} \int_{t=0}^{t_f} [Y(t) - T(x_{meas}, t; \mathbf{P}^{k+1})]^2 dt \qquad \text{(N5.2.1)}$$

From the iterative procedure of the conjugate gradient method, we have

$$\mathbf{P}^{k+1} = \mathbf{P}^k - \beta^k \mathbf{d}^k \tag{N5.2.2}$$

By substituting $\mathbf{P}^{k+1}$ into equation (N5.2.1), we obtain

$$\min_{\beta^k} S(\mathbf{P}^{k+1}) = \min_{\beta^k} \int_{t=0}^{t_f} [Y(t) - T(x_{meas}, t; \mathbf{P}^k - \beta^k \mathbf{d}^k)]^2 dt \tag{N5.2.3}$$

We now linearize $T(\mathbf{P}^k - \beta^k \mathbf{d}^k)$ by using a Taylor series expansion in the form

$$T(\mathbf{P}^k - \beta^k \mathbf{d}^k) = T[(P_1^k - \beta^k d_1^k),(P_2^k - \beta^k d_2^k),\dots,(P_N^k - \beta^k d_N^k)] \approx$$
$$\approx T(P_1^k, P_2^k, \dots, P_N^k) - \beta^k \frac{\partial T}{\partial P_1^k} d_1^k - \beta^k \frac{\partial T}{\partial P_2^k} d_2^k - \dots - \beta^k \frac{\partial T}{\partial P_N^k} d_N^k \tag{N5.2.4}$$

By making

$$d_1^k = \Delta P_1^k \quad , \quad d_2^k = \Delta P_2^k \quad , \quad \cdots \quad , \quad d_N^k = \Delta P_N^k \tag{N5.2.5}$$

the equation above becomes

$$T(\mathbf{P}^k - \beta^k \mathbf{d}^k) \approx T(P_1^k, P_2^k, \dots, P_N^k) - \beta^k \sum_{j=1}^{N} \frac{\partial T}{\partial P_j^k} \Delta P_j^k \tag{N5.2.6}$$

where N is the number of parameters.

Let
$$\Delta T(\mathbf{d}^k) = \sum_{j=1}^{N} \frac{\partial T}{\partial P_j^k} \Delta P_j^k \tag{N5.2.7}$$

Then equation (N5.2.6) can be written as

$$T(\mathbf{P}^k - \beta^k \mathbf{d}^k) \approx T(\mathbf{P}^k) - \beta^k \Delta T(\mathbf{d}^k) \tag{N5.2.8}$$

By substituting equation (N5.2.8) into equation (N5.2.3), we obtain

$$\min_{\beta^k} S(\mathbf{P}^{k+1}) = \min_{\beta^k} \int_{t=0}^{t_f} [Y(t) - T(x_{meas},t;\mathbf{P}^k) + \beta^k \Delta T(x_{meas},t;\mathbf{d}^k)]^2 dt \quad \text{(N5.2.9)}$$

which is then minimized with respect to β^k to yield

$$\beta^k = \frac{\int_{t=0}^{t_f} [T(x_{meas},t;\mathbf{P}^k) - Y(t)]\, \Delta T(x_{meas},t;\mathbf{d}^k)\, dt}{\int_{t=0}^{t_f} [\Delta T(x_{meas},t;\mathbf{d}^k)]^2\, dt} \quad \text{(N5.2.10)}$$

where $\Delta T(x_{meas},t;\mathbf{d}^k)$ is the solution of the sensitivity problem given by equations (2.3.4), obtained by setting $\Delta P_j = d_j^k$, for j = 1, ..., N, in the computation of

$$\Delta g_p(t) = \sum_{j=1}^{N} \Delta P_j\, C_j(t) \quad \text{(N5.2.11)}$$

NOTE 6. HILBERT SPACES

We present in this note some definitions and properties regarding Hilbert spaces. For further details on the subject, the reader should consult references [14,27].

A *Hilbert space* is a Banach space in which the norm is given by an inner (or scalar) product $<\cdot\,,\cdot>$, that is,

$$\|u\| = \langle u, u \rangle^{1/2} \quad \text{(N6.2.1)}$$

where $\|\cdot\|$ designates the norm in the space.

For u belonging to a linear space V, a *norm* on this space is a mapping from V into the non-negative real axis, $[0,\infty)$, satisfying the following properties:

(i) $\|u\| = 0$ if and only if $u = 0$;
(ii) $\|\lambda u\| = |\lambda|\, \|u\|$, where λ is a scalar;
(iii) $\|u + v\| \le \|u\| + \|v\|$, for any u and v in V.

Property (iii) is the so-called *triangle inequality*.

In a Hilbert space V, the *inner product* is given by the following symmetric bilinear form:

$$\langle u,v\rangle \equiv \frac{1}{4}(\|u+v\|^2 - \|u-v\|^2) \quad , \ \forall u,v \in V \tag{N6.2.2}$$

The *vector space* R^N with the Euclidean norm

$$\|\mathbf{P}\| = \left(\sum_{j=1}^{N} P_j^2\right)^{1/2} \tag{N6.2.3.a}$$

is a Hilbert space, with inner product given by

$$\langle \mathbf{P},\mathbf{R}\rangle = \sum_{j=1}^{N} P_j R_j = \mathbf{P}^T\mathbf{R} \tag{N6.2.3.b}$$

where $\quad \mathbf{P}^T = [P_1, P_2, \ldots, P_N] \qquad \mathbf{R}^T = [R_1, R_2, \ldots, R_N]$

and the superscript T denotes transpose.

Similarly, the *space of square-integrable real valued functions in a domain* Ω, $L_2(\Omega)$, satisfying

$$\int_{\Omega} [f(w)]^2 \, dw < \infty \quad \text{for} \ \ w \ \text{in} \, \Omega$$

is a Hilbert space with norm

$$\|f(w)\| = \left\{ \int_{\Omega} [f(w)]^2 \, dw \right\}^{1/2} \tag{N6.2.4.a}$$

and inner product

$$\langle f(w), g(w)\rangle = \int_{\Omega} f(w) g(w) \, dw, \quad \text{for} \ \ f(w) \, \text{and} \, g(w) \in L_2(\Omega) \tag{N6.2.4.b}$$

If Ω refers to the time domain, $0 < t < t_f$, equations (N6.2.4) become respectively

$$\|f(t)\| = \left\{ \int_{t=0}^{t_f} [f(t)]^2 \, dt \right\}^{1/2} \tag{N6.2.5.a}$$

and

$$\langle f(t),g(t)\rangle = \int_{t=0}^{t_f} f(t)g(t)\,dt, \quad \text{for} \quad f(t) \text{ and } g(t) \text{ in } L_2(0,t_f) \tag{N6.2.5.b}$$

Similarly, if Ω refers to the joint time and spatial domains, $0 < t < t_f$ and $0 < x < 1$, equations (N6.2.4) can be written respectively as:

$$\|f(x,t)\| = \left\{ \int_{x=0}^{1} \int_{t=0}^{t_f} [f(x,t)]^2\,dt\,dx \right\}^{1/2} \tag{N6.2.6.a}$$

and

$$\langle f(x,t),g(x,t)\rangle = \int_{x=0}^{1} \int_{t=0}^{t_f} f(x,t)g(x,t)\,dt\,dx, \quad \text{for} \quad f(x,t) \text{ and } g(x,t) \text{ in } L_2[(0,t_f)\text{x}(0,1)] \tag{N6.2.6.b}$$

Other expressions for the norm and inner product can be developed from equations (N6.2.4) for various domains Ω of interest.

The reader should note that the expressions for the conjugation coefficients (2.2.4) and (2.4.9) are analogous. They are given by inner products in the R^N and $L_2(0,t_f)$ spaces, equations (N6.2.3.b) and (N6.2.5.b), respectively. Also, expressions (2.3.13) and (2.4.5.b) are inner products of the gradient direction with the direction of perturbed parameters $\Delta\mathbf{P}$ in R^N, and with the direction of the perturbed function $\Delta g_p(t)$ in $L_2(0,t_f)$, respectively. Therefore, they give the directional derivative of $S(\mathbf{P})$ and $S[g_p(t)]$, respectively, in the direction of the perturbed unknown quantities. Note that $S(\mathbf{P})$ and $S[g_p(t)]$, given by equations (2.1.3.a) and (2.4.2), are the squares of the norms in the R^N and $L_2(0,t_f)$ spaces, respectively.

NOTE 7. SEARCH STEP-SIZE FOR TECHNIQUE IV

Similarly to Techniques II and III, the search step size for Technique IV is obtained as the one that minimizes the objective functional given by equation (2.4.2) at each iteration, that is,

$$\min_{\beta^k} S[g_p^{k+1}(t)] = \min_{\beta^k} \int_{t=0}^{t_f} \left\{ Y(t) - T\left[x_{meas}, t; g_p^{k+1}(t)\right]\right\}^2 dt \tag{N7.2.1}$$

The iterative procedure of the conjugate gradient method for function estimation is given by

$$g_p^{k+1}(t)=g_p^k(t)-\beta^k d^k(t) \tag{N7.2.2}$$

Thus, equation (N7.2.1) can be written as

$$\min_{\beta^k} S[g_p^{k+1}(t)]=\min_{\beta^k}\int_{t=0}^{t_f}\left\{Y(t)-T\left[x_{meas},t;g_p^k(t)-\beta^k d^k(t)\right]\right\}^2 dt \tag{N7.2.3}$$

By linearizing $T[g_p^k(t)-\beta^k d^k(t)]$ and making

$$d^k(t)=\Delta g_p^k(t) \tag{N7.2.4}$$

we obtain

$$T[g_p^k(t)-\beta^k d^k(t)]\approx T[g_p^k(t)]-\beta^k\frac{\partial T}{\partial g_p^k}\Delta g_p^k(t) \tag{N7.2.5}$$

Let

$$\Delta T[d^k(t)]=\frac{\partial T}{\partial g_p^k}\Delta g_p^k(t) \tag{N7.2.6}$$

and then equation (N7.2.5) can be written as

$$T[g_p^k(t)-\beta^k d^k(t)]\approx T[g_p^k(t)]-\beta^k\Delta T[d^k(t)] \tag{N7.2.7}$$

By substituting equation (N7.2.7) into equation (N7.2.3), we obtain

$$\min_{\beta^k} S[g_p^{k+1}(t)]=\min_{\beta^k}\int_{t=0}^{t_f}\{Y(t)-T[x_{meas},t;g_p^k(t)]+\beta^k\Delta T[x_{meas},t;d^k(t)]\}^2 dt \tag{N7.2.8}$$

By performing the minimization above, we find the following expression for the search step size for Technique IV:

$$\beta^k=\frac{\int_{t=0}^{t_f}\{T[x_{meas},t;g_p^k(t)]-Y(t)\}\Delta T[x_{meas},t;d^k(t)]\,dt}{\int_{t=0}^{t_f}\{\Delta T[x_{meas},t;d^k(t)]\}^2 dt} \tag{N7.2.9}$$

where $\Delta T\ [x_{meas},t;d^k(t)]$ is the solution of the sensitivity problem given by equations (2.3.4), obtained by setting $\Delta g_p^k(t)=d^k(t)$.

Part TWO

APPLICATIONS

Chapter 3

INVERSE CONDUCTION

In the previous chapter we presented four powerful methodologies for solving inverse heat transfer problems. This chapter is devoted to the application of these techniques to the solution of *Inverse Heat Conduction* problems. The specific examples considered here include the solution of the following problems:

- Estimation of constant thermal conductivity components of an orthotropic solid [1-3]
- Estimation of initial condition [5]
- Estimation of timewise variation of the strength of a line heat source [6]
- Estimation of timewise and spacewise variation of the strength of a volumetric heat source [7]
- Estimation of temperature-dependent properties and reaction function [8,9]
- Estimation of thermal diffusivity and relaxation time for a hyperbolic heat conduction model [13]
- Estimation of contact conductance between periodically contacting surfaces [21]
- Estimation of contact conductance between a solidifying metal and a metal mold [26,27]

3-1 ESTIMATION OF CONSTANT THERMAL CONDUCTIVITY COMPONENTS OF AN ORTHOTROPIC SOLID [1-3]

In nature, several materials have direction-dependent thermal conductivities including, among others, woods, rocks and crystals. This is also the case for some man-made materials, for example, composites.

Such kind of materials is denoted *anisotropic*, as an opposition to *isotropic* materials, in which the thermal conductivity does not vary with direction. A special case of anisotropic materials involve those where a thermal conductivity component can be identified along the three mutually orthogonal directions. They are referred to as *orthotropic* materials.

In this section we illustrate the application of **Technique I**, the Levenberg-Marquardt Method, to the estimation of the three thermal conductivity components of orthotropic solids.

Basic steps of Technique I, including the definitions of the direct and inverse problems, are presented below. The reader should refer to section 2.1 for the iterative procedure, stopping criteria and computational algorithm of Technique I, which are not repeated here.

Direct Problem

The physical problem considered here involves the three-dimensional linear heat conduction in an orthotropic solid, with thermal conductivity components k_1^*, k_2^* and k_3^* in the x^*, y^* and z^* directions, respectively. The solid is considered to be a parallelepiped with sides a^*, b^* and c^*, initially at the uniform temperature T_0^*. For times $t^*>0$, uniform heat fluxes q_1^*, q_2^* and q_3^* are supplied at the surfaces $x^* = a^*, y^* = b^*$ and $z^* = c^*$, respectively, while the other three remaining boundaries at $x^* = 0$, $y^* = 0$ and $z^* = 0$ are supposed insulated. The mathematical formulation of such physical problem is given in dimensionless form by

$$k_1 \frac{\partial^2 T}{\partial x^2} + k_2 \frac{\partial^2 T}{\partial y^2} + k_3 \frac{\partial^2 T}{\partial z^2} = \frac{\partial T}{\partial t} \quad \text{in } 0<x<a,\ 0<y<b,\ 0<z<c;\ t>0 \tag{3.1.1.a}$$

$$\frac{\partial T}{\partial x} = 0 \text{ at } x=0 \quad ; \quad k_1 \frac{\partial T}{\partial x} = q_1 \quad \text{at } x=a \quad , \text{ for } t>0 \tag{3.1.1.b,c}$$

$$\frac{\partial T}{\partial y} = 0 \text{ at } y=0 \quad ; \quad k_2 \frac{\partial T}{\partial y} = q_2 \quad \text{at } y=b \quad , \text{ for } t>0 \tag{3.1.1.d,e}$$

$$\frac{\partial T}{\partial z} = 0 \text{ at } z=0 \quad ; \quad k_3 \frac{\partial T}{\partial z} = q_3 \quad \text{at } z=c \quad , \text{ for } t>0 \tag{3.1.1.f,g}$$

$$T=0 \qquad \text{for } t=0 \ ; \ \text{in } 0<x<a,\ 0<y<b,\ 0<z<c \tag{3.1.1.h}$$

The superscript * denotes dimensional variables and the following dimensionless groups were introduced:

$$t = \frac{k^*_{ref}\, t^*}{\rho^* C^* L^{*2}} \quad T = \frac{T^* - T_0}{\dfrac{q^*_{ref} L^*}{k^*_{ref}}} \quad x = \frac{x^*}{L^*} \quad y = \frac{y^*}{L^*} \quad z = \frac{z^*}{L^*} \tag{3.1.2.a-e}$$

$$k_1 = \frac{k^*_1}{k^*_{ref}} \quad k_2 = \frac{k^*_2}{k^*_{ref}} \quad k_3 = \frac{k^*_3}{k^*_{ref}} \tag{3.1.2.f-h}$$

$$q_1 = \frac{q^*_1}{q^*_{ref}} \quad q_2 = \frac{q^*_2}{q^*_{ref}} \quad q_3 = \frac{q^*_3}{q^*_{ref}} \tag{3.1.2.i-k}$$

where L^* is a characteristic length, while q^*_{ref} and k^*_{ref} are reference values for heat flux and thermal conductivity, respectively.

In the *direct problem* associated with the physical problem described above, the three thermal conductivity components k_1, k_2 and k_3, as well as the solid geometry, initial and boundary conditions, are known. The objective of the direct problem is then to determine the transient temperature field in the body.

The solution of the direct problem (3.1.1) can be obtained analytically as a superposition of three one-dimensional solutions in the form [2,3]:

$$T(x,y,z,t) = \frac{q_1 a}{k_1}\theta\left(\frac{x}{a}, \frac{k_1 t}{a^2}\right) + \frac{q_2 b}{k_2}\theta\left(\frac{y}{b}, \frac{k_2 t}{b^2}\right) + \frac{q_3 c}{k_3}\theta\left(\frac{z}{c}, \frac{k_3 t}{c^2}\right) \tag{3.1.3.a}$$

where

$$\theta(\xi,\tau) = -\frac{1}{6} + \frac{\xi^2}{2} + \tau + \sum_{i=1}^{\infty} (-1)^{(i+1)} \frac{2}{(i\pi)^2} \cos(i\pi\xi)\exp[-\tau(i\pi)^2] \tag{3.1.3.b}$$

Inverse Problem

For the *inverse problem* considered here, the thermal conductivity components k_1, k_2 and k_3 are regarded as unknown, while the other quantities appearing in the formulation of the direct problem described above are assumed to be known with sufficient degree of accuracy.

For the estimation of the vector of unknown parameters $\mathbf{P}^T = [k_1, k_2, k_3]$ we assume available the readings of three temperature sensors. Since it is desirable to have a non-intrusive experiment, we consider each of the sensors to be located at the insulated surfaces x=0, y=0 and z=0.

The solution of the present parameter estimation problem is obtained through the minimization of the least-square norm.

$$S(\mathbf{P}) = [\mathbf{Y} - \mathbf{T}(\mathbf{P})]^T [\mathbf{Y} - \mathbf{T}(\mathbf{P})] \tag{3.1.4}$$

where, for the case involving multiple sensors (see equation 2.1.32.a), we have

$$[\mathbf{Y} - \mathbf{T}(\mathbf{P})]^T = [\vec{Y}_1 - \vec{T}_1(\mathbf{P}), \vec{Y}_2 - \vec{T}_2(\mathbf{P}), \ldots, \vec{Y}_I - \vec{T}_I(\mathbf{P})] \tag{3.1.5}$$

Each element $[\vec{Y}_i - \vec{T}_i(\mathbf{P})]$ is a vector of length equal to the number of sensors, M. In the present case M=3, so that we can write (see equation 2.1.32.b):

$$[\vec{Y}_i - \vec{T}_i(\mathbf{P})] = [Y_{i1} - T_{i1}(\mathbf{P}), Y_{i2} - T_{i2}(\mathbf{P}), Y_{i3} - T_{i3}(\mathbf{P})] \quad \text{for } i = 1, \ldots, I \tag{3.1.6}$$

In equation (3.1.6), Y_{im}, $i = 1, \ldots, I$, $m = 1, 2, 3$, are the measured temperatures of the sensor m at time t_i. The estimated temperatures $T_{im}(\mathbf{P})$ are obtained from the solution of the direct problem given equation (3.1.3.a), by using the current available estimate for the vector of unknown parameters $\mathbf{P}^T = [k_1, k_2, k_3]$.

The least-squares norm (3.1.4) is minimized here by using Technique I.

Results

We use simulated measurements in the form given by equation (2.5.2) in order to examine the accuracy of Technique I, as applied to the estimation of the unknown thermal conductivity components. The simulated measurements were generated by solving the direct problem with the exact values $k_1 = 1$, $k_2 = 2$ and $k_3 = 3$. In this case we considered the solid to be a cube with sides $a = b = c = 1$, with unitary heat fluxes supplied at the boundaries $x = a = 1$, $y = b = 1$ and $z = c = 1$, that is, $q_1 = q_2 = q_3 = 1$. However, before proceeding to the solution of the present parameter estimation problem, we perform an analysis of the sensitivity coefficients and choose experimental variables based on the D-optimal design, as discussed in Note 2 of Chapter 2.

Since the unknown parameters can assume different values, the analysis of the sensitivity coefficients is much simplified by using their relative versions defined by equation (2.1.26). In the present case, the *relative sensitivity coefficients* with respect to k_1, k_2 and k_3 are given respectively by:

$$J_1 = k_1 \frac{\partial T}{\partial k_1} \quad , \quad J_2 = k_2 \frac{\partial T}{\partial k_2} \quad , \quad J_3 = k_3 \frac{\partial T}{\partial k_3} \tag{3.1.7.a-c}$$

Due to the analytical nature of the solution of the direct problem given by equation (3.1.3.a), analytical expressions can also be obtained for the relative sensitivity coefficients. We note in equation (3.1.3.a) that, since the solution of the direct problem is obtained as a superposition of three one-dimensional solutions in the x, y and z directions, the relative sensitivity coefficient J_1 is a function of x, but not of y and z. The analytical expression for the relative sensitivity coefficient with respect to k_1 is given by:

$$J_1 = -\frac{q_1\, a}{k_1}\theta\left(\frac{x}{a}, \frac{k_1\, t}{a^2}\right) + \frac{q_1\, t}{a}\left\{1 - 2\sum_{i=1}^{\infty}(-1)^{1+i}\cos\left(\frac{i\pi\, x}{a}\right)\exp\left[\frac{(-k_1\, t)(i\,\pi)^2}{a^2}\right]\right\} \tag{3.1.8}$$

where the function $\theta(\xi,\tau)$ is obtained from equation (3.1.3.b). Analogous expressions can be obtained for the sensitivity coefficients J_2 and J_3, by making appropriate substitutions in equation (3.1.8). We note that the present estimation problem is non-linear, since the sensitivity coefficients are functions of the unknown parameters.

Figures 3.1.1.a-c present the transient behaviour of the relative sensitivity coefficients for sensors located at (0, 0.9, 0.9), (0.9, 0, 0.9) and (0.9, 0.9, 0), respectively. We note in these figures that, for each sensor, the sensitivity coefficient for the thermal conductivity in the direction normal to the surface where the sensor is located is positive, while the other sensitivity coefficients are negative (with the exception for very small times). Such figures show that the measurements are immediately affected by the thermal conductivities in the directions not normal to the surface where the sensor is located; but a lagging is observed in the sensors' response with respect to changes in the thermal conductivity in the other direction. As expected, this lagging is reduced as the value of such thermal conductivity is increased, which can be clearly noticed by comparing the curve for J_1 in figure 3.1.1.a, with the curve for J_2 in figure 3.1.1.b and with the curve for J_3 in figure 3.1.1.c (recall that the values k_1=1, k_2=2 and k_3=3 were used to generate the curves for the sensitivity coefficients). For each sensor location, the sensitivity coefficients with respect to the thermal conductivities in the directions not normal to the plane where the sensor is located tend to be linearly-dependent. The sensitivity coefficient for the thermal conductivity in the other direction does not seem to be linearly-dependent to the others. We also notice in figures 3.1.1.a-c that, if we consider a pair of sensors, the sensitivity coefficients are identical for the thermal conductivity in the direction parallel to the surfaces where they are located. Take as an example the sensors at (0.9, 0, 0.9) and (0.9, 0.9, 0), as shown in figures 3.1.1.b and 3.1.1.c, respectively. We notice that the curves for the sensitivity coefficient J_1 are identical for these sensors. The reason for this behaviour is because the sensitivity coefficient J_1 is a function of x, but not of y and z, as can be observed with the analysis of equation (3.1.8).

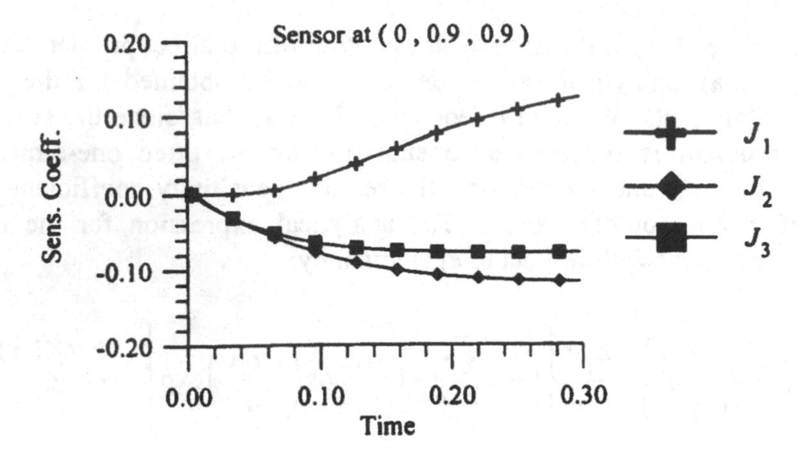

Figure 3.1.1.a – Relative Sensitivity Coefficients for a sensor located at (0, 0.9, 0.9).

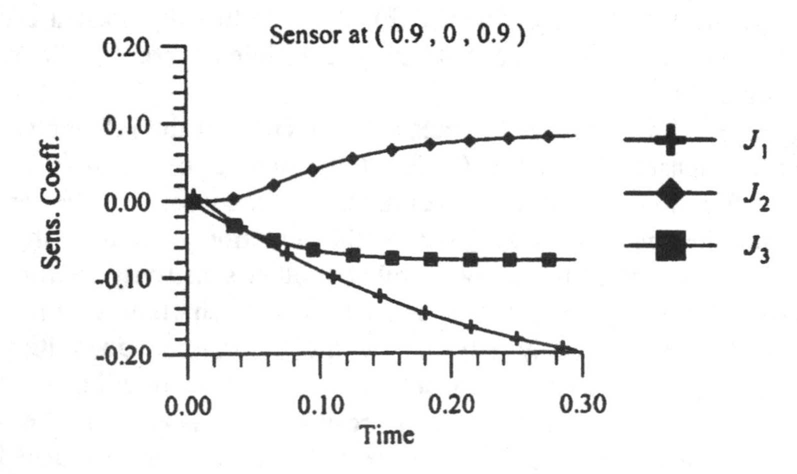

Figure 3.1.1.b - Relative Sensitivity Coefficients for a sensor located at (0.9, 0, 0.9).

The linear-dependence of two sensitivity coefficients at each sensor location makes impossible the estimation of the parameters by using the measurements of a single sensor, since two columns of the sensitivity matrix become linearly-dependent. In fact, difficulties were observed in the convergence of Technique I when the measurements of only one sensor were used in the analysis. However, the estimation of the three thermal conductivity components is possible if the measurements of more than one sensor, located at the positions shown in figures 3.1.1.a-c, are utilized. Such is the case because each row of the sensitivity matrix would have two columns proportional; but the proportional columns alternate for the rows corresponding to different sensors. Hence, the columns of the sensitivity matrix are not linearly-dependent.

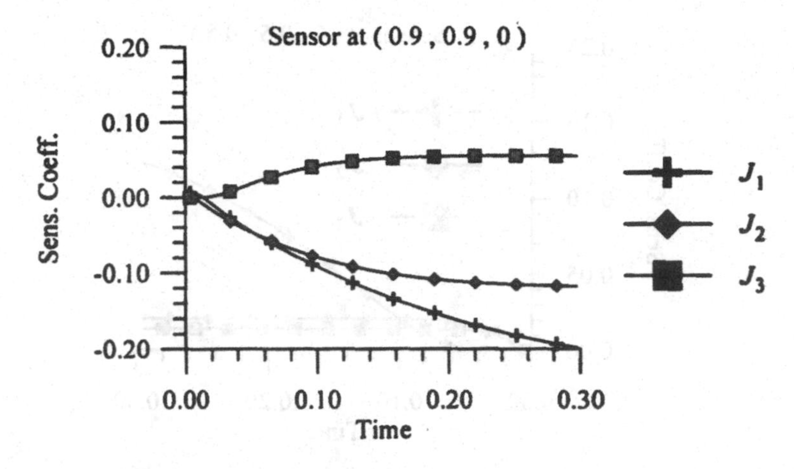

Figure 3.1.1.c - Relative Sensitivity Coefficients for a sensor located at (0.9, 0.9, 0).

Figures 3.1.2.a-c present the transient variation of the sensitivity coefficients for sensors located at the positions (0, 0.5, 0.5), (0.5, 0, 0.5) and (0.5, 0.5, 0), respectively. By comparing figures 3.1.1.a (sensor at 0, 0.9, 0.9) and 3.1.2.a (sensor at 0, 0.5, 0.5) we notice that the curves for J_1 are identical for these two different sensor locations. This is because the x position of the two sensors are the same (see equation 3.1.8). Similar behaviors are noticed for J_2 and J_3, as can be observed in figures 3.1.1.b and 3.1.2.b, as well as in figures 3.1.1.c and 3.1.2.c, respectively. We notice in figures 3.1.2.a-c that the sensitivity coefficients are positive for the thermal conductivities in the directions not normal to the surfaces where the sensors are located, while in figures 3.1.1.a-c such sensitivity coefficients are negative. This is in accordance with the physics of the problem, since an increase in the thermal conductivities tends to decrease the temperature in regions closer to the hottest point in the solid (point 1, 1, 1), but tends to increase the temperatures in regions far from such point. At each sensor location of figures 3.1.2.a-c, the sensitivity coefficients tend to be more linearly-dependent than those of figures 3.1.1.a-c. Also, the sensitivity coefficients for the thermal conductivities in the directions not normal to the surfaces where the sensors are located attain smaller absolute values in figures 3.1.2.a-c than in figures 3.1.1.a-c. As a result, the estimation with sensors located at (0, 0.5, 0.5), (0.5, 0, 0.5) and (0.5, 0.5, 0) is more difficult and not as accurate as the estimation with sensors located at (0, 0.9, 0.9), (0.9, 0, 0.9) and (0.9, 0.9, 0). This fact will be apparent later in the analysis of the results.

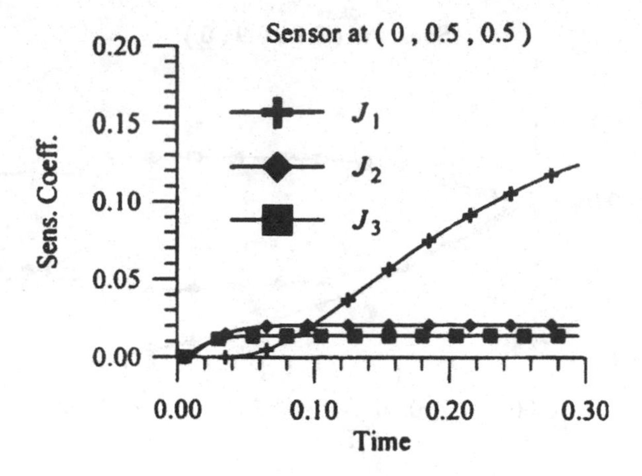

Figure 3.1.2.a – Relative Sensitivity Coefficients for a sensor located at (0, 0.5, 0.5).

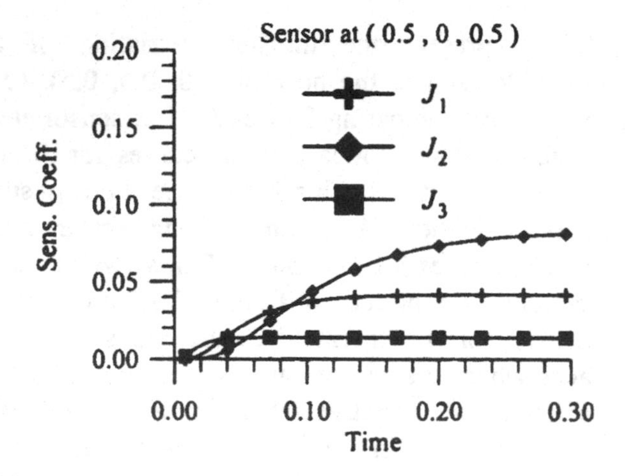

Figure 3.1.2.b – Relative Sensitivity Coefficients for a sensor located at (0.5, 0, 0.5).

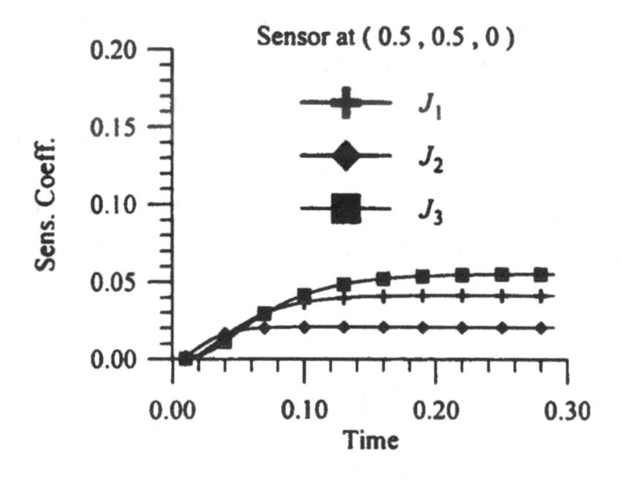

Figure 3.1.2.c – Relative Sensitivity Coefficients for a sensor located at (0.5, 0.5, 0).

Based on the concepts described in Note 2 of Chapter 2, we choose the optimal duration of the experiment by considering available for the inverse analysis a large but fixed number of measurements, of three sensors located at (0, 0.9, 0.9), (0.9, 0, 0.9) and (0.9, 0.9, 0). We also take into account the maximum temperature in the region, T_{max}, which is obtained from equation (3.1.3.a) for the point $x = y = z = 1$ at each final time considered. Hence, we choose to maximize the determinant of the matrix $\mathbf{F}_I^*$, the elements of which are defined by (see equation N2.2.5):

$$\left[\mathbf{F}_I^*\right]_{p,q} = \frac{1}{3t_f}\sum_{m=1}^{3}\int_{t=0}^{t_f}\left(k_p\frac{\partial T_m}{\partial k_p}\right)\left(k_q\frac{\partial T_m}{\partial k_q}\right)\left(\frac{1}{T_{max}}\right)^2 dt \tag{3.1.9}$$

where the subscripts p and q refer to the matrix row and column, respectively ($p,q = 1, 2, 3$).

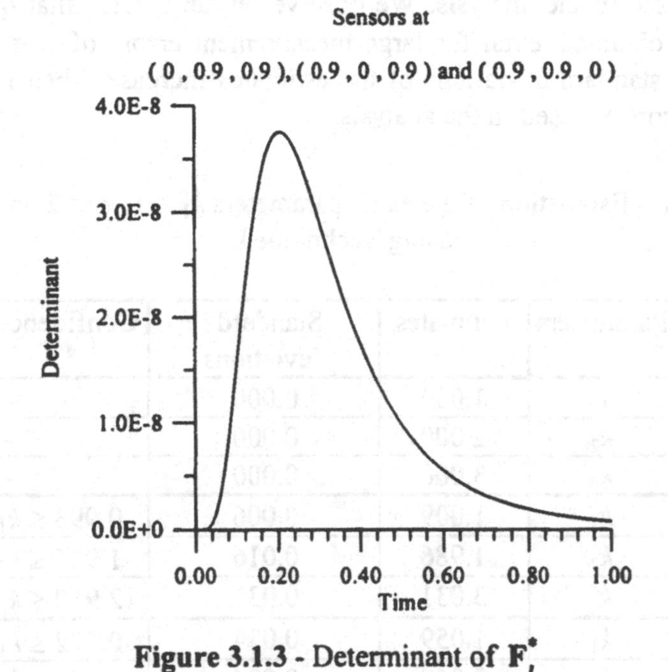

Figure 3.1.3 - Determinant of $\mathbf{F}_I^*$

Figure 3.1.3 shows the variation of the determinant of $\mathbf{F}_I^*$ with time. An analysis of this figure reveals that, for three sensors located at (0, 0.9, 0.9), (0.9, 0, 0.9) and (0.9, 0.9, 0), the duration of the experiment should be taken as $t_f = 0.22$ where such determinant is maximum, so that the confidence region of the estimated parameters is minimized. A similar analysis involving three sensors located at (0, 0.5, 0.5), (0.5, 0, 0.5) and (0.5, 0.5, 0) yields a maximum determinant of 7×10^{-11} for $t_f = 0.3$. Such a value for the determinant is about three orders of magnitude smaller than the maximum determinant of figure 3.1.3. Similarly to the analysis of the sensitivity coefficients, this gives also an indication that the measurements of sensors located at (0, 0.9, 0.9), (0.9, 0, 0.9) and (0.9, 0.9, 0) provide more accurate estimates than the measurements of sensors located at (0, 0.5, 0.5), (0.5, 0, 0.5) and (0.5, 0.5, 0).

We now present the results obtained with the estimation procedure of Technique I, by using in the analysis 100 transient measurements of three sensors located at (0, 0.9, 0.9), (0.9, 0, 0.9) and (0.9, 0.9, 0). The duration of the experiment was taken as $t_f = 0.22$, in accordance with the analysis of figure 3.1.3. The IMSL [4] version of Technique I in the form of subroutine DBCLSJ was used for the estimation of the thermal conductivity components k_1, k_2 and k_3. For the results presented below, we have used as initial guesses in the iterative procedure of Technique I the values $k_1^0 = k_2^0 = k_3^0 = 0.1$.

Table 3.1.1 illustrates the results obtained for the estimated parameters, standard deviations and 99% confidence intervals, for different levels of measurement errors, including $\sigma = 0$, $\sigma = 0.01Y_{max}$ and $\sigma = 0.05Y_{max}$, where Y_{max} is the maximum measured temperature. This table shows that the exact values $k_1 = 1$, $k_2 = 2$ and $k_3 = 3$ are perfectly recovered when errorless measurements ($\sigma = 0$) are used in the analysis. We observe on table 3.1.1 that quite accurate estimates are obtained, even for large measurement errors of $\sigma = 0.05Y_{max}$. As expected, the standard-deviations of the estimates increase when measurements with larger errors are used in the analysis.

Table 3.1.1 – Estimation of the exact parameters $k_1 = 1$, $k_2 = 2$ and $k_3 = 3$ by using Technique I.

σ	Parameters	Estimates	Standard-deviations	Confidence Intervals
0	k_1	1.000	0.000	-
	k_2	2.000	0.000	-
	k_3	3.000	0.000	-
$0.01Y_{max}$	k_1	1.009	0.006	$0.993 \le k_1 \le 1.026$
	k_2	1.986	0.016	$1.945 \le k_2 \le 2.027$
	k_3	3.031	0.031	$2.950 \le k_3 \le 3.111$
$0.05Y_{max}$	k_1	1.059	0.034	$0.972 \le k_1 \le 1.146$
	k_2	2.060	0.084	$1.845 \le k_2 \le 2.276$
	k_3	2.961	0.150	$2.574 \le k_3 \le 3.348$

The standard-deviations and confidence intervals presented in table 3.1.1 were computed in accordance with the concepts described in Note 1 of Chapter 2. Based on such concepts, we can also obtain expressions for the confidence regions at the 99% confidence level. For measurements involving errors of $\sigma = 0.01Y_{max}$ and $\sigma = 0.05Y_{max}$, the confidence regions are given respectively by (see equation N1.2.4 in Chapter 2):

$$46936+24465\,k_1^2+4034.82\,k_2^2-k_2(14685.3+399.28\,k_3)-$$
$$k_1(48383.4+128.615\,k_2+245.66\,k_3)-5240.9\,k_3+1036.41\,k_3^2\leq 0 \tag{3.1.10.a}$$

$$1822.8+873.653\,k_1^2+144.429\,k_2^2-k_2(535.801+16.0212\,k_3)-$$
$$k_1(1791.01+11.1704\,k_2+12.1191\,k_3)-219.778\,k_3+44.8506\,k_3^2\leq 0 \tag{3.1.10.b}$$

We note that the exact values $k_1 = 1$, $k_2 = 2$ and $k_3 = 3$ fall inside the confidence regions given by equations (3.1.10).

Finally, let us consider in the analysis 100 transient measurements of three sensors located at (0, 0.5, 0.5), (0.5, 0, 0.5) and (0.5, 0.5, 0), instead of (0, 0.9, 0.9), (0.9, 0, 0.9) and (0.9, 0.9, 0). In this case, the duration of the experiment was chosen as t_f=0.3, which yielded the maximum determinant of the matrix $\mathbf{F}_1^*$ for the new locations for the sensors. For measurements with σ=0.01Y_{max}, the estimated parameters were $k_1 = 1.007$, $k_2 = 2.006$ and $k_3 = 2.873$, with standard-deviations of $\sigma_{k1} = 0.018$, $\sigma_{k2} = 0.044$ and $\sigma_{k3} = 0.078$, respectively. As expected from the analyses of the sensitivity coefficients and of the maximum determinant of $\mathbf{F}_1^*$, we note that the parameters estimated with the sensors located at (0, 0.5, 0.5), (0.5, 0, 0.5) and (0.5, 0.5, 0) are not as accurate as those shown in table 3.1.1, which were estimated with the sensors located at (0, 0.9, 0.9), (0.9, 0, 0.9) and (0.9, 0.9, 0).

3-2 ESTIMATION OF INITIAL CONDITION [5]

In this section we discuss the inverse problem of estimating the unknown initial condition in a slab of finite thickness by using **Technique IV**, the conjugate gradient method with adjoint problem for function estimation. The solution of inverse problems with Technique IV consists of the following basic steps: direct problem, inverse problem, sensitivity problem, adjoint problem, gradient equation, iterative procedure, stopping criterion and computational algorithm.

The details of such steps, as applied to the inverse problem considered here, are described below.

Direct Problem

The mathematical formulation in dimensionless form of the physical problem considered here, is given by:

$$\frac{\partial^2\theta(X,\tau)}{\partial X^2}=\frac{\partial\theta(X,\tau)}{\partial\tau} \qquad \text{in } 0<X<1, \;\; \tau>0 \tag{3.2.1.a}$$

$$\frac{\partial\theta}{\partial X}=0 \qquad \text{at } X=0 \text{ and } X=1, \text{ for } \tau>0 \tag{3.2.1.b,c}$$

$$\theta(X,0)=F(X) \qquad \text{in } 0<X<1, \text{ for } \tau=0 \tag{3.2.1.d}$$

The direct problem is concerned with the determination of the temperature field $\theta(X,\tau)$, when the initial condition $F(X)$ is known.

Inverse Problem

In the inverse problem, the initial condition $F(X)$ is regarded as unknown and is to be estimated by using the transient measurements of two sensors, located at the boundaries $X = 0$ and $X = 1$, respectively. Figure 3.2.1 shows the geometry, coordinates and the locations of the temperature sensors.

The solution of this inverse heat transfer problem involves the minimization of the following functional:

$$S[F(X)]=\int_{\tau=0}^{\tau_f}\left\{[Z(0,\tau)-\theta(0,\tau)]^2+[Z(1,\tau)-\theta(1,\tau)]^2\right\}d\tau \tag{3.2.2}$$

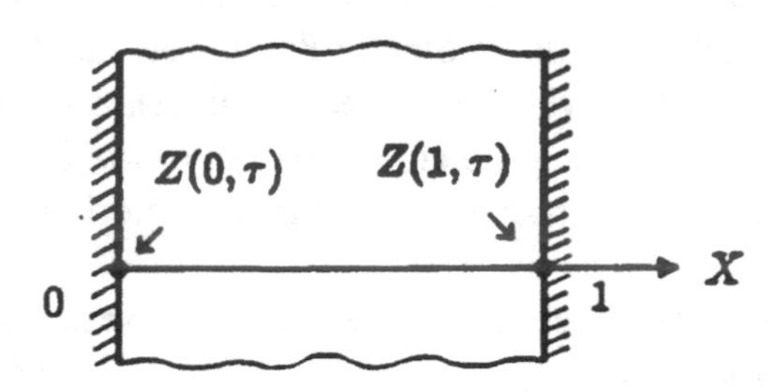

Figure 3.2.1 - Geometry and sensor locations.

where $Z(X,\tau)$ and $\theta(X,\tau)$ are the measured and estimated temperatures, respectively. In order to apply the conjugate gradient method for the minimization of the functional (3.2.2), we need to develop two auxiliary problems, called the *sensitivity* and *adjoint problems*, as described next.

Sensitivity Problem

This problem is obtained by replacing in the above direct problem (3.2.1) $\theta(X,\tau)$ by $[\theta(X,\tau)+\Delta\theta(X,\tau)]$ and $F(X)$ by $[F(X)+\Delta F(X)]$, and by subtracting from the resulting expressions the original direct problem, where $\Delta\theta(X,\tau)$ and $\Delta F(X)$ are small perturbations. We find:

$$\frac{\partial^2 \Delta\theta(X,\tau)}{\partial X^2}=\frac{\partial \Delta\theta(X,\tau)}{\partial \tau}, \quad \text{in } 0<X<1,\ \tau>0 \tag{3.2.3.a}$$

$$\frac{\partial \Delta\theta}{\partial X}=0 \quad \text{at } X=0 \text{ and } X=1, \text{for } \tau>0 \tag{3.2.3.b,c}$$

$$\Delta\theta(X,0)=\Delta F(X) \quad \text{for } \tau=0,\ \text{in } 0<X<1 \tag{3.2.3.d}$$

Adjoint Problem

The adjoint problem is obtained by multiplying equation (3.2.1.a) by the Lagrange Multiplier $\lambda(X,\tau)$, integrating the resulting expression over time and space domains and adding the result to the functional given by equation (3.2.2). We obtain:

$$S[F(X)]=\int_{\tau=0}^{\tau_f}[\theta(0,\tau)-Z(0,\tau)]^2 d\tau+\int_{\tau=0}^{\tau_f}[\theta(1,\tau)-Z(1,\tau)]^2 d\tau+ \\ \int_{\tau=0}^{\tau_f}\int_{X=0}^{1}\lambda(X,\tau)\left[\frac{\partial^2\theta}{\partial X^2}-\frac{\partial\theta}{\partial\tau}\right]dX\,d\tau \tag{3.2.4}$$

Then, the variation $\Delta S[F(X)]$ of the functional $S[F(X)]$ is obtained by perturbing $F(X)$ by $\Delta F(X)$ and $\theta(X,\tau)$ by $\Delta\theta(X,\tau)$, performing integration by parts and utilizing the boundary and initial conditions of the sensitivity problem. Then, by requiring that the coefficients of $\Delta\theta(X,\tau)$ vanish, the following adjoint problem is obtained

$$\frac{\partial^2\lambda}{\partial X^2}+2\{[\theta(0,\tau)-Z(0,\tau)]\delta(X-0)+[\theta(1,\tau)-Z(1,\tau)]\delta(X-1)\}=-\frac{\partial\lambda}{\partial\tau}$$

$$\text{in } 0<X<1,\quad \text{for } 0<\tau<\tau_f \tag{3.2.5.a}$$

$$\frac{\partial \lambda}{\partial X} = 0 \qquad \text{at } X = 0 \text{ and } X = 1, \quad \text{for } 0 < \tau < \tau_f \tag{3.2.5.b,c}$$

$$\lambda = 0 \qquad \text{in } 0 < X < 1, \quad \text{for } \tau = \tau_f \tag{3.2.5.d}$$

Gradient Equation

In the process of obtaining the adjoint problem, the following integral term is left:

$$\Delta S[F(X)] = \int_{X=0}^{1} \lambda(X,0)\, \Delta F(X)\, dX \tag{3.2.6.a}$$

By using the assumption that $F(X)$ belongs to the space of square integrable functions in the domain $0 < X < 1$, we can write:

$$\Delta S[F(X)] = \int_{X=0}^{1} \nabla S[F(X)]\, \Delta F(X)\, dX \tag{3.2.6.b}$$

Thus, by comparing equations (3.2.6.a,b), we obtain the gradient equation for the functional as

$$\nabla S[F(X)] = \lambda(X,0) \tag{3.2.7}$$

Iterative Procedure

The iterative procedure of the conjugate gradient method, as applied to the estimation of the initial function $F(X)$ is given by:

$$F^{k+1}(X) = F^{k}(X) - \beta^{k} d^{k}(X) \tag{3.2.8.a}$$

where the superscript k refers to the number of iterations. The direction of descent is obtained as:

$$d^{k}(X) = \nabla S[F^{k}(X)] + \gamma^{k} d^{k-1}(X) \tag{3.2.8.b}$$

and the conjugation coefficient used here is given by the Fletcher-Reeves expression as:

$$\gamma^k = \frac{\int_{X=0}^{1}\left\{\nabla S[F^k(X)]\right\}^2 dX}{\int_{X=0}^{1}\left\{\nabla S[F^{k-1}(X)]\right\}^2 dX} \quad \text{for } k = 1, 2, \ldots \quad \text{with } \gamma^0 = 0 \ , \ \text{for } k = 0$$

(3.2.8.c)

By applying the procedure outlined in Note 7 of chapter 2, the search step-size is obtained as:

$$\beta^k = \frac{\int_{\tau=0}^{\tau_f}\left\{[\theta(0,\tau)-Z(0,\tau)]\,\Delta\theta(0,\tau)+[\theta(1,\tau)-Z(1,\tau)]\,\Delta\theta(1,\tau)\right\}d\tau}{\int_{\tau=0}^{\tau_f}\left\{[\Delta\theta(0,\tau)]^2+[\Delta\theta(1,\tau)]^2\right\}d\tau} \tag{3.2.9}$$

where $\Delta\theta\ (X,\tau)$ is the solution of the sensitivity problem, equations (3.2.3), obtained by setting $\Delta F(X) = d(X)$.

Stopping Criterion

The conjugate gradient method requires the stopping criterion based on the *Discrepancy Principle* in order to pursue an iterative regularization character, as discussed in Chapter 2. In the case of the present estimation problem, the stopping criterion is given by:

$$S[F(X)] < \varepsilon \tag{3.2.10}$$

where $S[F(x)]$ is obtained from equation (3.2.2).

In order to obtain the tolerance ε, we assume

$$|Z(\tau)-\theta(\tau)| \approx \sigma \tag{3.2.11}$$

where σ is the constant standard deviation of the measurements. Thus, ε is obtained from equation (3.2.2) as:

$$\varepsilon = 2\sigma^2\tau_f \tag{3.2.12}$$

The *computational algorithm* of Technique IV, as applied to the estimation of the unknown initial condition $F(X)$, is quite similar to the one presented in section 2-4, and is not repeated here for the sake of brevity.

Results

The accuracy of the inverse analysis for estimating the initial condition was examined by using simulated temperature readings. The simulated temperature data containing measurement errors, Z, were generated by solving the direct problem for a specified initial condition $F(X)$ and by adding to it an error term, as outlined in section 2-5.

Figures 3.2.2.a and 3.2.2.b show the estimated functions for an exact sine variation used in the direct problem to generate the measurements, and for the final experimental time of $\tau_f = 0.024$ and $\tau_f = 0.040$, respectively. The standard deviation of measurement errors was taken as $\sigma = 0.04$, representing an error of up to 10% in the input data. An examination of figures 3.2.2.a,b reveals that the accuracy of estimation improves as the final time increases. The reason for this behaviour is that as time elapses, more information reaches the boundaries allowing better estimations.

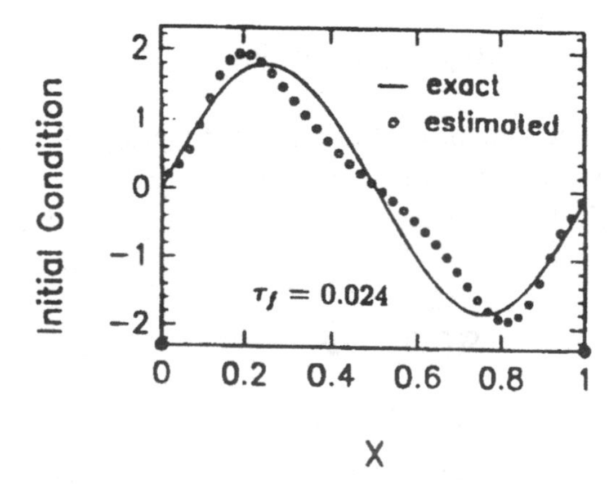

Figure 3.2.2.a - Estimated initial condition for final experimental time $\tau_f = 0.024$.

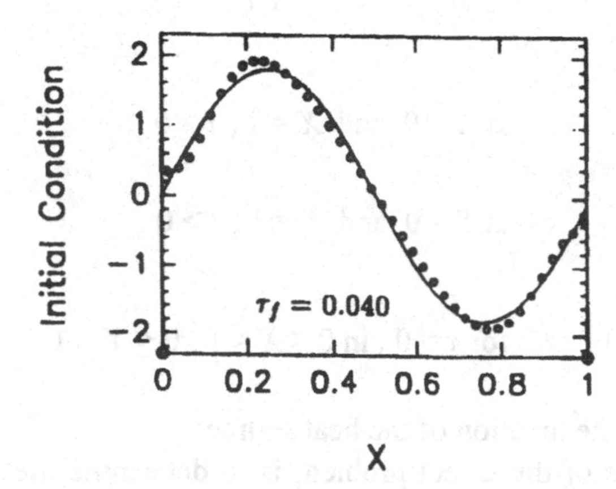

Figure 3.2.2.b - Estimated initial condition for final experimental time $\tau_f = 0.040$.

3-3 ESTIMATION OF TIMEWISE VARIATION OF THE STRENGTH OF A LINE HEAT SOURCE [6]

In this section we illustrate the application of **Technique IV**, the conjugate gradient method with adjoint problem, for the estimation of the timewise-varying strength of a line-heat source, in a two-dimensional inverse heat conduction problem. We assume that no *a priori* information is available on the functional form of such variation of the heat source. The basic steps in the analysis include: direct problem, inverse problem, sensitivity problem, adjoint problem, gradient equation, iterative procedure, stopping criterion and computational algorithm. We present below the details of such basic steps, except for the stopping criterion and the computational algorithm. They are very similar to those presented in section 2-4 and are not repeated here for the sake of brevity.

Direct Problem

The physical problem considered here involves two dimensional transient heat conduction in a dimensionless square domain $0 < X < 1$, $0 < Y < 1$, initially at zero temperature. For times $\tau > 0$, a line heat source of strength $G(\tau)$ is activated to generate energy while the lateral surfaces are kept insulated.

The mathematical formulation of this heat conduction problem in dimensionless form is given by

$$\frac{\partial^2\theta}{\partial X^2}+\frac{\partial^2\theta}{\partial Y^2}+G(\tau)\delta(X-X^*)\delta(Y-Y^*)=\frac{\partial\theta}{\partial\tau} \qquad \text{in } 0<X<1\,,\,0<Y<1\,,\,\tau>0 \tag{3.3.1.a}$$

$$\frac{\partial\theta}{\partial X}=0 \qquad \text{at } X=0 \text{ and } X=1\,,\,\tau>0 \tag{3.3.1.b,c}$$

$$\frac{\partial\theta}{\partial Y}=0 \qquad \text{at } Y=0 \text{ and } Y=1\,,\,\tau>0 \tag{3.3.1.d,e}$$

$$\theta(X,Y,\tau)=0 \qquad \text{for } \tau=0\,,\text{ in } 0<X<1\,,\,0<Y<1 \tag{3.3.1.f}$$

where X^* and Y^* is the location of the heat source.

The objective of the direct problem is to determine the temperature field $\theta(X,Y,\tau)$ in the medium, where the strength $G(\tau)$ of the heat source is known.

Inverse Problem

The inverse problem is concerned with the estimation of the unknown timewise varying strength, $G(\tau)$, of the line heat source, by utilizing the transient temperature readings of a temperature sensor. Figures (3.3.1.a,b) show the locations of the source and sensor, in two different configurations tested.

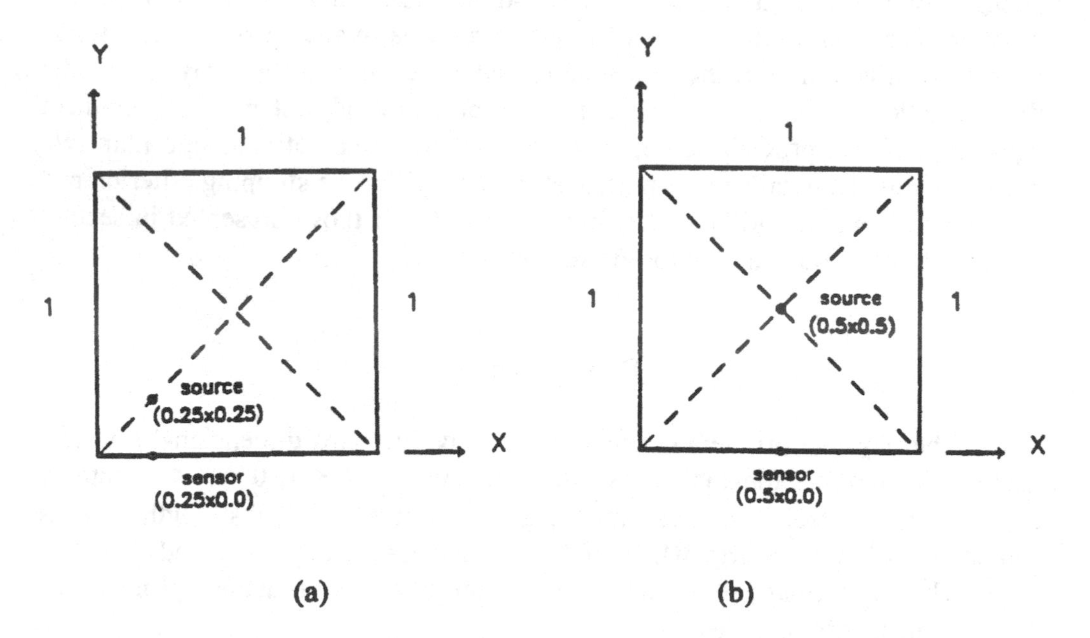

Figure 3.3.1 - Locations of the line heat source and temperature sensor.

The inverse problem is solved here by the application of Technique IV, since no *a priori* information is available on the functional form of the timewise variation of the heat source.

The inverse problem is solved so that the following functional is minimized.

$$S[G(\tau)] = \sum_{m=1}^{M} \int_{\tau=0}^{\tau_f} [Z_m(\tau) - \theta_m(\tau)]^2 d\tau \tag{3.3.2}$$

where $Z_m(\tau)$ and $\theta_m(\tau)$ are the measured and estimated temperatures, respectively, while M is the total number of sensors, assumed greater than one in the formulation for the sake of generality.

In order to implement the iterative algorithm of the conjugate gradient method, we need to develop the *sensitivity* and *adjoint problems*, as described below.

Sensitivity Problem

Suppose that the energy generation rate $G(\tau)$ is perturbed by a small amount $\Delta G(\tau)$; it results in a small change in temperature by an amount $\Delta\theta(X,Y,\tau)$. Then, the sensitivity problem governing $\Delta\theta(X,Y,\tau)$ is obtained by replacing in the direct problem (3.3.1), $\theta(X,Y,\tau)$ by $[\theta(X,Y,\tau) + \Delta\theta(X,Y,\tau)]$, $G(\tau)$ by $[G(\tau) + \Delta G(\tau)]$ and subtracting from the resulting expressions the original direct problem. We obtain:

$$\nabla^2[\Delta\theta(X,Y,\tau)] + \Delta G(\tau)\delta(X - X^*)\delta(Y - Y^*) = \frac{\partial[\Delta\theta(X,Y,\tau)]}{\partial\tau}$$

$$\text{in } 0 < X < 1,\ 0 < Y < 1, \tau > 0 \tag{3.3.3.a}$$

$$\frac{\partial(\Delta\theta)}{\partial X} = 0 \qquad \text{at } X = 0 \text{ and } X = 1, \text{ for } \tau > 0 \tag{3.3.3.b,c}$$

$$\frac{\partial(\Delta\theta)}{\partial Y} = 0 \qquad \text{at } Y = 0 \text{ and } Y = 1, \text{ for } \tau > 0 \tag{3.3.3.d,e}$$

$$\Delta\theta(X,Y,\tau) = 0 \qquad \text{for } \tau = 0, \text{ in } 0 < X < 1, 0 < Y < 1 \tag{3.3.3.f}$$

where ∇^2 is the Laplacian in rectangular coordinates. Equations (3.3.3) give the sensitivity problem for the determination of the sensitivity function $\Delta\theta(X,Y,\tau)$.

Adjoint Problem

The adjoint problem is developed by multiplying equation (3.3.1.a) by the Lagrange multiplier $\lambda(X,Y,\tau)$, integrating the resulting expression over the time and spatial domains and then adding this result to the functional given by equation (3.3.2). We obtain:

$$S[G(\tau)] = \sum_{m=1}^{M} \int_{\tau=0}^{\tau_f} [\theta_m(\tau) - Z_m(\tau)]^2 d\tau +$$

$$\int_{\tau=0}^{\tau_f} \int_{Y=0}^{1} \int_{X=0}^{1} \lambda(X,Y,\tau) \left[\nabla^2 \theta + G(\tau)\delta(X - X^*)\delta(Y - Y^*) - \frac{\partial \theta}{\partial \tau} \right] dX\, dY\, d\tau \tag{3.3.4}$$

The variation $\Delta S[G(\tau)]$ of the functional $S[G(\tau)]$ is obtained by perturbing $G(\tau)$ by $\Delta G(\tau)$ and $\theta(X,Y,\tau)$ by $\Delta\theta(X,Y,\tau)$ in equation (3.3.4) and subtracting from it the original equation (3.3.4). By neglecting the second order terms, performing integrations by parts and using the boundary and initial conditions of the sensitivity problem, we obtain the following adjoint problem for the determination of the Lagrange Multiplier $\lambda(X,Y,\tau)$:

$$\nabla^2 \lambda + \sum_{m=1}^{M} 2\left[\theta_m(\tau) - Z_m(\tau)\right] \delta(X - X_m)\, \delta(Y - Y_m) = -\frac{\partial \lambda}{\partial \tau}$$

$$\text{in } 0 < X < 1,\ 0 < Y < 1, \text{ for } 0 < \tau < \tau_f \tag{3.3.5.a}$$

$$\frac{\partial \lambda}{\partial X} = 0 \qquad \text{at } X = 0 \text{ and } X = 1, \text{ for } 0 < \tau < \tau_f \tag{3.3.5.b,c}$$

$$\frac{\partial \lambda}{\partial Y} = 0 \qquad \text{at } Y = 0 \text{ and } Y = 1, \text{ for } 0 < \tau < \tau_f \tag{3.3.5.d,e}$$

$$\lambda = 0 \qquad \text{for } \tau = \tau_f,\ \text{in } 0 < X < 1,\ 0 < Y < 1 \tag{3.3.5.f}$$

where the points (X_m, Y_m), $m = 1,2,...M$ are the locations of the sensors.

Gradient Equation

In the process of obtaining the above adjoint problem, the expression for the variation of the functional $\Delta S[G(\tau)]$ reduces to

$$\Delta S[G(\tau)] = \int_{\tau=0}^{\tau_f} \lambda(X^*,Y^*,\tau)\,\Delta G(\tau)\,d\tau \tag{3.3.6.a}$$

By assuming that $G(\tau)$ belongs to the space of square integrable functions in $0 < \tau < \tau_f$, we can write

$$\Delta S[G(\tau)] = \int_{\tau=0}^{\tau_f} \nabla S[G(\tau)]\,\Delta G(\tau)\,d\tau \tag{3.3.6.b}$$

Then, by comparing equations (3.3.6.a,b) we find the *gradient equation* as

$$\nabla S[G(\tau)] = \lambda(X^*,Y^*,\tau) \tag{3.3.7}$$

Iterative Procedure

The conjugate gradient method of minimization, as applied to the estimation of the unknown function $G(\tau)$, is written as

$$G^{k+1}(\tau) = G^k(\tau) - \beta^k d^k(\tau) \tag{3.3.8.a}$$

where the superscript k refers to the number of iterations, and the direction of descent is taken as

$$d^k(\tau) = \nabla S[G^k(\tau)] + \gamma^k d^{k-1}(\tau) \tag{3.3.8.b}$$

The conjugation coefficient γ^k is given by the Fletcher-Reeves expression as

$$\gamma^k = \frac{\int_{\tau=0}^{\tau_f} \{\nabla S[G^k(\tau)]\}^2 d\tau}{\int_{\tau=0}^{\tau_f} \{\nabla S[G^{k-1}(\tau)]\}^2 d\tau} \qquad \text{for } k = 1,2,\ldots \quad \text{with } \gamma^0 = 0 \tag{3.3.8.c}$$

and the search step-size β^k is determined by the minimization of the objective function (3.3.2) as (see Note 7 in Chapter 2):

$$\beta^k = \frac{\sum_{m=1}^{M} \int_{\tau=0}^{\tau_f} \left[\theta_m(\tau) - Z_m(\tau)\right] \Delta\theta(d^k)\, d\tau}{\sum_{m=1}^{M} \int_{\tau=0}^{\tau_f} \left[\Delta\theta(d^k)\right]^2 d\tau} \tag{3.3.9}$$

where $\Delta\theta(d^k)$ is the solution of the sensitivity problem (3.3.3), obtained by setting $\Delta G(\tau) = d^k(\tau)$. Note that the only difference between β^k, as given by the above expression (3.3.9) and that given by equation (N7.2.9) in Note 7 in Chapter 2, is that the former contains a summation term due to the presence of multiple sensors.

Results

The accuracy of the inverse analysis for estimating the timewise varying strength of an unknown line heat source $G(\tau)$, located at a specified position (X^*, Y^*), is now examined by using simulated measured data. Several test cases have been run with simulated test data $Z_m(\tau)$ and the estimated values were compared with the exact results.

For all the cases considered here, the stopping criterion given by the discrepancy principle was used to stop the iterations. The functions exhibiting a step change or a sharp corner are generally the most difficult cases to be recovered by inverse analysis. In order to perform the tests under most strict conditions, functions involving abrupt changes in the form of step and triangular variations were considered for $G(\tau)$.

A finite difference mesh of 25 x 25 nodes was used for the spatial discretization and the value of the dimensionless final time, $\tau_f = 6.9 \times 10^{-2}$, was divided into 280 time steps for all the results presented here. For a 10 cm thick region, this value of final dimensionless time corresponds to a physical final time $t_f = 69$ seconds for a material having thermal diffusivity $\alpha = 10^{-5}$ m^2/s and to a physical time $t_f = 6900$ seconds for an insulating material having $\alpha = 10^{-7}$ m^2/s.

By examining equations (3.3.5.f) and (3.3.7), we note that the gradient equation is null at the final time τ_f. Therefore, the initial guess used for $G(\tau)$ is never changed by the iterative procedure of the conjugate gradient method. In order to avoid such difficulty, the calculations were repeated few times by using for the initial guess, previously estimated values for $G(\tau)$ at a time τ in the neighbourhood of τ_f.

Two different locations of the source and sensor considered in the present study included the cases:

(a) The source $G(\tau)$ at (0.25, 0.25) and the sensor at (0.25, 0.0), as shown in figure 3.3.1.a

(b) The source $G(\tau)$ at (0.5, 0.5) and the sensor at (0.5, 0.0), as illustrated in figure 3.3.1.b

Figure 3.3.2 presents the exact and estimated strength $G(\tau)$, for the configuration shown in figure 3.3.1.a and for measured data involving a standard deviation $\sigma = 0.0025$. The agreement between exact and estimated functions is quite good.

Figures 3.3.3 illustrate the effect on the inverse problem solution, of locating a sensor farther from the source. It can be clearly noticed that the estimation deteriorates for the case involving the configuration shown in figure 3.3.1.b, where the source is located at the center of the region, as compared to that obtained with the configuration of figure 3.3.1.a.

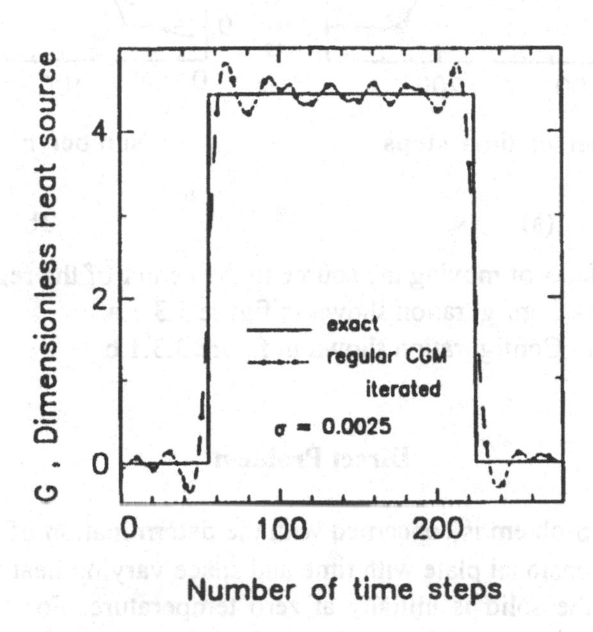

Figure 3.3.2 - The estimation of the strength of a line heat source varying with time as a step function.

3-4 ESTIMATION OF TIMEWISE AND SPACEWISE VARIATIONS OF THE STRENGTH OF A VOLUMETRIC HEAT SOURCE [7]

In the previous section we presented the estimation of the timewise varying strength of a line heat source by the application of **Technique IV**. We now apply this technique for estimating the timewise and spacewise varying strength of a volumetric heat source, $G(X, \tau)$, in a plate. The solution technique follows the methodology described previously, which includes the following basic steps: direct problem, inverse problem, sensitivity problem, adjoint problem, gradient equation, iterative procedure, stopping criterion and computational algorithm.

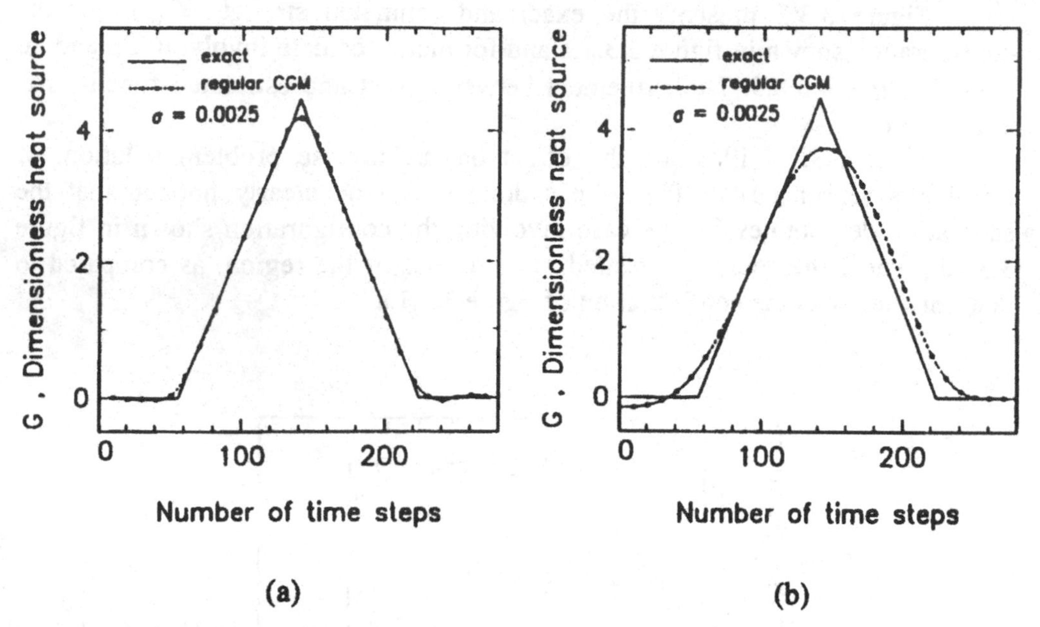

Figure 3.3.3 - Effects of moving the source to the center of the region.
(a) - Configuration shown in figure 3.3.1.a.
(b) - Configuration shown in figure 3.3.1.b.

Direct Problem

The direct problem is concerned with the determination of the temperature field in a one-dimensional plate with time and space varying heat source, $G(X, \tau)$. We assume that the solid is initially at zero temperature. For times $t > 0$, the energy source is activated while the boundaries at $X = 0$ and $X = 1$ are insulated. The mathematical formulation of this direct problem in dimensionless form is given by

$$\frac{\partial^2 \theta(X,\tau)}{\partial X^2} + G(X,\tau) = \frac{\partial \theta(X,\tau)}{\partial \tau} \quad \text{in } 0 < X < 1, \text{for } \tau > 0 \tag{3.4.1.a}$$

$$\frac{\partial \theta}{\partial X} = 0 \quad \text{at } X = 0, \quad \text{for } \tau > 0 \tag{3.4.1.b}$$

$$\frac{\partial \theta}{\partial X} = 0 \quad \text{at } X = 1, \quad \text{for } \tau > 0 \tag{3.4.1.c}$$

$$\theta(X,\tau) = 0 \quad \text{for } \tau = 0, \quad \text{in } 0 < X < 1 \tag{3.4.1.d}$$

Inverse Problem

For the inverse problem, the source function $G(X,\tau)$ is regarded as unknown. In order to estimate $G(X,\tau)$, we consider available the transient readings of M temperature sensors in the region and choose to minimize the following functional:

$$S[G(X,\tau)]=\sum_{m=1}^{M}\int_{\tau=0}^{\tau_f}[Z_m(\tau)-\theta_m(\tau)]^2 d\tau \tag{3.4.2}$$

where $Z_m(\tau)$ is the measured temperature of sensor m (m = 1, ..., M), while $\theta_m(\tau)$ is the estimated temperature at the sensor location, which is obtained from the solution of the direct problem by using an estimate for $G(X,\tau)$.

For the solution of the present inverse problem via the conjugate gradient method of function estimation, we need to develop the sensitivity and adjoint problems as described next.

Sensitivity Problem

The sensitivity problem is obtained by replacing in the direct problem (3.4.1), $\theta(X,\tau)$ by $[\theta(X,\tau)+\Delta\theta(X,\tau)]$, $G(X,\tau)$ by $[G(X,\tau)+\Delta G(X,\tau)]$ and subtracting from the resulting expression the original direct problem, where $\Delta\theta(X,\tau)$ and $\Delta G(X,\tau)$ are small perturbations. We find:

$$\frac{\partial^2[\Delta\theta(X,\tau)]}{\partial X^2}+\Delta G(X,\tau)=\frac{\partial[\Delta\theta(X,\tau)]}{\partial\tau} \quad \text{in } 0<X<1, \text{ for } \tau>0 \tag{3.4.3.a}$$

$$\frac{\partial(\Delta\theta)}{\partial X}=0 \quad \text{at } X=0, \text{ for } \tau>0 \tag{3.4.3.b}$$

$$\frac{\partial(\Delta\theta)}{\partial X}=0 \quad \text{at } X=1, \text{ for } \tau>0 \tag{3.4.3.c}$$

$$\Delta\theta(X,\tau)=0 \quad \text{for } \tau=0, \text{ in } 0<X<1 \tag{3.4.3.d}$$

Adjoint Problem

The adjoint problem is developed by multiplying equation (3.4.1.a) by the Lagrange multiplier $\lambda(X,\tau)$, integrating the resulting expression over time and space domains and then adding the result to the functional given by equation (3.4.2). We obtain:

$$S[G(X,\tau)] = \sum_{m=1}^{M} \int_{\tau=0}^{\tau_f} [\theta_m(\tau) - Z_m(\tau)]^2 d\tau + \int_{\tau=0}^{\tau_f} \int_{X=0}^{1} \lambda(X,\tau) \left[\frac{\partial^2 \theta}{\partial X^2} + G(X,\tau) - \frac{\partial \theta}{\partial \tau} \right] dX\, d\tau \tag{3.4.4}$$

The variation $\Delta S[G(X,\tau)]$ of the functional $S[G(X,\tau)]$ is obtained by perturbing $G(X,\tau)$ by $\Delta G(X,\tau)$ and $\theta(X,\tau)$ by $\Delta\theta(X,\tau)$ in equation (3.4.4) and subtracting from it the original equation (3.4.4). By neglecting second-order terms, performing integration by parts and using the boundary and initial conditions of the sensitivity problem, we obtain after some manipulations the following adjoint problem:

$$\frac{\partial^2 \lambda(X,\tau)}{\partial X^2} + \sum_{m=1}^{M} 2\left[\theta_m(\tau) - Z_m(\tau)\right] \delta(X - X_m) = -\frac{\partial \lambda(X,\tau)}{\partial \tau} \quad \text{in } 0 < X < 1, \quad \text{for } 0 < \tau < \tau_f \tag{3.4.5.a}$$

$$\frac{\partial \lambda}{\partial X} = 0 \quad \text{at } X = 0, \quad \text{for } 0 < \tau < \tau_f \tag{3.4.5.b}$$

$$\frac{\partial \lambda}{\partial X} = 0 \quad \text{at } X = 1, \quad \text{for } 0 < \tau < \tau_f \tag{3.4.5.c}$$

$$\lambda = 0 \quad \text{for } \tau = \tau_f, \quad \text{in } 0 < X < 1 \tag{3.4.5.d}$$

Gradient Equation

In the process used to obtain the adjoint problem, the following integral term is left:

$$\Delta S[G(X,\tau)] = \int_{\tau=0}^{\tau_f} \int_{X=0}^{1} \lambda(X,\tau)\, \Delta G(X,\tau) dX\, d\tau \tag{3.4.6.a}$$

By using the hypothesis that $G(X, \tau)$ belongs to the space of square integrable functions in the domain $0 < \tau < \tau_f$ and $0 < X < 1$, we can write

$$\Delta S[G(X,\tau)] = \int_{\tau=0}^{\tau_f} \int_{X=0}^{1} \nabla S[G(X,\tau)] \Delta G(X,\tau)\, dX\, d\tau \tag{3.4.6.b}$$

Thus, by comparing equations (3.4.6.a,b) we obtain the gradient equation as

$$\nabla S[G(X,\tau)] = \lambda(X,\tau) \tag{3.4.7}$$

Iterative Procedure

The iterative procedure of Technique IV, as applied to the estimation of the function $G(X, \tau)$, is given by:

$$G^{k+1}(X,\tau) = G^k(X,\tau) - \beta^k d^k(X,\tau) \tag{3.4.8.a}$$

where the direction of descent at iteration k is obtained as a conjugation of the gradient direction and of the previous direction of descent, in the form

$$d^k(X,\tau) = \nabla S[G(X,\tau)] + \gamma^k d^{k-1}(X,\tau) \tag{3.4.8.b}$$

The conjugation coefficient is obtained from the Fletcher-Reeves expression as

$$\gamma^k = \frac{\int_{\tau=0}^{\tau_f} \int_{X=0}^{1} \left\{\nabla S[G^k(X,\tau)]\right\}^2 dX\, d\tau}{\int_{\tau=0}^{\tau_f} \int_{X=0}^{1} \left\{\nabla S[G^{k-1}(X,\tau)]\right\}^2 dX\, d\tau} \quad \text{for } k = 1,2,\ldots \text{ with } \gamma^0 = 0 \tag{3.4.8.c}$$

and the search step size is determined as (see Note 7 in Chapter 2)

$$\beta^k = \frac{\sum_{m=1}^{M} \int_{\tau=0}^{\tau_f} \left[\theta_m(\tau) - Z_m(\tau)\right] \Delta\theta(d^k)\, d\tau}{\sum_{m=1}^{M} \int_{\tau=0}^{\tau_f} \left[\Delta\theta(d^k)\right]^2 d\tau} \tag{3.4.8.d}$$

where $\Delta\theta(d^k)$ is the solution of the sensitivity problem given by equations (3.4.3), obtained by setting $\Delta G(X, \tau) = d^k(X, \tau)$.

Results

The accuracy of Technique IV, as applied to the estimation of $G(X,\tau)$ is examined by using simulated measured temperature data. The stopping criterion was based on the ***Discrepancy Principle*** as described in section 2-4. Similarly, the computational algorithm presented in section 2-4 can be applied to the present estimation problem with few modifications. Hence, they are not repeated here.

Figures 3.4.1.a,b show the estimated function $G(X,\tau)$ by using errorless measurements ($\sigma = 0$) taken by seven equally spaced temperature sensors. Figure 3.4.1.a shows G as a function of position at different dimensionless times (i.e., $\tau/\tau_f = 0.1$, 0.3 and 0.5), while figure 3.4.1.b shows G as a function of time τ/τ_f at three different locations (ie., $X = 0.13$, 0.25 and 0.5). Similar results, obtained by using 9 equally spaced temperature sensors containing measurement error ($\sigma = 0.05$), are shown in figure 3.4.2.a,b. The results were good, showing the feasibility of such estimates. Reasonably accurate estimates were obtained for $G(X,\tau)$ with standard deviation $\sigma = 0.05$, corresponding to an error of up to 13%.

3-5 ESTIMATION OF TEMPERATURE-DEPENDENT PROPERTIES AND REACTION FUNCTION [8,9]

In the previous sections of this chapter, we considered inverse problems involving linear heat conduction. In this section, we illustrate the solution of the inverse problems of estimating the temperature-dependent thermal conductivity, heat capacity or reaction function. The reaction-diffusion type of problems considered here are found in nonlinear heat conduction, chemical reactor analysis, combustion, enzyme kinetics, population dynamics and many other practical applications.

Inverse problems of estimating temperature-dependent properties and reaction function have been generally solved by using Technique III [10-12]. However, in situations where no information is available on the functional form of the unknown quantity, the inverse problem can be recast as a function estimation problem. Here we apply **Technique IV**, ***the conjugate gradient method with adjoint problem for function estimation***, to solve such classes of inverse problems. Details on the basic steps of Technique IV are described below. Also, a comparison of Techniques III and IV is presented for the case of estimating the reaction function.

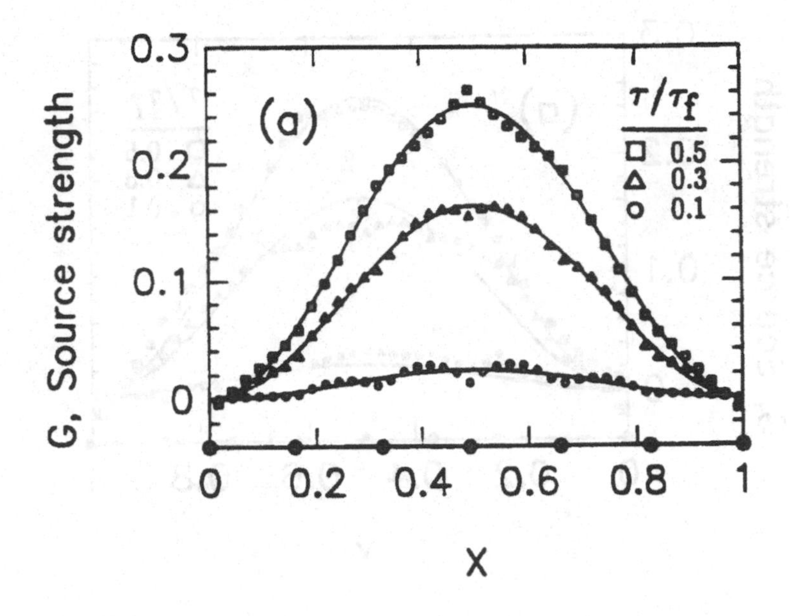

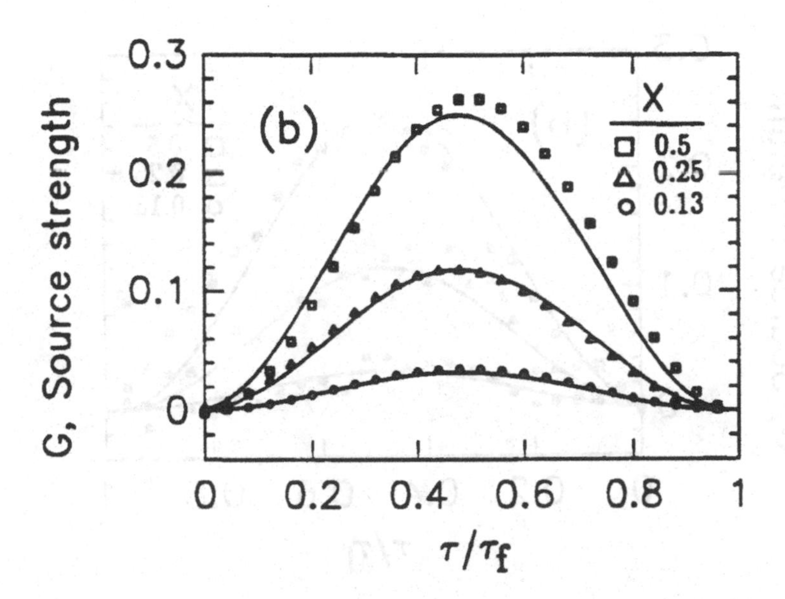

Figure 3.4.1 - The estimation of a space and time dependent volumetric heat source by using 7 temperature sensors and $\sigma = 0$.
(a) Spatial variation. (b) Timewise variation.

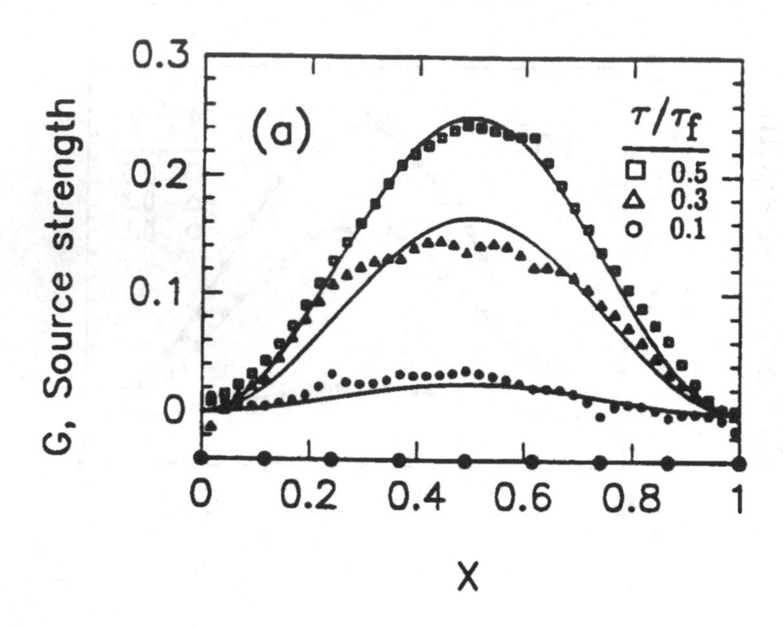

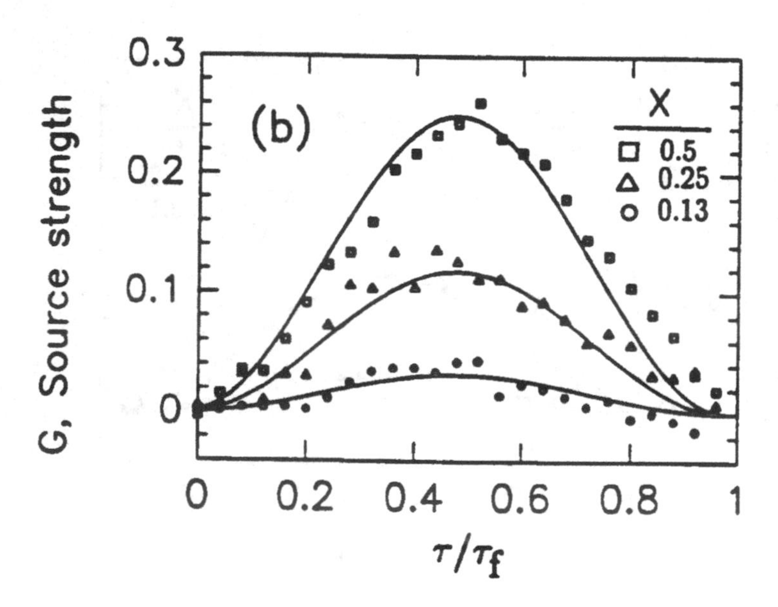

Figure 3.4.2 - The estimation of a space and time dependent volumetric heat source by using 9 temperature sensors and $\sigma = 0.05$.
(a) Spatial variation. (b) Timewise variation.

Direct Problem

For the present study, we consider the following nonlinear, one-dimensional heat conduction problem with temperature dependent properties and reaction function:

$$C(T)\frac{\partial T(x,t)}{\partial t}-\frac{\partial}{\partial x}\left[k(T)\frac{\partial T}{\partial x}\right]-g(T)=0 \quad \text{in } 0<x<L\text{, for } t>0 \tag{3.5.1.a}$$

$$\frac{\partial T}{\partial x}=0 \quad \text{at } x=0\text{, for } t>0 \tag{3.5.1.b}$$

$$k(T)\frac{\partial T}{\partial x}=\phi_L(t) \quad \text{at } x=L\text{, for } t>0 \tag{3.5.1.c}$$

$$T(x,0)=F(x) \quad \text{for } t=0\text{, in } 0<x<L \tag{3.5.1.d}$$

The *direct problem* defined above by equations (3.5.1) is concerned with the determination of the temperature distribution $T(x,t)$ in the medium, when the physical properties $C(T)$ and $k(T)$, the boundary and initial conditions, and the reaction function $g(T)$ are known.

Inverse Problem

Consider the following three different inverse problems of estimating:

(i) $g(T)$ unknown, but $k(T)$ and $C(T)$ known
(ii) $k(T)$ unknown, but $C(T)$ and $g(T)$ known
(iii) $C(T)$ unknown, but $k(T)$ and $g(T)$ known

where $g(T)$ is the energy generation rate (reaction-function), $k(T)$ is the thermal conductivity and $C(T)$ is the heat capacity.

For the solution of each of these inverse problems, we consider transient temperature readings available from M temperature sensors at the positions x_m, m = 1, 2, ..., M. To solve these inverse problems, one needs to minimize the following objective functional $S[P(T)]$ defined as

$$S[P(T)]=\sum_{m=1}^{M}\int_{t=0}^{t_f}\{Y_m(t)-T[x_m,t;P(T)]\}^2\,dt \tag{3.5.2}$$

where $P(T) \equiv g(T)$, $k(T)$ or $C(T)$, unknown quantities,
$Y_m(t)$ is the measured temperature, and

$T[x_m, t; P(T)]$ is the estimated temperature.

The estimated temperatures are obtained from the solution of the direct problem by using an estimate for the unknown quantity $P(T)$.

The development of the sensitivity and adjoint problems, required for the implementation of Technique IV, are described next.

Sensitivity Problem

In order to develop the sensitivity problem we assume that the unknown quantity $P(T)$ is perturbed by an amount $\varepsilon\Delta P(T)$. Thus, the temperature $T(x,t)$ undergoes a variation $\varepsilon\Delta T(x,t)$, that is,

$$T_\varepsilon(x,t) = T(x,t) + \varepsilon\Delta T(x,t) \tag{3.5.3.a}$$

where ε is a real number and, as a subscript, ε refers to a perturbed variable.

Due to the nonlinear character of the problem, the perturbation of temperature causes variations on the temperature-dependent properties, as well as on the reaction function. The resulting perturbed quantities are linearized as:

$$k_\varepsilon(T_\varepsilon) = k(T+\varepsilon\Delta T) + \varepsilon\Delta k(T) \approx k(T) + \left(\frac{dk}{dT}\right)\varepsilon\Delta T + \varepsilon\Delta k(T) \tag{3.5.3.b}$$

$$C_\varepsilon(T_\varepsilon) = C(T+\varepsilon\Delta T) + \varepsilon\Delta C(T) \approx C(T) + \left(\frac{dC}{dT}\right)\varepsilon\Delta T + \varepsilon\Delta C(T) \tag{3.5.3.c}$$

$$g_\varepsilon(T_\varepsilon) = g(T+\varepsilon\Delta T) + \varepsilon\Delta g(T) \approx g(T) + \left(\frac{dg}{dT}\right)\varepsilon\Delta T + \varepsilon\Delta g(T) \tag{3.5.3.d}$$

where $\Delta k(T) = \Delta C(T) = 0$ for $P(T) = g(T)$ unknown,
$\Delta C(T) = \Delta g(T) = 0$ for $P(T) = k(T)$ unknown, and
$\Delta k(T) = \Delta g(T) = 0$ for $P(T) = C(T)$ unknown.

For convenience in the subsequent analysis, the differential equation (3.5.1.a) of the direct problem is written as

$$D(T) \equiv C(T)\frac{\partial T(x,t)}{\partial t} - \frac{\partial}{\partial x}\left[k(T)\frac{\partial T}{\partial x}\right] - g(T) = 0 \tag{3.5.4.a}$$

and the perturbed form of this equation becomes

$$D_\varepsilon(T_\varepsilon) \equiv C_\varepsilon(T_\varepsilon)\frac{\partial T_\varepsilon(x,t)}{\partial t} - \frac{\partial}{\partial x}\left[k_\varepsilon(T_\varepsilon)\frac{\partial T_\varepsilon}{\partial x}\right] - g_\varepsilon(T_\varepsilon) = 0 \tag{3.5.4.b}$$

To develop the sensitivity problem we apply a limiting process to the differential equations (3.5.4.a,b) in the form:

$$\lim_{\varepsilon \to 0} \frac{D_{\varepsilon}(T_{\varepsilon}) - D(T)}{\varepsilon} = 0 \tag{3.5.5}$$

and similar limiting processes are applied for the boundary and initial conditions of the direct problem. After some manipulations, the following sensitivity problem results for the determination of the sensitivity function $\Delta T(x,t)$:

$$\frac{\partial (C\Delta T)}{\partial t} - \frac{\partial^2 (k\Delta T)}{\partial x^2} - \frac{dg}{dT}\Delta T - \frac{\partial}{\partial x}\left(\Delta k \frac{\partial T}{\partial x}\right) + \Delta C \frac{\partial T}{\partial t} - \Delta g = 0$$

$$\text{in } 0 < x < L, \text{ for } t > 0 \tag{3.5.6.a}$$

$$\frac{\partial (k\Delta T)}{\partial x} = 0 \qquad \text{at } x = 0, \text{ for } t > 0 \tag{3.5.6.b}$$

$$\frac{\partial (k\Delta T)}{\partial x} = -\frac{\Delta k}{k}\phi_L \qquad \text{at } x = L, \text{ for } t > 0 \tag{3.5.6.c}$$

$$\Delta T = 0 \qquad \text{for } t = 0, \text{ in } 0 < x < L \tag{3.5.6.d}$$

where $\Delta T \equiv \Delta T(x,t)$, $C \equiv C(T)$, $k \equiv k(T)$, $g \equiv g(T)$, $\Delta k \equiv \Delta k(T)$, $\Delta C \equiv \Delta C(T)$ and $\Delta g \equiv \Delta g(T)$.

The procedure used here to develop the sensitivity problem for the nonlinear case is more general than that given in Chapter 2 for the linear case. Since the original problem involves temperature-dependent quantities, it is more convenient to use here the limiting process given by equation (3.5.5). A similar approach is used for the derivation of the adjoint problem, as described next.

Adjoint Problem

In order to derive the adjoint problem and the gradient equation, we multiply equation (3.5.1.a) by the Lagrange multiplier $\lambda(x,t)$ and integrate over the time and space domains. The resulting expression is then added to the functional given by equation (3.5.2) to obtain:

$$S[P(T)] = \int_{x=0}^{L}\int_{t=0}^{t_f}\sum_{m=1}^{M}(T-Y)^2\delta(x-x_m)\,dt\,dx + \int_{x=0}^{L}\int_{t=0}^{t_f}\left\{C(T)\frac{\partial T(x,t)}{\partial t} - \frac{\partial}{\partial x}\left[k(T)\frac{\partial T}{\partial x}\right] - g(T)\right\}\lambda(x,t)\,dt\,dx \tag{3.5.7}$$

where $\delta(.)$ is the Dirac delta function.

The above extended functional $S[P(T)]$ undergoes a variation $\Delta S[P(T)]$ when the unknown quantity and the temperature undergo variations $\varepsilon\Delta P(T)$ and $\varepsilon\Delta T(x, t)$, respectively. The variation $\Delta S[P(T)]$ can be conveniently obtained by applying the following limiting process:

$$\Delta S[P(T)] = \lim_{\varepsilon\to 0}\frac{S[P_\varepsilon(T_\varepsilon)] - S[P(T)]}{\varepsilon} \tag{3.5.8}$$

where the term $S[P_\varepsilon(T_\varepsilon)]$ is obtained by writing equation (3.5.7) for the perturbed quantities given by equations (3.5.3). We obtain

$$\begin{aligned}\Delta S[P(T)] = {} & 2\int_{t=0}^{t_f}\int_{x=0}^{L}\sum_{m=1}^{M}(T-Y)\delta(x-x_m)\Delta T\,dx\,dt + \int_{x=0}^{L}\int_{t=0}^{t_f}\frac{\partial(C\Delta T)}{\partial t}\lambda(x,t)\,dt\,dx - \\ & -\int_{t=0}^{t_f}\int_{x=0}^{L}\frac{\partial^2(k\Delta T)}{\partial x^2}\lambda(x,t)\,dx\,dt - \int_{t=0}^{t_f}\int_{x=0}^{L}\frac{\partial}{\partial x}\left(\Delta k\frac{\partial T}{\partial x}\right)\lambda(x,t)\,dx\,dt - \\ & -\int_{x=0}^{L}\int_{t=0}^{t_f}\frac{dg}{dT}\Delta T\,\lambda(x,t)\,dt\,dx + \int_{x=0}^{L}\int_{t=0}^{t_f}\frac{\partial T}{\partial t}\Delta C\lambda(x,t)\,dt\,dx - \\ & -\int_{x=0}^{L}\int_{t=0}^{t_f}\lambda(x,t)\Delta g\,dt\,dx\end{aligned} \tag{3.5.9}$$

The inner integrals in the second, third and fourth terms of equation (3.5.9) are integrated by parts and the boundary and initial conditions of the sensitivity problem are utilized. In the resulting expression, the terms containing $\Delta T(x,t)$ are then allowed to go to zero to obtain the following adjoint problem:

$$-C\frac{\partial\lambda}{\partial t} - k\frac{\partial^2\lambda}{\partial x^2} - \frac{dg}{dT}\lambda + 2\sum_{m=1}^{M}(T-Y)\delta(x-x_m) = 0$$

$$\text{in } 0 < x < L,\ \text{for } 0 < t < t_f \tag{3.5.10.a}$$

$$\frac{\partial \lambda}{\partial x} = 0 \qquad \text{at } x = 0, \text{for } 0 < t < t_f \tag{3.5.10.b}$$

$$\frac{\partial \lambda}{\partial x} = 0 \qquad \text{at } x = L, \text{for } 0 < t < t_f \tag{3.5.10.c}$$

$$\lambda = 0 \qquad \text{for } t = t_f, \text{ in } 0 < x < L \tag{3.5.10.d}$$

Gradient Equation

In the process used to obtain the adjoint problem (3.5.10), equation (3.5.9) reduces to

$$\Delta S[P(T)] = \int_{x=0}^{L}\int_{t=0}^{t_f} \frac{\partial T}{\partial x}\frac{\partial \lambda}{\partial x}\Delta k(T)\,dt\,dx + \int_{x=0}^{L}\int_{t=0}^{t_f} \frac{\partial T}{\partial t}\lambda(x,t)\Delta C(T)\,dt\,dx - \\ - \int_{x=0}^{L}\int_{t=0}^{t_f} \lambda(x,t)\Delta g(T)\,dt\,dx \tag{3.5.11}$$

For a function $P(x,t)$ belonging to the space of square integrable functions in the domain $(0,t_f)$ x $(0,L)$, we can write:

$$\Delta S[P(x,t)] = \int_{x=0}^{L}\int_{t=0}^{t_f} \nabla S[P(x,t)]\,\Delta P(x,t)\,dt\,dx \tag{3.5.12}$$

By assuming that there exists one-to-one correspondence between the temperature T and the pair (x,t), that is, $P(T) \equiv P(x,t)$ and $\Delta P(T) \equiv \Delta P(x,t)$, we can transform the minimization of the functional given by equation (3.5.2) from the temperature space to the (x,t) space. Therefore, we can compare equations (3.5.11) and (3.5.12) to obtain the gradient equations for the cases of unknown reaction function, thermal conductivity and volumetric heat capacity, respectively as

$$\nabla S[g(T)] = -\lambda(x,t) \quad , \quad \text{for } P(T) = g(T) \tag{3.5.13.a}$$

$$\nabla S[k(T)] = \frac{\partial T}{\partial x}\frac{\partial \lambda}{\partial x} \quad , \quad \text{for } P(T) = k(T) \tag{3.5.13.b}$$

$$\nabla S[C(T)] = \frac{\partial T}{\partial t}\lambda(x,t) \quad , \quad \text{for } P(T) = C(T) \tag{3.5.13.c}$$

We note that the sensitivity and adjoint problems given by equations (3.5.6) and (3.5.10), respectively, are linear, although the direct problem given by equations (3.5.1) is nonlinear.

Iterative Procedure

The following iterative procedure based on the conjugate gradient method is applied for the estimation of $P(T)$:

$$P^{i+1}(T) = P^i(T) - \beta^i d^i(T) \tag{3.5.14.a}$$

where the superscript i denotes the number of iterations and the direction of descent $d^i(T)$ is given by:

$$d^i(T) = \nabla S\left[T; P^i(T)\right] + \gamma^i d^{i-1}(T) \tag{3.5.14.b}$$

The expression of Polak and Ribiere is used here for the conjugation coefficient γ^i:

$$\gamma^i = \frac{\int_{x=0}^{L}\int_{t=0}^{t_f}\left\{\nabla S\left[T;P^i(T)\right] - \nabla S\left[T;P^{i-1}(T)\right]\right\}\nabla S\left[T;P^i(T)\right]dt\,dx}{\int_{x=0}^{L}\int_{t=0}^{t_f}\left\{\nabla S\left[T;P^{i-1}(T)\right]\right\}^2 dt\,dx} \tag{3.5.14.c}$$

for $i = 1,2,\ldots$ with $\gamma^0 = 0$

The search step-size β^i is obtained by minimizing the functional given by equation (3.5.2) with respect to β^i. The following expression results (see Note 7 in Chapter 2):

$$\beta^i = \frac{\int_{t=0}^{t_f}\sum_{m=1}^{M}\left[T(x_m,t;P^i) - Y_m(t)\right]\Delta T(x_m,t;d^i)\,dt}{\int_{t=0}^{t_f}\sum_{m=1}^{M}\left[\Delta T(x_m,t;d^i)\right]^2 dt} \tag{3.5.14.d}$$

where $\Delta T(x_m, t; d^i)$ is the solution of the sensitivity problem at position x_m and time t, which is obtained from equations (3.5.6) by setting $\Delta P(T) = d^i(T)$.

Once $d^i(T)$ is computed from equation (3.5.14.b) and β^i from equation (3.5.14.d), the iterative process given equation (3.5.14.a) can be applied to determine $P^{i+1}(T)$, until a specified stopping criterion based on the discrepancy principle is satisfied, as described in Chapter 2.

Results

In order to examine the accuracy of Technique IV, as applied to the analysis of the inverse problems previously described, we studied test cases by using simulated measured temperatures as the input data for the inverse analysis. To generate the simulated measurements, the direct problem given by equations (3.5.1) was expressed in dimensionless form by introducing the following dimensionless variables:

$$\theta = \frac{T - T_0}{\frac{\phi_L}{k_0} L} \; ; \; \Gamma = \frac{g(T)L}{\phi_L} \; ; \; \tau = \frac{k_0 t}{L^2 C_0} \; ; \; x = \frac{x}{L} \; ; \; \phi = \frac{q}{\phi_L} \qquad (3.5.15.a\text{-}e)$$

and by taking the coefficients $k(T)$ and $C(T)$ in the form

$$k(T) = k_0 \kappa(\theta) \quad \text{and} \quad C(T) = C_0 \chi(\theta) \qquad (3.5.16.a,b)$$

where k_0 and C_0 are constants with units of $k(T)$ and $C(T)$, respectively; $\kappa(\theta)$ and $\chi(\theta)$ are dimensionless functions of θ; T_0 is the initial temperature in the medium which is assumed to be uniform; and ϕ_L is the heat flux applied at the boundary $x = L$, which is assumed to be constant.

The direct, sensitivity and adjoint problems were solved by using finite differences with 51 mesh points and 100 time steps. These values were chosen by comparing the numerical solution of the direct problem with a known analytic solution. The agreement between the two solutions was better than 1%.

The accuracy of the present method of inverse analysis was verified under strict conditions by using the measurements of a single sensor. In such a case, the requirement of one-to-one correspondence between the temperature T and the pair (x,t), used to derive the gradient equations (3.5.13), is automatically satisfied.

Consider initially the inverse problem of estimating the reaction function, i.e. $P(T) = g(T)$, with $k(T)$ and $C(T)$ known. For simplicity, we have assumed $K(\theta) = \chi(\theta) = 1$. Figures 3.5.1-3 present the results for exponential, triangular and step variations for the dimensionless reaction function, respectively, obtained with errorless measurements ($\sigma = 0$) and measurements with random error, $\sigma = 0.01\ \theta_{max}$, where θ_{max} is the maximum measured temperature. Note in these figures that very accurate results are obtained, even for functions containing sharp corners and discontinuities, which are the most difficult to be recovered by an inverse analysis.

In order to compare the present function estimation approach of Technique IV with the traditional approach of Technique III, we also solved the inverse problem of estimating the reaction function parameterized with B-Splines trial functions in the form:

$$g(T) = \sum_{j=1}^{N} P_j B_j(T) \tag{3.5.17}$$

Thus, the inverse problem of estimating the reaction function reduces to the problem of estimating the unknown parameters P_j, $j = 1, ..., N$, where $B_j(T)$ are the known B-Splines. The number N of trial functions used in the parameterization is also considered known.

The iterative procedure of Technique III can be found in section 2-3 and is not repeated here. In order to implement such a procedure, the sensitivity and adjoint problems given by equations (3.5.6) and (3.5.10) are also required. The gradient vector components are shown to be given by:

$$[\nabla S(\mathbf{P})]_j = -\int_{x=0}^{L} \int_{t=0}^{t_f} \lambda(x,t) B_j[T(x,t)] dt\, dx \tag{3.5.18}$$

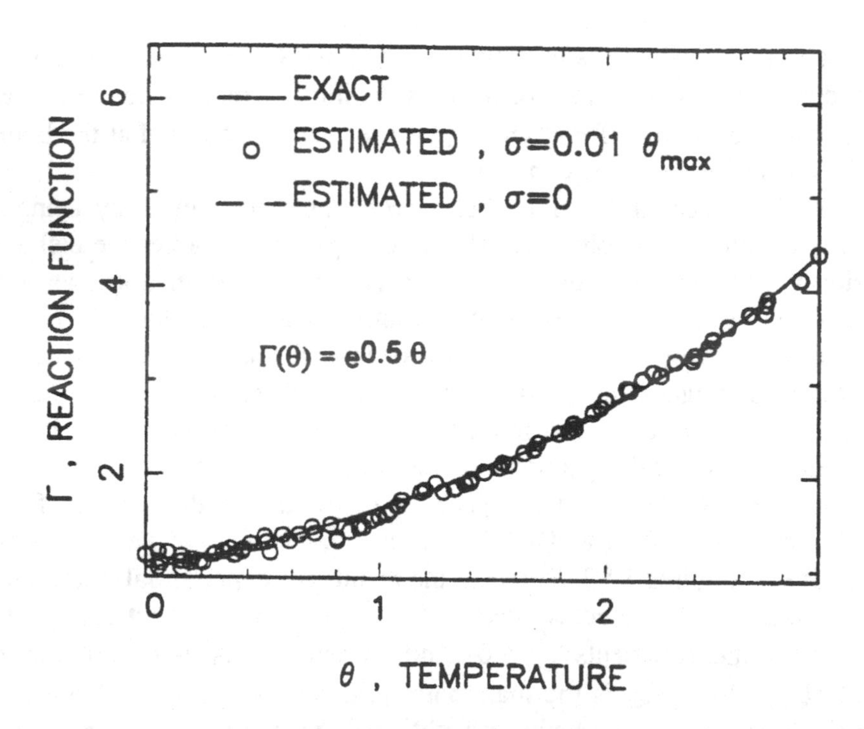

Figure 3.5.1 - Inverse solution with exponential variation for the reaction function in the form $\Gamma(\theta) = e^{0.5\theta}$.

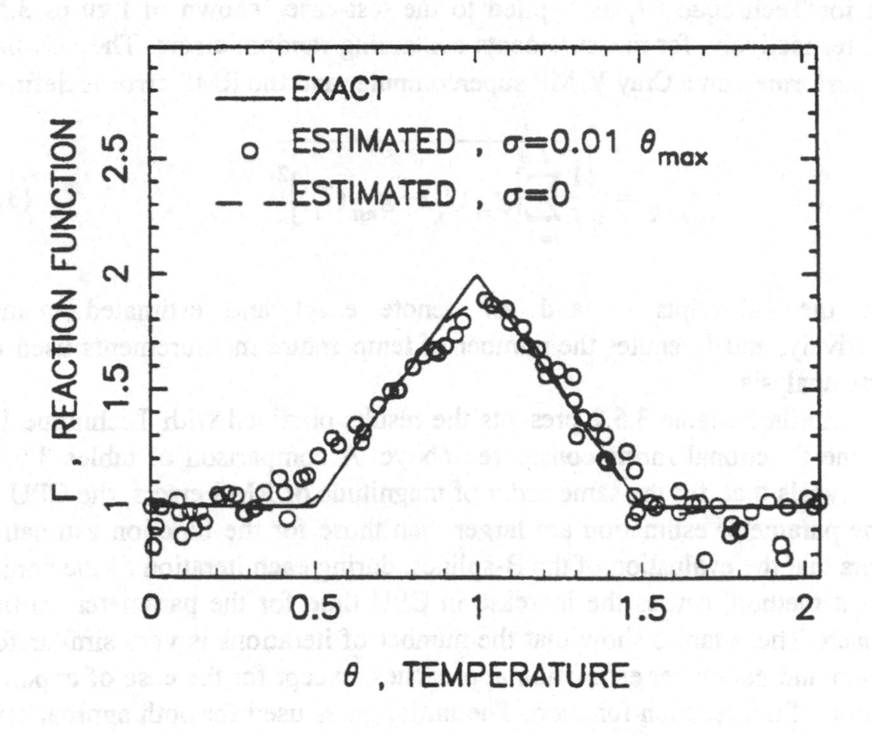

Figure 3.5.2 - Inverse solution with triangular variation for the reaction function.

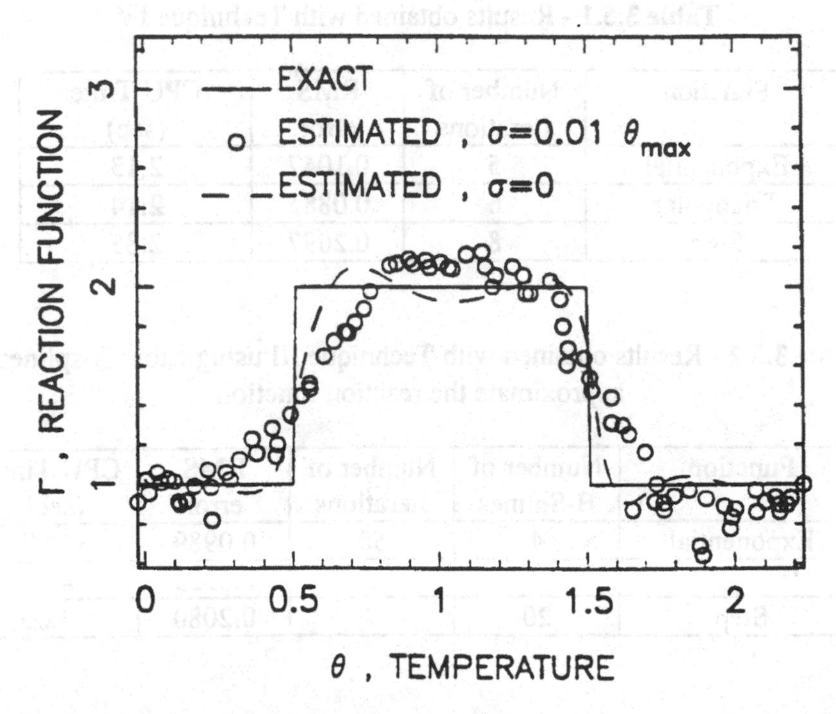

Figure 3.5.3 - Inverse solution with step variation for the reaction function.

Table 3.5.1 presents the number of iterations, the RMS error and the CPU times for Technique IV, as applied to the test-cases shown in figures 3.5.1 to 3.5.3, respectively, for measurements containing random errors. The calculations were performed on a Cray Y-MP supercomputer and the RMS error is defined as:

$$e_{RMS} = \sqrt{\frac{1}{I}\sum_{i=1}^{I}\left[g_{ex}(T_i) - g_{est}(T_i)\right]^2} \tag{3.5.19}$$

where the subscripts *ex* and *est* denote exact and estimated quantities, respectively, and I denotes the number of temperature measurements used in the inverse analysis.

Similarly, table 3.5.2 presents the results obtained with Technique III for the same functional forms considered above. A comparison of tables 3.5.1 and 3.5.2 reveals that, for the same order of magnitude of RMS errors, the CPU times for the parameter estimation are larger than those for the function estimation. It appears that the evaluation of the B-splines, during each iteration of the conjugate gradient method, causes the increase in CPU time for the parameter estimation approach. These tables show that the number of iterations is very similar for the function and parameter estimation approaches, except for the case of exponential variation of the reaction function. The initial guess used for both approaches was the exact value of the reaction function at the final temperature measured by the sensor, so that the instabilities inherent of Technique IV at the final temperature value could be avoided.

Table 3.5.1 - Results obtained with Technique IV

Function	Number of Iterations	RMS error	CPU Time (sec)
Exponential	5	0.1047	2.13
Triangular	6	0.0885	2.14
Step	8	0.2097	2.85

Table 3.5.2 - Results obtained with Technique III using cubic B-splines to approximate the reaction function

Function	Number of B-Splines	Number of Iterations	RMS error	CPU Time (sec)
Exponential	4	50	0.0989	31.07
Triangular	15	7	0.0923	4.52
Step	20	7	0.2080	4.60

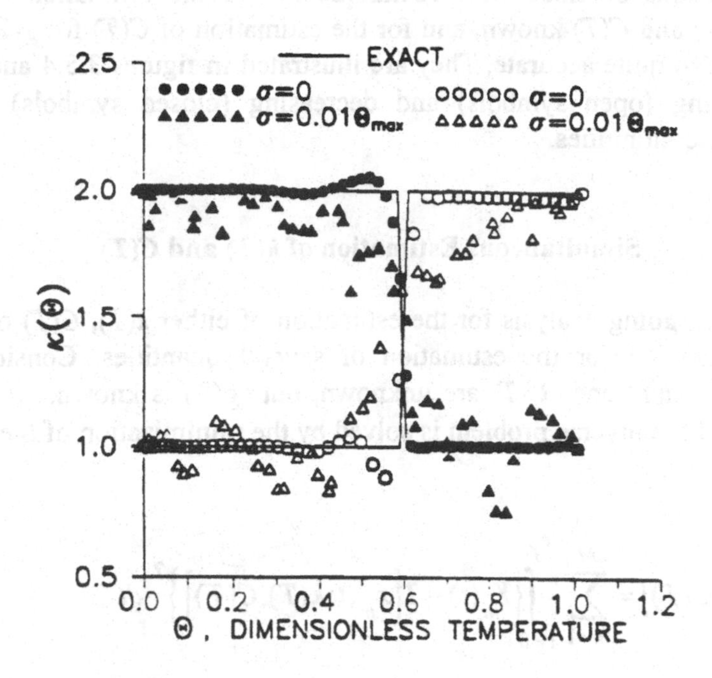

Figure 3.5.4 - Estimation of the dimensionless thermal conductivity $\kappa(\theta)$. Functional forms containing discontinuities.

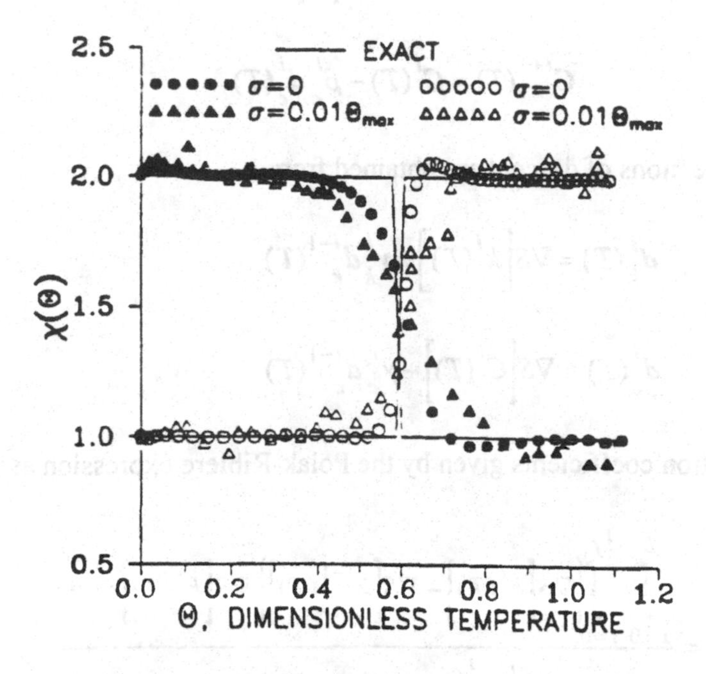

Figure 3.5.5 - Estimation of the dimensionless volumetric heat capacity $\chi(\theta)$. Functional forms containing discontinuities.

The results obtained with Technique IV for the estimation of $k(T)$, by assuming $g(T)$ and $C(T)$ known, and for the estimation of $C(T)$ for $g(T)$ and $k(T)$ known, are also quite accurate. They are illustrated in figures 3.5.4 and 3.5.5 for both increasing (open symbols) and decreasing (closed symbols) functions, containing discontinuities.

Simultaneous Estimation of $k(T)$ and $C(T)$

The foregoing analysis for the estimation of either $k(T)$, $C(T)$ or $g(T)$ can be easily extended for the estimation of several quantities. Consider, as an example, that $k(T)$ and $C(T)$ are unknown, but $g(T)$ is known. In this case, $\Delta g(T) = 0$ and the inverse problem is solved by the minimization of the following functional.

$$S[k(T),C(T)] = \sum_{m=1}^{M} \int_{t=0}^{t_f} \{Y_m(t) - T[x_m,t;k(T),C(T)]\}^2 \, dt \tag{3.5.20}$$

The iterative procedures of the conjugate gradient method, for the simultaneous estimation of thermal conductivity and volumetric heat capacity, are given by

$$k^{i+1}(T) = k^i(T) - \beta_k^i d_k^i(T) \tag{3.5.21.a}$$

$$C^{i+1}(T) = C^i(T) - \beta_c^i d_c^i(T) \tag{3.5.21.b}$$

where the directions of descent are obtained from

$$d_k^i(T) = \nabla S\left[k^i(T)\right] + \gamma_k^i d_k^{i-1}(T) \tag{3.5.22.a}$$

$$d_c^i(T) = \nabla S\left[C^i(T)\right] + \gamma_c^i d_c^{i-1}(T) \tag{3.5.22.b}$$

with conjugation coefficients given by the Polak-Ribiere expression as

$$\gamma_k^i = \frac{\int_{x=0}^{L} \int_{t=0}^{t_f} \{\nabla S[k^i(T)] - \nabla S[k^{i-1}(T)]\} \nabla S[k^i(T)] \, dt \, dx}{\int_{x=0}^{L} \int_{t=0}^{t_f} \{\nabla S[k^{i-1}(T)]\}^2 \, dt \, dx} \tag{3.5.23.a}$$

$$\gamma_c^i = \frac{\int_{x=0}^{L}\int_{t=0}^{t_f}\left\{\nabla S\left[C^i(T)\right]-\nabla S\left[C^{i-1}(T)\right]\right\}\nabla S\left[C^i(T)\right]dt\,dx}{\int_{x=0}^{L}\int_{t=0}^{t_f}\left\{\nabla S\left[C^{i-1}(T)\right]\right\}^2 dt\,dx} \tag{3.5.23.b}$$

with $\gamma_C^0 = \gamma_k^0 = 0$ for $i = 0$.

The gradient directions for thermal conductivity, $\nabla S[k(T)]$, and heat capacity, $\nabla S[C(T)]$, are given by equations (3.5.13.b,c), respectively. The direct, sensitivity and adjoint problems are not changed for the simultaneous estimation of $k(T)$ and $C(T)$, and they are given by equations (3.5.1), (3.5.6) and (3.5.10), respectively.

The search step sizes β_k^i and β_c^i are obtained by minimizing the functional (3.5.20) with respect to these two quantities. By using equations (3.5.21), we can write equation (3.5.20) as

$$S\left[k^{i+1},C^{i+1}\right] = \sum_{m=1}^{M}\int_{t=0}^{t_f}\left\{Y_m - T_m\left(k^i - \beta_k^i d_k^i, C^i - \beta_c^i d_c^i\right)\right\}^2 dt \tag{3.5.24}$$

where the functional dependence of several quantities were omitted above for simplicity.

The estimated temperature $T_m\left(k^i - \beta_k^i d_k^i, C^i - \beta_c^i d_c^i\right)$ is linearized by a Taylor series expansion in the form:

$$T_m\left(k^i - \beta_k^i d_k^i, C^i - \beta_c^i d_c^i\right) \approx T_m(k^i, C^i) - \beta_k^i \frac{\partial T_m}{\partial k^i} d_k^i - \beta_c^i \frac{\partial T_m}{\partial C^i} d_c^i \tag{3.5.25}$$

Let $\qquad d_k^i = \Delta k^i \qquad$ (3.5.26.a)

and $\qquad d_c^i = \Delta C^i \qquad$ (3.5.26.b)

Then, equation (3.5.25) can be written as (see Note 7 in Chapter 2)

$$T_m\left(k^i - \beta_k^i d_k^i, C^i - \beta_c^i d_c^i\right) \approx T_m(k^i, C^i) - \beta_k^i \Delta T_{k,m}^i - \beta_c^i \Delta T_{c,m}^i \tag{3.5.27}$$

where $\Delta T^i_{k,m}$ and $\Delta T^i_{c,m}$ are the solutions of the sensitivity problem, equations (3.5.6), at the measurement locations x_m, $m = 1, ..., M$, obtained by setting

$$\Delta k^i = d^i_k \quad , \quad \Delta C^i = \Delta g^i = 0$$

and
$$\Delta C^i = d^i_c \quad , \quad \Delta k^i = \Delta g^i = 0 \quad , \quad \text{respectively.}$$

By substituting equation (3.5.27), we can write the functional (3.5.24) as

$$S\left[k^{i+1}, C^{i+1}\right] = \sum_{m=1}^{M} \int_{t=0}^{t_f} \left\{ Y_m - T_m\left(k^i, C^i\right) + \beta^i_k \, \Delta T^i_{k,m} + \beta^i_c \, \Delta T^i_{c,m} \right\}^2 dt$$

(3.5.28)

The above equation is minimized with respect to β^i_k and β^i_c to obtain the following expressions for the search step sizes:

$$\beta^i_k = \frac{F_1 A_{22} - F_2 A_{12}}{A_{11} A_{22} - A_{12}^2} \qquad \beta^i_c = \frac{F_2 A_{11} - F_1 A_{12}}{A_{11} A_{22} - A_{12}^2} \tag{3.5.29.a,b}$$

where

$$A_{11} = \sum_{m=1}^{M} \int_{t=0}^{t_f} \left(\Delta T^i_{k,m}\right)^2 dt \qquad A_{22} = \sum_{m=1}^{M} \int_{t=0}^{t_f} \left(\Delta T^i_{c,m}\right)^2 dt$$

$$A_{12} = \sum_{m=1}^{M} \int_{t=0}^{t_f} \Delta T^i_{k,m} \Delta T^i_{c,m} \, dt$$

$$F_1 = \sum_{m=1}^{M} \int_{t=0}^{t_f} \left(T^i_m - Y^i_m\right) \Delta T^i_{k,m} \, dt \qquad F_2 = \sum_{m=1}^{M} \int_{t=0}^{t_f} \left(T^i_m - Y^i_m\right) \Delta T^i_{c,m} \, dt$$

(3.5.30.a-e)

After developing expressions for the directions of descent, equations (3.5.22), and for the search step-sizes, equations (3.5.29), the iterative procedure of the conjugate gradient method given by equations (3.5.21) can be applied for the simultaneous estimation of $k(T)$ and $C(T)$.

3-6 ESTIMATION OF THERMAL DIFFUSIVITY AND RELAXATION TIME WITH A HYPERBOLIC HEAT CONDUCTION MODEL [13]

So far in this chapter, we considered inverse heat conduction problems mathematically modeled by the parabolic heat conduction equation. The Fourier's Law serves as the constitutive equation relating the heat flux to the temperature gradient in the classical theory of diffusion, based on the parabolic heat conduction model. In accordance with Fourier's Law, heat propagates with an infinite speed in a conducting medium, that is, the effect of a thermal disturbance is felt instantaneously, although not homogeneously, in all parts of the medium. Despite such an unacceptable notion of energy transport in solids, Fourier's Law is accurate in describing heat conduction in most engineering situations encountered in daily life. However, there are practical situations in which the effects of the finite speed of heat propagation become important. For such situations, a constitutive equation which allows a time lag between the heat flux vector and the temperature gradient is given by [14]

$$q(\mathbf{r},t)+\tau\frac{\partial q(\mathbf{r},t)}{\partial t}=-k\nabla T(\mathbf{r},t) \tag{3.6.1}$$

where τ is the *relaxation time*, an intrinsic property of the medium. This equation, when combined with the energy equation

$$-\nabla\cdot q(\mathbf{r},t)+g(\mathbf{r},t)=\rho c_p\frac{\partial T(\mathbf{r},t)}{\partial t} \tag{3.6.2}$$

yields the following *hyperbolic equation* for heat conduction in the medium [15,16]

$$\frac{\partial T(\mathbf{r},t)}{\partial t}+\tau\frac{\partial^2 T}{\partial t^2}=\alpha\nabla^2 T(\mathbf{r},t)+\frac{1}{\rho c_p}\left[g(\mathbf{r},t)+\tau\frac{\partial g(\mathbf{r},t)}{\partial t}\right] \tag{3.6.3}$$

Equation (3.6.3) predicts a wave behavior for the heat propagation, where the *thermal wave speed*, *C*, is related to the relaxation time and to the thermal diffusivity by

$$C=\sqrt{\frac{\alpha}{\tau}} \tag{3.6.4}$$

Equation (3.6.3) has been applied on the modeling of physical processes dealing with extremely short time responses, extremely high-rate change of temperature and heat flux, initial conditions involving the time-rate change of temperature ($\partial T/\partial t$) and temperatures approaching the absolute zero [15-17]. In experiments on the propagation of heat waves in liquid and solid helium, as well as in dielectric crystals at cryogenic temperatures, values of relaxation time of the order of 10^{-6} seconds and of thermal diffusivity of the order of 10 m^2/sec were reported [17].

In this section we present an inverse analysis for the simultaneous estimation of *the thermal diffusivity α and the relaxation time τ* for a hyperbolic heat conduction model, by using transient temperature measurements taken in a semi-infinite region. The resulting parameter estimation problem is solved with **Technique I**, Levenberg-Marquardt Method, and an analysis of the sensitivity coefficients permits the design of an optimum experiment with respect to the heat flux boundary condition at the surface of the semi-infinite medium.

Direct Problem

The direct problem is concerned with the determination of the temperature field in the medium, when the physical properties, the initial and the boundary conditions are known.

Here we consider a semi-infinite medium with no energy generation, subjected to a time-dependent heat flux at the boundary $x = 0$ and to equilibrium initial conditions. The mathematical formulation of this problem is given by:

$$\frac{\partial T}{\partial t} + \tau \frac{\partial^2 T}{\partial t^2} = \alpha \frac{\partial^2 T}{\partial x^2} \qquad \text{for } x > 0,\ t > 0 \qquad (3.6.5.a)$$

$$T(x,0) = 0 \qquad \text{for } t = 0,\ x > 0 \qquad (3.6.5.b)$$

$$\frac{\partial T}{\partial t} = 0 \qquad \text{for } t = 0,\ x > 0 \qquad (3.6.5.c)$$

$$q(0,t) = q_0(t) \qquad \text{for } x = 0\ \ t > 0 \qquad (3.6.5.d)$$

where

$$q(x,t) = -k\frac{\partial T}{\partial x} - \tau \frac{\partial q}{\partial t} \qquad (3.6.5.e)$$

The solution of problem (3.6.5) is obtained by the application of Duhamel's Theorem [18,19]. For the case of $q_0(t)$ containing N jump discontinuities, it is given by

$$T(x,t) = \int_{t'=0}^{t} \phi(x,t-t')\frac{dq_0(t')}{dt'}dt' + \sum_{i=0}^{N-1} \phi(x,t-\lambda_i)\Delta q_i H(t-\lambda_i) \qquad (3.6.6)$$

where $H(.)$ is the Heaviside step function, Δq_i is the magnitude of the step change in the surface heat flux at time λ_i, and $\phi(x,t)$ is the solution of problem (3.6.5) for $q_0(t) = 1$ W/m^2.

Here we assume that $q_0(t)$ is constant in each of the N intervals $\lambda_i < t < \lambda_{i+1}$, for $i = 0, 1, ..., N$-1. Thus, the solution of problem (3.6.5) is obtained as [19]:

$$T(x,t) = \sum_{i=0}^{N-1} \frac{\Delta q_i}{k} \sqrt{\alpha\tau} H\left(t - \lambda_i - \sqrt{\frac{\tau}{\alpha}} x\right)$$

$$\left\{ e^{-(t-\lambda_i)/2\tau} I_0\left[\frac{1}{2\tau}\sqrt{(t-\lambda_i)^2 - \frac{\tau}{\alpha}x^2}\right] + \frac{1}{\tau}\int_{t'=\sqrt{\frac{\tau}{\alpha}}x}^{t-\lambda_i} e^{-t'/2\tau} I_0\left[\frac{1}{2\tau}\sqrt{t'^2 - \frac{\tau}{\alpha}x^2}\right] dt' \right\} \tag{3.6.7}$$

The corresponding parabolic solution of problem (3.6.5) for $\tau = 0$ is also obtained via Duhamel's Theorem and is given by [19]:

$$T(x,t) = \sum_{i=0}^{N-1} \frac{\Delta q_i}{k} \sqrt{4\alpha(t-\lambda_i)}\, ierfc\left(\frac{x}{\sqrt{4\alpha(t-\lambda_i)}}\right) H(t-\lambda_i) \tag{3.6.8.a}$$

where

$$ierfc(z) = \frac{1}{\sqrt{\pi}} e^{-z^2} - z\, erfc(z) \tag{3.6.8.b}$$

Inverse Problem

The inverse problem is concerned with the simultaneous estimation of thermal diffusivity and relaxation time, from the knowledge of transient temperature measurements taken with a single sensor in the medium. The boundary and initial conditions of problem (3.6.5), as well as the sensor location, are assumed to be known exactly, but the temperature measured data may contain random errors.

The solution of such an inverse problem is obtained so that the least squares norm is minimized with respect to each of the unknown parameters. The least squares norm is written in matrix form as

$$S(\mathbf{P}) = [\mathbf{Y} - \mathbf{T}(\mathbf{P})]^T [\mathbf{Y} - \mathbf{T}(\mathbf{P})] \tag{3.6.9.a}$$

where

$$[\mathbf{Y}-\mathbf{T}(\mathbf{P})]^T = [Y(t_1)-T(t_1;\mathbf{P}),...,Y(t_I)-T(t_I;\mathbf{P})] \tag{3.6.9.b}$$

$$\mathbf{P}^T = [\alpha,\tau] \tag{3.6.9.c}$$

$Y(t_i)$ is the measured temperature at time t_i and I is the total number of measurements. The estimated temperature $T(t_i;\mathbf{P})$ at time t_i and at the measurement location is obtained from the solution of problem (3.6.5) by using an estimate for the unknown vector **P**.

The Technique I, Levenberg-Marquardt method, was chosen for the minimization of the least squares norm (3.6.9.a). Such method requires the computation of the sensitivity matrix, which for the present case involving the estimation of the thermal diffusivity α and relaxation time τ, is given by:

$$\mathbf{J}(\mathbf{P}) = \begin{bmatrix} J_{1\alpha} & J_{1\tau} \\ J_{2\alpha} & J_{2\tau} \\ \vdots & \vdots \\ J_{I\alpha} & J_{I\tau} \end{bmatrix} \tag{3.6.10.a}$$

where the sensitivity coefficients $J_{i\alpha}$ and $J_{i\tau}$ are given as

$$J_{i\alpha} = \frac{\partial T(t_i;\mathbf{P})}{\partial \alpha} \quad \text{and} \quad J_{i\tau} = \frac{\partial T(t_i;\mathbf{P})}{\partial \tau} \qquad \text{for } i=1,...,I \tag{3.6.10.b,c}$$

Details of the Levenberg-Marquardt method are omitted here for the sake of brevity, but they can be found in section 2-1 in Chapter 2. The subroutine DBCLSJ of the IMSL [4], based on this method, was used here to obtain estimates for α and τ, by using simulated experimental data, as described next.

Results

Before we proceed to the examination of the accuracy of Technique I, as applied to the present inverse problem, we shall determine the timewise variation of the heat flux at the boundary $x = 0$ which provides the most meaningful temperature measurements for the estimation of thermal diffusivity and relaxation time. This is accomplished by an analysis of the sensitivity coefficients $J_{i\alpha}$ and $J_{i\tau}$, defined by equations (3.6.10.b,c). The sensitivity coefficients represent the changes in the temperature $T(t_i;\,\mathbf{P})$ with respect to the unknown parameters α and τ. It is desirable to have large, linearly independent sensitivity coefficients and the sensors should be placed at locations where the temperature readings are most sensitive to changes in the values of the unknown parameters.

Here we consider the following three different timewise variations for the boundary heat flux in an experiment of duration t_f:

(i) *A constant heat flux of strength* $q_0(t) = q_c$: For this case we have $N = 1$ and $\Delta q_0(t) = q_c$ at $\lambda_0 = 0$.

(ii) *On-Off Heat Flux with period P*: The step changes in the heat flux are

$$\Delta q_i = (-1)^i 2q_c \qquad \text{for } i = 0,...,N-1 \tag{3.6.11.a}$$

and the times when the changes Δq_i occur are given by:

$$\lambda_i = (i/2)P \qquad \text{for } i = 0,...,N-1 \tag{3.6.11.b}$$

where $N = (2t_f/P)$. The durations of nonzero and zero heat fluxes were considered to be equal.

(iii) *Single-Pulse Heat Flux of duration* λ_1: For this case we have $N = 2$ with

$$\Delta q_0 = \frac{t_f}{\lambda_1} q_c \text{ at } \lambda_0 = 0 \text{ and } \Delta q_1 = -\frac{t_f}{\lambda_1} q_c \text{ at } \lambda_1 \tag{3.6.11.c}$$

We note that the magnitude of the nonzero heat flux was chosen so that the total energy input during the experiment would be the same for the three cases considered.

For the sake of generality and simplicity in the comparison, the sensitivity coefficients are determined in dimensionless form by introducing the following dimensionless variables:

$$\theta = \frac{T(x,t)}{q_c \frac{L}{k}} \quad ; \quad Q = \frac{q}{q_c} \quad ; \quad \eta = \frac{1}{2}\frac{x}{L} \tag{3.6.12.a-c}$$

$$\xi = \frac{1}{2}\frac{\alpha t}{L^2} \quad ; \quad \Omega_\alpha = \frac{\alpha}{|\Delta q| \frac{L}{k}} J_\alpha \quad ; \quad \Omega_\tau = \frac{\tau}{|\Delta q| \frac{L}{k}} J_\tau \tag{3.6.12.d-f}$$

where the characteristic length L is defined as

$$L = \sqrt{\alpha \tau} \tag{3.6.12.g}$$

Figures 3.6.1.a-c present the dimensionless temperature variation for constant, on-off and single-pulse boundary heat fluxes, respectively, at a position

$\eta = 0.5$ and for an experiment of duration $\xi_f = 1$. For sake of comparison, the dimensionless temperatures obtained from the solution of the parabolic problem are also included in these figures.

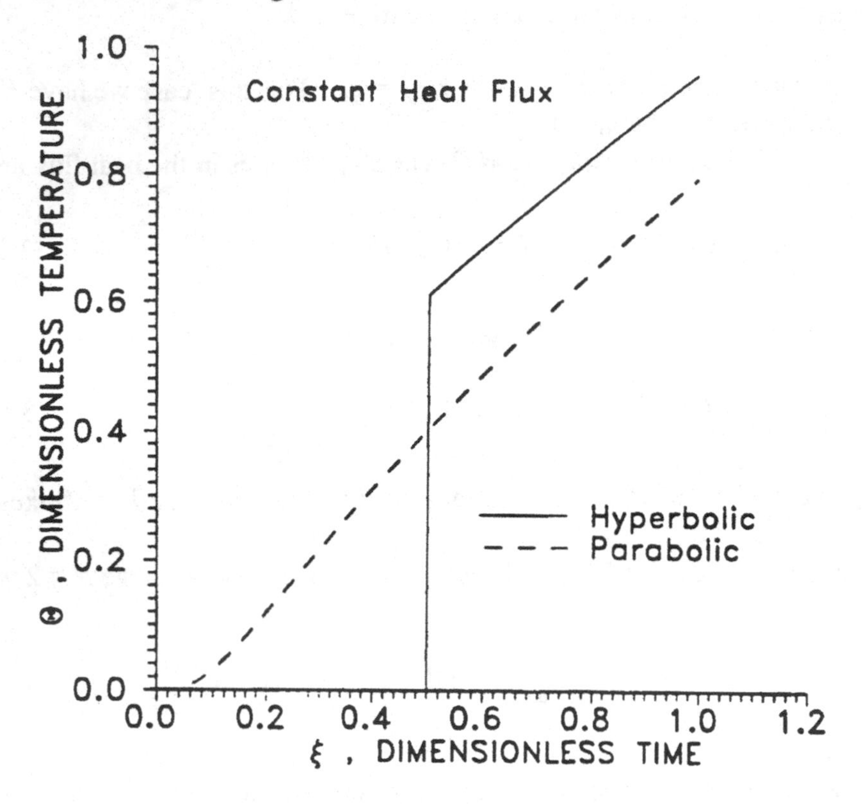

Figure 3.6.1.a - Temperature distribution at position $\eta = 0.5$ for a constant heat flux applied at $\eta = 0$.

Figures 3.6.2.a-c present the dimensionless sensitivity coefficients corresponding to the three cases shown in figures 3.6.1.a-c, respectively. The behaviors of the sensitivity coefficients for the hyperbolic problem are quite different for each case considered; they look like their corresponding temperature profiles shown in figures 3.6.1.a-c. For times $\xi < 0.5$, the sensitivity coefficients are zero because the thermal wave has not yet reached the point $\eta = 0.5$. For the cases of on-off and single-pulse heat fluxes and for $\xi > 0.5$, the sensitivity coefficients become very small and practically linearly-dependent during those periods that correspond to a zero boundary heat flux, as seen in figure 3.6.2.b,c. Such fact indicates that the simultaneous estimation of α and τ is very difficult for these two cases. For the on-off heat flux, the measurements would have to be synchronized with the nonzero boundary heat flux, and for the single-pulse heat flux, all the measurements would have to be taken during the very short period when the temperature wave passes through the measurement point. On the other hand, for the constant heat flux boundary condition for times $\xi > 0.5$, the sensitivity coefficients are not linearly dependent and attain relatively large values, as seen in figure 3.6.2.a. Therefore, the foregoing analysis of the sensitivity coefficients reveals that constant heat flux is the best boundary

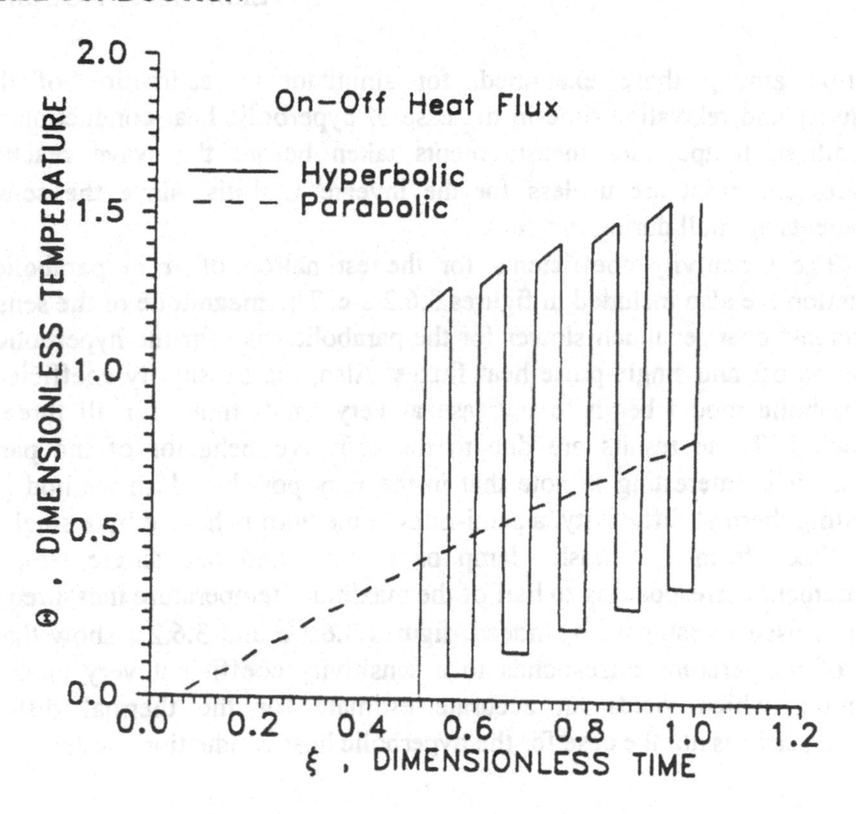

Figure 3.6.1.b - Temperature distribution at position $\eta = 0.5$ for on-off heat flux of period $P = 0.1$ applied at $\eta = 0$.

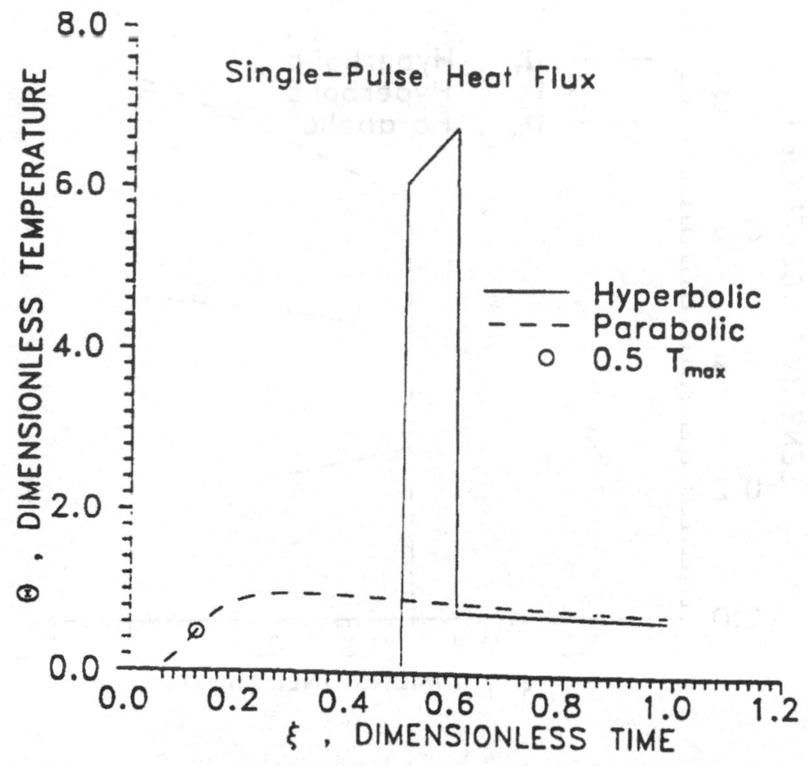

Figure 3.6.1.c - Temperature distribution at position $\eta = 0.5$ for a single-pulse heat flux of duration $\lambda_1 = 0.1$ applied at $\eta = 0$.

condition among those examined, for simultaneous estimation of thermal diffusivity and relaxation time in the case of hyperbolic heat conduction model. In addition, temperature measurements taken before the wave reaches the measurement point are useless for the inverse analysis, since the sensitivity coefficients are null during this time.

The sensitivity coefficients for the estimation of α in parabolic heat conduction are also included in figures 3.6.2.a-c. The magnitude of the sensitivity coefficients changes much slower for the parabolic than for the hyperbolic case, for the on-off and single-pulse heat fluxes. Also, the sensitivity coefficients for the parabolic model begin to increase at very small times for all three cases considered. These results are due to the diffusive behavior of the parabolic solution. It is interesting to note that in the very popular Flash method [20] of estimating thermal diffusivity, a semi-infinite medium is heated by a single-pulse heat flux from a flash lamp or a laser, and one single temperature measurement corresponding to half of the maximum temperature measured by the sensor is used to estimate α. Indeed, figures 3.6.1.c and 3.6.2.c show that such value of temperature corresponds to a sensitivity coefficient very close to its maximum, which yields an accurate estimate for the thermal diffusivity. However, such is not the case for the hyperbolic heat conduction model.

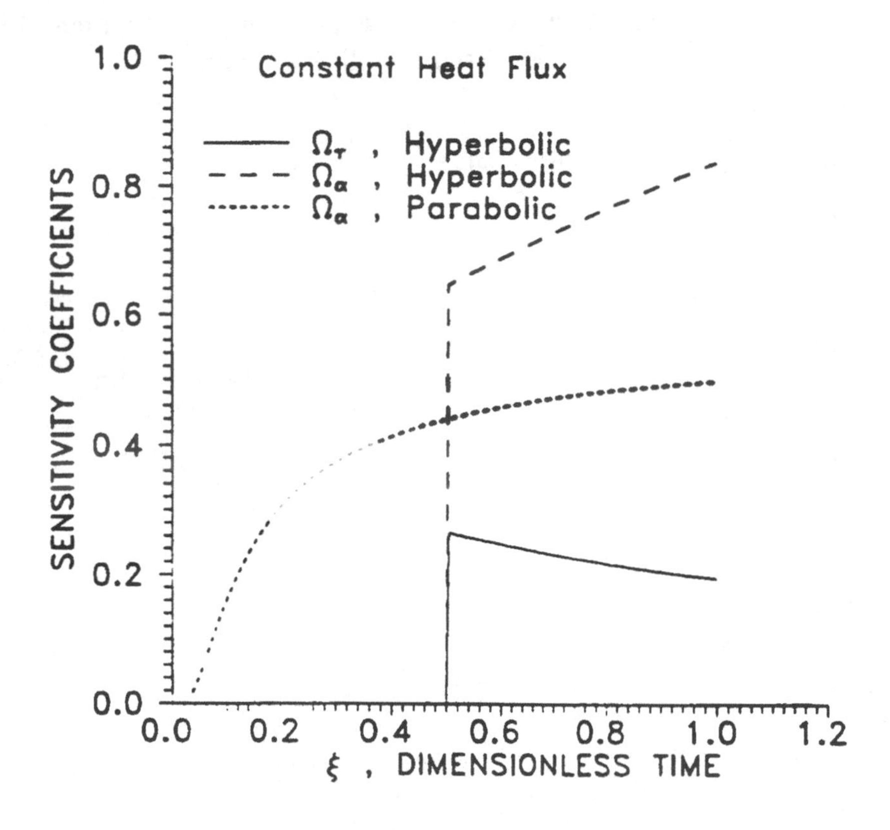

Figure 3.6.2.a - Dimensionless sensitivity coefficients for a sensor at $\eta = 0.5$ for a constant heat flux applied at $\eta = 0$.

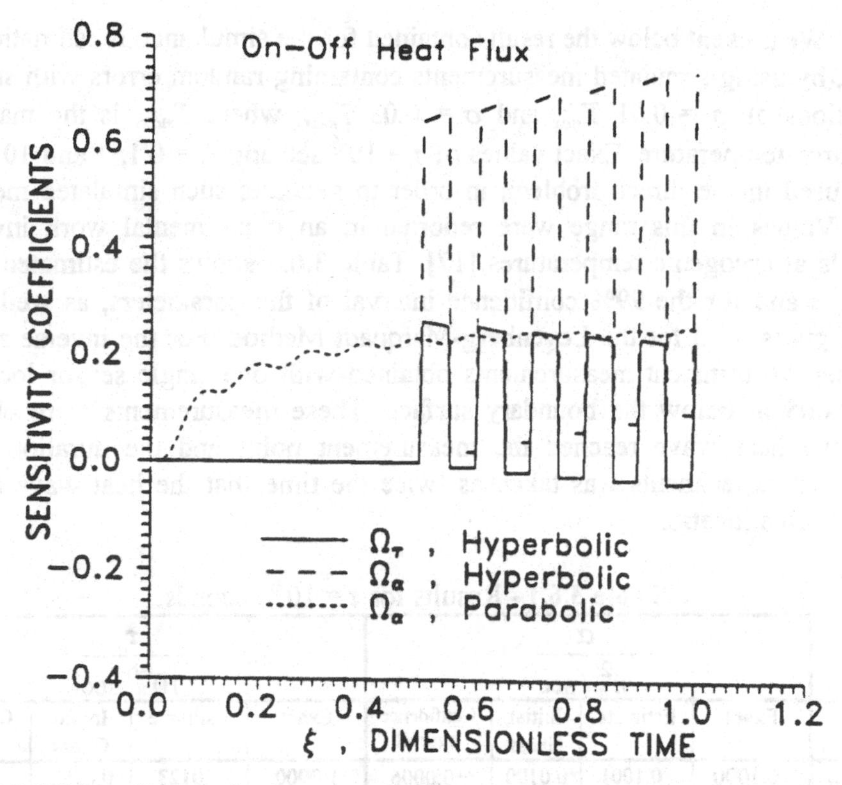

Figure 3.6.2.b - Dimensionless sensitivity coefficients for a sensor at $\eta = 0.5$ for on-off heat flux of period $P = 0.1$ applied at $\eta = 0$.

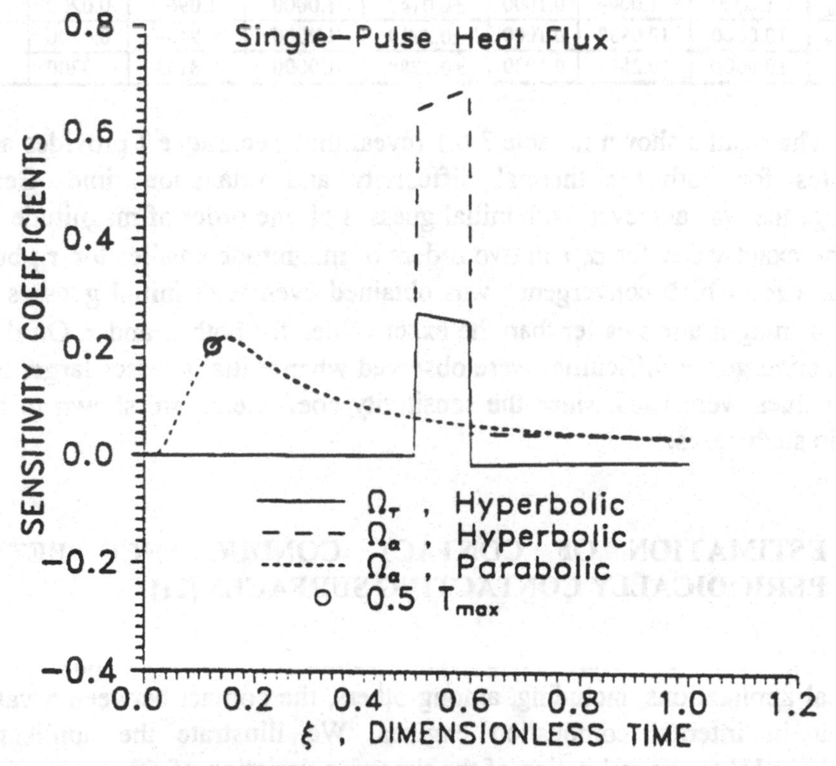

Figure 3.6.2.c - Dimensionless sensitivity coefficients for a sensor at $\eta = 0.5$ for a single-pulse heat flux of duration $\lambda_1 = 0.1$ applied at $\eta = 0$.

We present below the results obtained for the simultaneous estimation of α and τ, by using simulated measurements containing random errors with standard deviations of $\sigma = 0.01\ T_{max}$ and $\sigma = 0.03\ T_{max}$, where T_{max}, is the maximum measured temperature. Exact values of $\tau = 10^{-6}$ sec and $\alpha = 0.1$, 1 and 10 m²/sec were used in the direct problem, in order to generate such simulated measured data. Values in this range were reported in an experimental work involving crystals at cryogenic temperatures [17]. Table 3.6.1 shows the estimated values for α, τ and for the 99% confidence interval of the parameters, as well as the initial guess used for the Levenberg-Marquadt Method. For the inverse analysis we used 46 transient measurements obtained with one single sensor located at $x = 0.005$ m below the boundary surface. These measurements were obtained after the heat wave reached the measurement point and the duration of the simulated experiments was taken as twice the time that the heat wave took to reach such a location.

Table 3.6.1 - Results for $\tau = 10^{-6}$ seconds.

	$\dfrac{\alpha}{m^2/\text{sec}}$				$\dfrac{\tau}{10^{-6}\ \text{sec}}$			
σ	Exact	Estimate	Initial Guess	Confidence Interval	Exact	Estimate	Initial Guess	Confidence Interval
$0.01T_{max}$	0.1000	0.1001	0.0100	±0.0006	1.0000	1.0123	0.0200	±0.0729
$0.03T_{max}$	0.1000	0.1004	0.0100	±0.0018	1.0000	1.0368	0.0200	±0.2161
$0.01T_{max}$	1.0000	1.0015	0.1000	±0.0049	1.0000	1.0322	0.0200	±0.0902
$0.03T_{max}$	1.0000	1.0046	0.1000	±0.0147	1.0000	1.0967	0.0200	±0.2704
$0.01T_{max}$	10.0000	10.0933	0.1000	±0.2759	1.0000	0.9409	0.0200	±0.1734
$0.03T_{max}$	10.0000	10.2871	0.1000	±0.8788	1.0000	0.8163	0.0200	±0.5653

The results shown in table 3.6.1 reveal that Technique I provides accurate estimates for both the thermal diffusivity and relaxation time. Generally, convergence was achieved with initial guesses of one order of magnitude smaller than the exact value for α, and two orders of magnitude smaller for τ; but there are cases for which convergence was obtained even with initial guesses of two orders of magnitude smaller than the exact values for both α and τ. On the other hand, convergence difficulties were observed when initial guesses larger than the exact values were used, since the sensitivity coefficients are shown to be very small in such cases.

3-7 ESTIMATION OF CONTACT CONDUCTANCE BETWEEN PERIODICALLY CONTACTING SURFACES [21]

Problems involving periodically contacting surfaces have different practical applications, including, among others, the contact between a valve and its seat in internal combustion engines. We illustrate the application of **Technique IV** to the estimation of the timewise variation of contact conductance between two one-dimensional solids with periodic contact. Small periods are usually the most difficult to perform an inverse analysis. The present approach is found to be accurate and stable, even for situations involving very small periods.

As discussed previously, the solution of this inverse problem by Technique IV requires the development of the direct, inverse, sensitivity and adjoint problems as well as the gradient equation. These basic steps to solve the problem are described next. Details on the other steps of Technique IV can be found in section 2-4.

Direct Problem

Figure 3.7.1 shows the geometry and the coordinates for the one-dimensional physical problem considered here. Two rods, referred to as regions 1 and 2, are contacting periodically with period τ and with a contact conductance $\bar{h}(\bar{t})$ at the interface. The non-contacting ends are kept at constant, but different temperatures $\bar{T}_{01}$ and $\bar{T}_{02}$. It is assumed that sufficient number of contacts has been made, so that the quasi-steady-state condition is established for the temperature distribution in the solids, that is, the temperature distribution in the regions during one period is identical to that in the following period.

The mathematical formulation of this heat conduction problem is given in *dimensionless form* as:

Region 1 ($0 \leq x \leq 1$):

$$\frac{\partial^2 T_1}{\partial x^2} = \frac{\partial T_1}{\partial t} \qquad \text{in } 0 < x < 1 \text{ , for } t > 0 \tag{3.7.1.a}$$

$$T_1 = 0 \qquad \text{at } x = 0 \text{ , for } t > 0 \tag{3.7.1.b}$$

$$-\frac{\partial T_1}{\partial x} = h(t)\left[T_1 - T_2\right] \qquad \text{at } x = 1 \text{ , for } t > 0 \tag{3.7.1.c}$$

$$T_1(x,0) = T_1(x,\tau) \tag{3.7.1.d}$$

Region 2 ($1 \leq x \leq 1+L$):

$$\frac{\partial^2 T_2}{\partial x^2} = \frac{1}{\alpha}\frac{\partial T_2}{\partial t} \qquad \text{in } 1 < x < 1+L \text{ , for } t > 0 \tag{3.7.2.a}$$

$$-k\frac{\partial T_2}{\partial x} = h(t)\left[T_1 - T_2\right] \qquad \text{at } x = 1 \text{ , for } t > 0 \tag{3.7.2.b}$$

$$T_2 = 1 \qquad \text{at } x = 1 + L \text{ , for } t > 0 \tag{3.7.2.c}$$

$$T_2(x,0) = T_2(x,\tau) \tag{3.7.2.d}$$

where the following dimensionless quantities were defined

$$x = \frac{\bar{x}}{\bar{L}_1} \quad h = \frac{\bar{h}\,\bar{L}_1}{\bar{k}_1} \quad t = \frac{\bar{\alpha}_1 \bar{t}}{\bar{L}_1^2} \quad \alpha = \frac{\bar{\alpha}_2}{\bar{\alpha}_1} \quad T = \frac{\bar{T} - \bar{T}_{01}}{\bar{T}_{02} - \bar{T}_{01}} \quad k = \frac{\bar{k}_2}{\bar{k}_1} \quad L = \frac{\bar{L}_2}{\bar{L}_1} \tag{3.7.3.a-g}$$

and the superscript "–" above indicates dimensional variables.

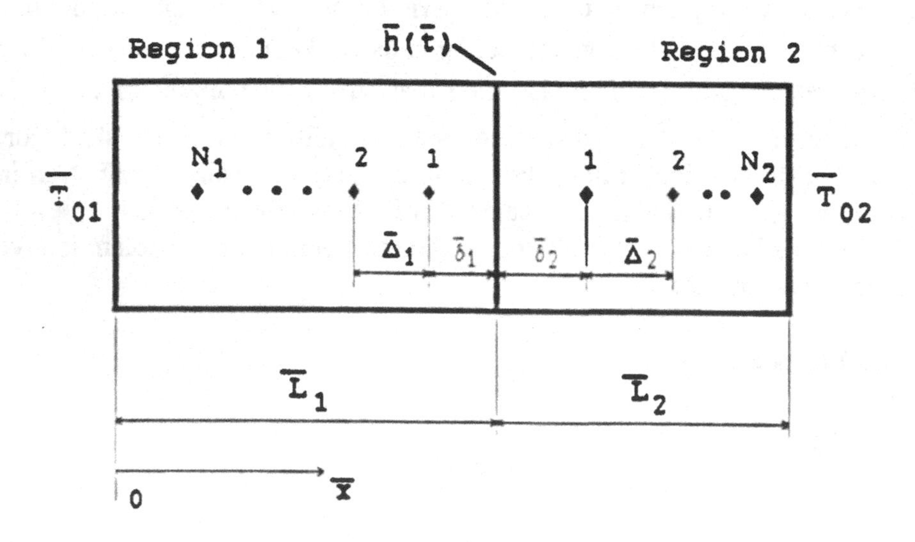

Figure 3.7.1 - Periodically contacting solids.

The direct problem considered here is concerned with the determination of the temperature field in the regions when the thermophysical properties, interface conductance, $h(t)$, and the boundary conditions at the outer ends of the regions are known.

Inverse Problem

For the inverse problem, the interface conductance, $h(t)$, is regarded as unknown, but everything else in the system of equations (3.7.1-2) is known and temperature readings taken at some appropriate locations within the medium are available.

Referring to the nomenclature shown in figure 3.7.1, we assume that N_1 sensors are located in Region 1 and N_2 sensors are located in Region 2. The first sensors are located at distances $\bar{\delta}_1$ and $\bar{\delta}_2$ from the interface, while the remaining sensors are located with equal spacing of $\bar{\Delta}_1$ in Region 1 and $\bar{\Delta}_2$ in Region 2.

Let the temperature recordings taken with these sensors over the period τ to be denoted by:

$$Y_{1i}(t) \equiv Y_{1i} \quad , \quad i = 1, 2, ..., N_1 \text{ in Region 1 and}$$
$$Y_{2j}(t) \equiv Y_{2j} \quad , \quad j = 1, 2, ..., N_2 \text{ in Region 2.}$$

Then, the *inverse problem* can be stated as: By utilizing the above mentioned measured temperature data Y_{1i} ($i = 1, 2,..., N_1$) and Y_{2j} ($j = 1, 2,..., N_2$), estimate the unknown interface conductance $h(t)$ over the period τ.

It is assumed that no prior information is available on the functional form of $h(t)$, except that the period τ is known. We are after the function $h(t)$ over the whole time domain (0, τ), with the assumption that $h(t)$ belongs to the space of square integrable functions in this domain, i.e.,

$$\int_{t=0}^{\tau} [h(t)]^2 dt < \infty$$

The solution of the present inverse problem is to be obtained in such a way that the following functional is minimized:

$$S[h(t)] = \int_{t=0}^{\tau} \left[\sum_{i=1}^{N_1} (T_{1i} - Y_{1i})^2 \right] dt + \int_{t=0}^{\tau} \left[\sum_{j=1}^{N_2} (T_{2j} - Y_{2j})^2 \right] dt \qquad (3.7.4)$$

where $T_{1i} \equiv T_{1i}(t)$ and $T_{2j} \equiv T_{2j}(t)$ are the estimated temperatures at the measurement locations in regions 1 and 2, respectively.

Sensitivity Problem

The sensitivity problem is obtained from the direct problem defined by equations (3.7.1) and (3.7.2) in the following manner. It is assumed that when $h(t)$ undergoes a variation $\Delta h(t)$, $T_1(x,t)$ is perturbed by $\Delta T_1(x,t)$ and $T_2(x,t)$ is perturbed by $\Delta T_2(x,t)$. Then, by replacing in the direct problem $h(t)$ by $[h(t) + \Delta h(t)]$, $T_1(x,t)$ by $[T_1(x,t) + \Delta T_1(x,t)]$ and $T_2(x,t)$ by $[T_2(x,t) + \Delta T_2(x,t)]$, subtracting from the resulting expressions the original direct problem and neglecting second-order terms, the following *Sensitivity Problem* for the sensitivity functions $\Delta T_1(x,t)$ and $\Delta T_2(x,t)$ is obtained:

Region 1 ($0 \leq x \leq 1$):

$$\frac{\partial^2 \Delta T_1}{\partial x^2} = \frac{\partial \Delta T_1}{\partial t} \qquad \text{in } 0 < x < 1 \text{ , for } t > 0 \tag{3.7.5.a}$$

$$\Delta T_1 = 0 \qquad \text{at } x = 0 \text{ , for } t > 0 \tag{3.7.5.b}$$

$$-\frac{\partial \Delta T_1}{\partial x} = \Delta h(t)\ \left[T_1 - T_2\right] + h(t)\ \left[\Delta T_1 - \Delta T_2\right] \qquad \text{at } x = 1 \text{ , for } t > 0 \tag{3.7.5.c}$$

$$\Delta T_1(x,0) = \Delta T_1(x,\tau) \tag{3.7.5.d}$$

Region 2 ($1 \leq x \leq 1 + L$):

$$\frac{\partial^2 \Delta T_2}{\partial x^2} = \frac{1}{\alpha}\frac{\partial \Delta T_2}{\partial t} \qquad \text{in } 1 < x < 1 + L \text{ , for } t > 0 \tag{3.7.6.a}$$

$$-k\frac{\partial \Delta T_2}{\partial x} = \Delta h(t)\ \left[T_1 - T_2\right] + h(t)\ \left[\Delta T_1 - \Delta T_2\right] \qquad \text{at } x = 1 \text{ , for } t > 0 \tag{3.7.6.b}$$

$$\Delta T_2 = 0 \qquad \text{at } x = 1 + L \text{ , for } t > 0 \tag{3.7.6.c}$$

$$\Delta T_2(x,0) = \Delta T_2(x,\tau) \tag{3.7.6.d}$$

Adjoint Problem

In the present inverse problem, the estimated temperatures need to satisfy two constraints, which are the heat conduction problems for regions 1 and 2, given by equations (3.7.1) and (3.7.2), respectively. Therefore, two Lagrange multipliers come into picture here. To obtain the adjoint problem, equation (3.7.1.a) is multiplied by the Lagrange multiplier $\lambda_1(x,t)$, equation (3.7.2.a) is multiplied by the Lagrange multiplier $\lambda_2(x,t)$ and the resulting expressions are integrated over the time and space domains. Then, the results are added to the right-hand side of equation (3.7.4) to yield the following expression for the functional $S[h(t)]$:

$$S[h(t)] = \int_{t=0}^{\tau}\left[\sum_{i=1}^{N_1}(T_{1i}-Y_{1i})^2\right]dt + \int_{t=0}^{\tau}\left[\sum_{j=1}^{N_2}(T_{2j}-Y_{2j})^2\right]dt +$$
$$+\int_{t=0}^{\tau}\int_{x=0}^{1}\lambda_1(x,t)\left[\frac{\partial^2 T_1}{\partial x^2}-\frac{\partial T_1}{\partial t}\right]dx\,dt + \int_{t=0}^{\tau}\int_{x=1}^{1+L}\lambda_2(x,t)\left[\frac{\partial^2 T_2}{\partial x^2}-\frac{1}{\alpha}\frac{\partial T_2}{\partial t}\right]dx\,dt \tag{3.7.7}$$

The variation $\Delta S[h(t)]$ is obtained by perturbing $T_1(x,t)$ by $\Delta T_1(x,t)$, $T_2(x,t)$ by $\Delta T_2(x,t)$ in equation (3.7.7), subtracting from the resulting expression the original equation (3.7.7) and neglecting second-order terms. We find

$$\Delta S[h(t)] = \int_{t=0}^{\tau}\sum_{i=1}^{N_1}2\,\Delta T_{1i}(T_{1i}-Y_{1i})\,dt + \int_{t=0}^{\tau}\sum_{j=1}^{N_2}2\,\Delta T_{2j}(T_{2j}-Y_{2j})\,dt +$$
$$+\int_{t=0}^{\tau}\int_{x=0}^{1}\lambda_1(x,t)\left[\frac{\partial^2 \Delta T_1}{\partial x^2}-\frac{\partial \Delta T_1}{\partial t}\right]dx\,dt + \tag{3.7.8}$$
$$+\int_{t=0}^{\tau}\int_{x=1}^{1+L}\lambda_2(x,t)\left[\frac{\partial^2 \Delta T_2}{\partial x^2}-\frac{1}{\alpha}\frac{\partial \Delta T_2}{\partial t}\right]dx\,dt$$

In equation (3.7.8), the last two integral terms are integrated by parts; the initial and boundary conditions of the sensitivity problem given by equations (3.7.5.b-d) and (3.7.6.b-d) are utilized and then $\Delta S[h(t)]$ is allowed to go to zero. The vanishing of the integrands containing $\Delta T_1(x,t)$ and $\Delta T_2(x,t)$ leads to the following *adjoint problem* for the determination of the Lagrange multipliers $\lambda_1(x,t)$ and $\lambda_2(x,t)$:

Region 1 ($0 \le x \le 1$):

$$\frac{\partial^2 \lambda_1}{\partial x^2}+\frac{\partial \lambda_1}{\partial t}+\sum_{i=1}^{N_1}2\left(T_{1i}-Y_{1i}\right)\delta\left(x-x_i\right)=0 \quad \text{in } 0<x<1\text{, for } t>0 \tag{3.7.9.a}$$

$$\lambda_1 = 0 \quad \text{at } x=0\text{, for } t>0 \tag{3.7.9.b}$$

$$\frac{\partial \lambda_1}{\partial x} = h(t)\left[\frac{\lambda_2}{k}-\lambda_1\right] \quad \text{at } x=1\text{, for } t>0 \tag{3.7.9.c}$$

$$\lambda_1(x,\tau)=\lambda_1(x,0) \tag{3.7.9.d}$$

Region 2 ($1 \leq x \leq 1 + L$):

$$\frac{\partial^2 \lambda_2}{\partial x^2} + \frac{1}{\alpha}\frac{\partial \lambda_2}{\partial t} + \sum_{j=1}^{N_2} 2\left(T_{2j} - Y_{2j}\right)\delta\left(x - x_j\right) = 0$$

$$\text{in } 1 < x < 1+L \text{ , for } t > 0 \tag{3.7.10.a}$$

$$\frac{\partial \lambda_2}{\partial x} = h(t)\left[\frac{\lambda_2}{k} - \lambda_1\right] \qquad \text{at } x = 1 \text{ , for } t > 0 \tag{3.7.10.b}$$

$$\lambda_2 = 0 \qquad \text{at } x = 1 + L \text{ , for } t > 0 \tag{3.7.10.c}$$

$$\lambda_2(x,\tau) = \lambda_2(x,0) \tag{3.7.10.d}$$

where $\delta(\cdot)$ is the Dirac delta function.

Gradient Equation

In the limiting process used to obtain the adjoint problem above, the following integral term is left:

$$\Delta S[h(t)] = \int_{t=0}^{\tau} \left\{\left[\frac{\lambda_2(1,t)}{k} - \lambda_1(1,t)\right]\left[T_1(1,t) - T_2(1,t)\right]\right\}\Delta h(t)\,dt \tag{3.7.11}$$

From the assumption that $h(t) \in L_2(0,\tau)$, $\Delta S[h(t)]$ is related to the gradient $\nabla S[h(t)]$ by:

$$\Delta S[h(t)] = \int_{t=0}^{\tau} \nabla S[h(t)]\Delta h(t)\,dt \tag{3.7.12}$$

Thus, a comparison of equations (3.7.11) and (3.7.12) leads to the following expression for the gradient $\nabla S[h(t)]$ of the functional $S[h(t)]$:

$$\nabla S[h(t)] = \left\{\left[\frac{\lambda_2(1,t)}{k} - \lambda_1(1,t)\right]\left[T_1(1,t) - T_2(1,t)\right]\right\} \tag{3.7.13}$$

Iterative Procedure

The following iterative procedure of Technique IV is used for the estimation of $h(t)$:

$$h^{k+1}(t) = h^{k}(t) - \beta^{k} d^{k}(t) \tag{3.7.14.a}$$

where β^k is the *Search Step Size* in going from iteration k to iteration k+1, and $d^k(t)$ is the *Direction of Descent* given by:

$$d^{k}(t) = \nabla S\left[h^{k}(t)\right] + \gamma^{k} d^{k-1}(t) \tag{3.7.14.b}$$

which is a conjugation of the gradient direction $\nabla S[h^k(t)]$ at iteration k and the direction of descent $d^{k-1}(t)$ at iteration k-1. The *Conjugation Coefficient* γ^k is determined from the Fletcher-Reeves expression as

$$\gamma^{k} = \frac{\int_{t=0}^{\tau}\left\{\nabla S\left[h^{k}(t)\right]\right\}^{2} dt}{\int_{t=0}^{\tau}\left\{\nabla S\left[h^{k-1}(t)\right]\right\}^{2} dt} \quad \text{for } k = 1,2,\ldots \text{ with } \gamma^{k} = 0 \ \text{ for } k = 0 \tag{3.7.14.c}$$

The step size β^k is determined by minimizing the functional $S[h(t)]$ defined by equation (3.7.4) in the following manner. By utilizing the expression for $h^{k+1}(t)$ given by equation (3.7.14.a), the functional given by equation (3.7.4) takes the form:

$$\begin{aligned} S\left[h^{k+1}(t)\right] = & \int_{t=0}^{\tau}\sum_{i=1}^{N_1}\left[T_{1i}\left(h^{k} - \beta^{k} d^{k}\right) - Y_{1i}\right]^{2} dt + \\ & + \int_{t=0}^{\tau}\sum_{j=1}^{N_2}\left[T_{2j}\left(h^{k} - \beta^{k} d^{k}\right) - Y_{2j}\right]^{2} dt \end{aligned} \tag{3.7.15}$$

By linearizing the temperatures $T_{1i}(h^k - \beta^k d^k)$ and $T_{2j}(h^k - \beta^k d^k)$ and minimizing the resulting expression with respect to β^k, we obtain the following expression for the search step size (see Note 7 in Chapter 2):

$$\beta^k = \frac{\int_{t=0}^{\tau_f} \left\{ \sum_{i=1}^{N_1} \left[T_{1i} - Y_{1i} \right] \Delta T_{1i} + \sum_{j=1}^{N_2} \left[T_{2j} - Y_{2j} \right] \Delta T_{2j} \right\} dt}{\int_{t=0}^{\tau_f} \left\{ \sum_{i=1}^{N_1} \left[\Delta T_{1i} \right]^2 + \sum_{j=1}^{N_2} \left[\Delta T_{2j} \right]^2 \right\} dt} \tag{3.7.16}$$

where T_{1i} and T_{2j} are the solutions of the direct problem (3.7.1) and (3.7.2), obtained by using the current estimate for $h(t)$; while the sensitivity functions ΔT_{1i} and ΔT_{2j} are the solutions of the sensitivity problem (3.7.5) and (3.7.6), obtained by setting $\Delta h(t) = d^k(t)$.

Results

The problems of periodically contacting surfaces involving very small periods are the most difficult to perform an inverse analysis. Therefore, to illustrate the accuracy of Technique IV under very strict conditions, we examine the problems for very small periods.

Consider two identical regions each of length $\bar{L}_1 = \bar{L}_2 = 0.1\,\text{m}$ and made of brass ($\bar{k}_1 = \bar{k}_2 = 106.1$ W/m K; $\bar{\alpha}_1 = \bar{\alpha}_2 = 3.4 \times 10^{-5}\ \text{m}^2/s$) studied experimentally in reference [22] and theoretically by solving the inverse problem using B-Splines in reference [23]. Each region contains four sensors and 18 measurements are made per sensor per period. Figure 3.7.1 shows the notation for the geometry, while Table 3.7.1 lists typical dimensional and dimensionless sensor locations, as well as periods of variation of $h(t)$.

Let $h(t)$ vary in the form

$$h(t) = \begin{cases} 2 & \text{for the contact period} \\ 0 & \text{for the non-contact period} \end{cases} \tag{3.7.17}$$

This dimensionless value of $h(t) = 2$ corresponds to a dimensional contact conductance of 2122 W/m^{2}K, which is encountered in the contact of metallic wavy surfaces [24]. Both exact and inexact simulated temperature measurements are considered, but all the other quantities used in the inverse analysis are assumed to be errorless.

Table 3.7.1 - Periods and sensor locations

	Dimensionless	Dimensional
Period	10^{-1}	29.41 s
	10^{-2}	2.94 s
	10^{-3}	0.29 s
Sensor Locations	$\delta = \Delta = 0.005$	$\bar{\delta} = \bar{\Delta} = 0.5$ mm
	$\delta = \Delta = 0.01$	$\bar{\delta} = \bar{\Delta} = 1$ mm
	$\delta = 0.05;\ \Delta = 0.1$	$\bar{\delta} = 5$ mm; $\bar{\Delta} = 10$ mm

Due to the periodic characteristic of the problem, it is shown [25] that under the quasi-steady-state condition, the temperature distribution in the regions vary only within a finite depth of δ_T below the surface. The temperature distributions in each region at the end of the contact ($T_{\tau/2}$) and non-contact (T_τ) periods are presented in figure 3.7.2 for $\tau = 10^{-1}$. Therefore, if the sensors are located outside this thermal layer δ_T, no difference can be detected between the temperature measurements for the contact and non-contact periods. Thus, to obtain meaningful results from the temperature measurements, the sensors must be located within the thermal layer δ_T. Here we define δ_T as the depth below the surface such that

$$\left| T_{\tau/2} - T_{\tau} \right| > \varepsilon_{\delta} \tag{3.7.18}$$

where ε_δ is a fixed tolerance.

Figure 3.7.3 shows the effects of the period τ and the contact conductance h during the contact period, on the dimensionless thickness δ_T of the thermal layer, for a tolerance $\varepsilon_\delta = 10^{-3}$. We note that the value of the contact conductance has negligible effect on δ_T; on the other hand, δ_T is strongly dependent on the period τ, such that δ_T becomes very small for short periods. As an example, for the physical case considered here, for $\tau = 10^{-4}$ (i.e., 0.029 s) the thermal layer is of the order of 10^{-3}, which corresponds to tenths of millimeter. The results presented in figure 3.7.3 are obviously dependent on the tolerance ε_δ, which is directly related to the accuracy of the sensors used. Therefore, these results are just a qualitative indication of the behavior of the thermal layer, with respect to variations in h and τ.

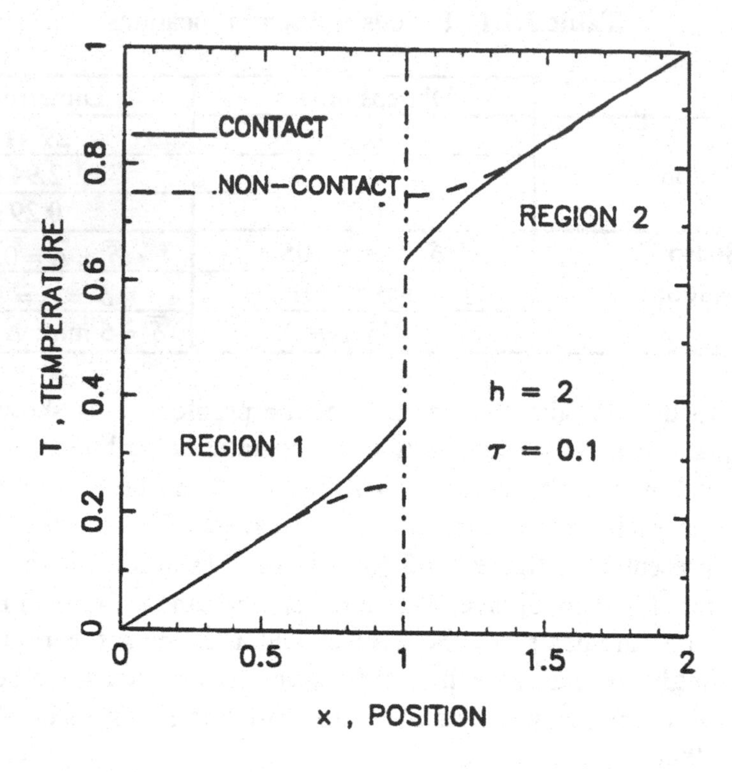

Figure 3.7.2 - Temperature distribution on the regions.

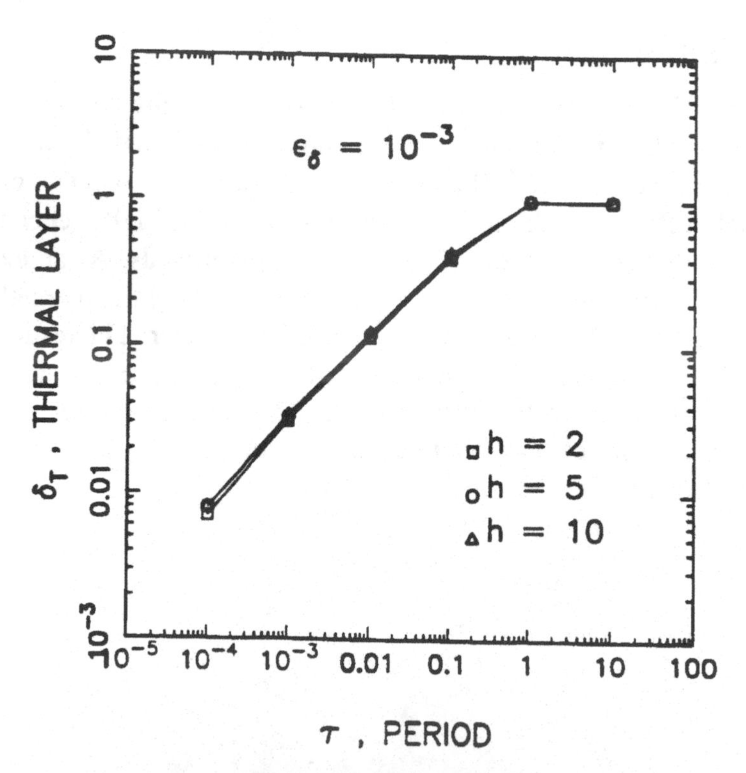

Figure 3.7.3 - Effects of τ and h on the thermal layer.

Figures 3.7.4 to 3.7.6 show the results obtained with simulated errorless measurements for the periodically varying contact conductance in the form of a step function. Note that for the 3 cases considered, the solutions obtained with the conjugate gradient method (CGM) are more accurate than those obtained with $h(t)$ parameterized with B-Splines in reference [23]. As a matter of fact, the B-Splines solutions exhibit oscillations near the discontinuities, which increase with decreasing period. On the other hand, the solutions with the Conjugate Gradient Method are very stable and do not exhibit oscillations, even for very small periods.

Similarly, the estimation of $h(t)$ with B-Splines become unstable when measurements with random errors are utilized in the analysis; but reasonably accurate results can be obtained with the conjugate gradient method of function estimation. This is illustrated in figure 3.7.7 for the period $\tau = 10^{-3}$ and for a standard-deviation of the measurement errors of $\sigma = 0.0065$.

We define the root mean square (RMS) error as

$$e_{RMS} = \sqrt{\frac{1}{I}\sum_{i=1}^{I}\left[h_{ex}(t) - h_{est}(t)\right]^2} \tag{3.7.19}$$

where I is the number of transient measurements per sensor, while the subscripts *ex* and *est* refer to the exact and estimated contact conductance, respectively.

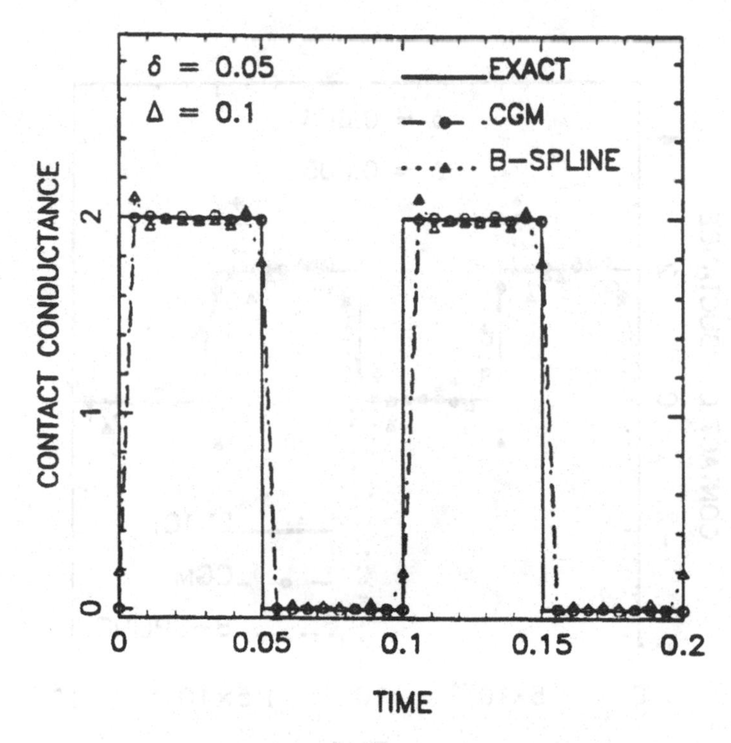

Figure 3.7.4 - Inverse solution for exact measurements and $\tau = 10^{-1}$.

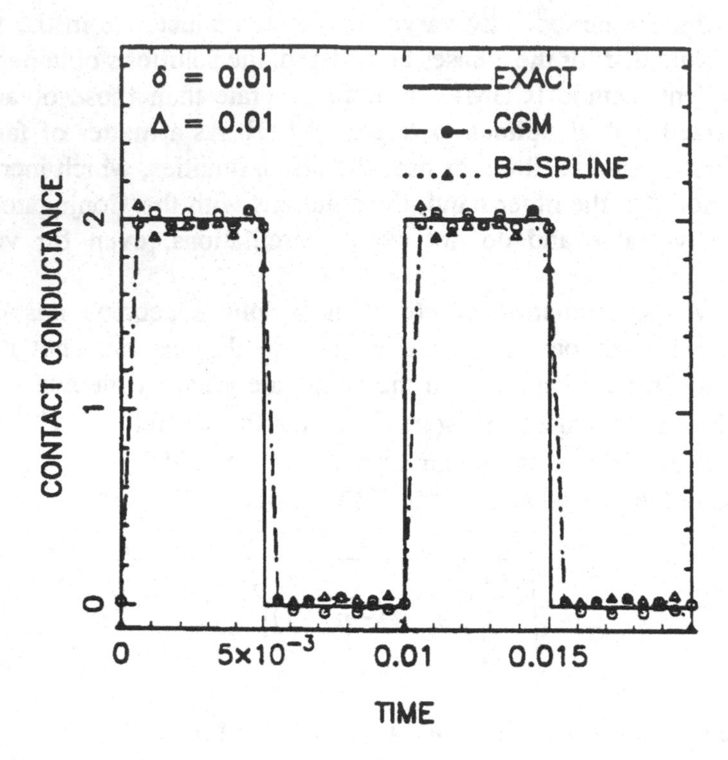

Figure 3.7.5 - Inverse solution for exact measurements and $\tau = 10^{-2}$.

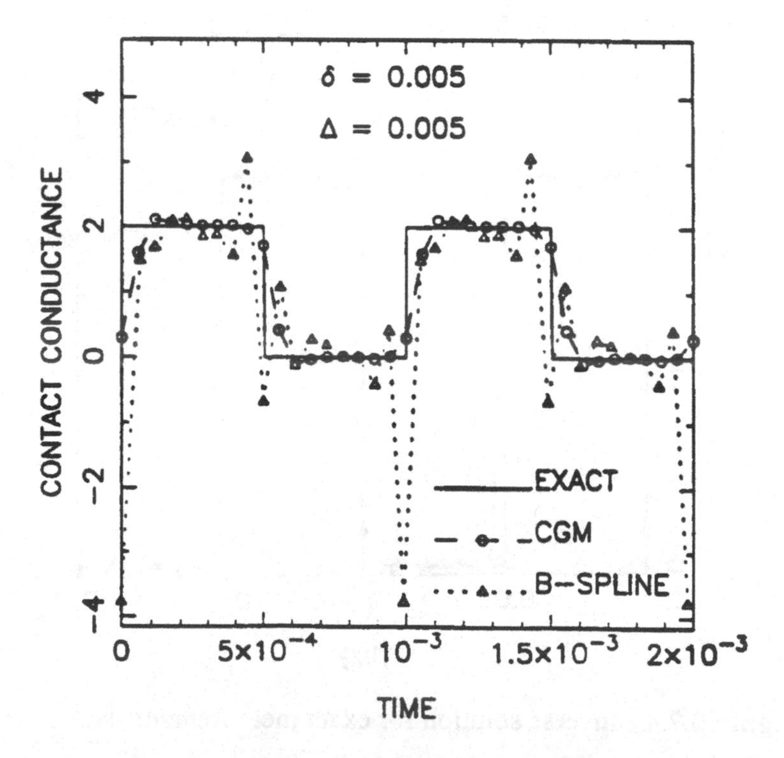

Figure 3.7.6 - Inverse solution for exact measurements and $\tau = 10^{-3}$.

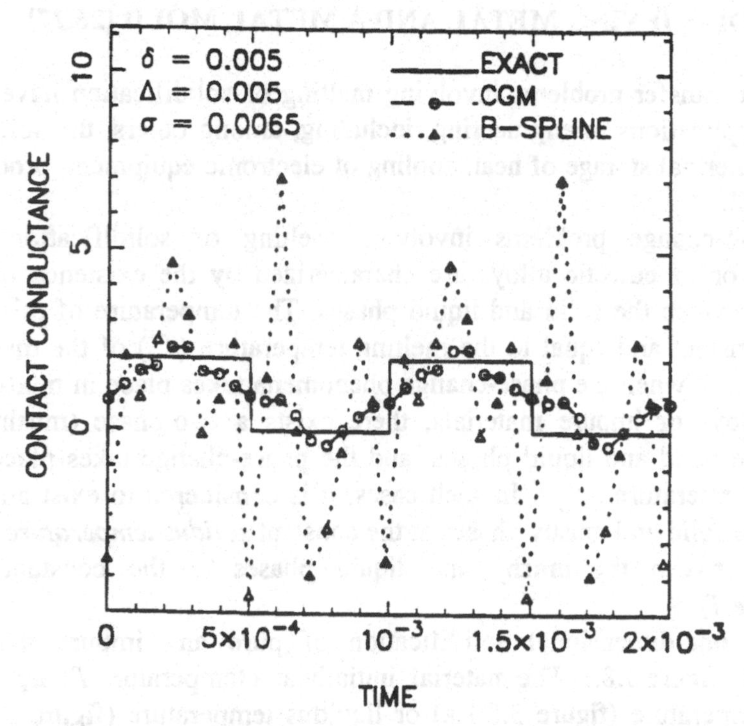

Figure 3.7.7 - Inverse solution for $\sigma = 0.0065$ and $\tau = 10^{-3}$.

Table 3.7.2 shows the values of e_{RMS} for cases involving measurements with standard-deviation $\sigma = 0.0065$. Clearly, the RMS error increases as the period decreases and, for $\tau = 10^{-3}$, the RMS error for the B-spline method is much higher than that for the conjugate gradient method. The increase in the RMS error with decreasing period is expected. This is due to the fact that for shorter periods, the measurement error increases relatively to the maximum temperature variation in the regions. This maximum variation occurs at the contacting interface (see figure 3.7.2) and gives the upper limit of the temperature variations the sensors will measure. The ratio between the measurement error e at the 99% confidence level for $\sigma = 0.0065$ and the maximum temperature variation in the regions T_{var} is presented in the last column of table 3.7.2.

Table 3.7.2 - Total rms error and relative error

Period	e_{RMS}		e/T_{var}
	B-spline	CGM	
10^{-1}	0.36	0.36	0.15
10^{-2}	0.60	0.43	0.42
10^{-3}	2.90	0.58	1.67

3-8 ESTIMATION OF THE CONTACT CONDUCTANCE BETWEEN A SOLIDIFYING METAL AND A METAL MOLD [26,27]

Heat transfer problems involving melting or solidification have different practical applications in engineering, including, among others, the solidification of metals, thermal storage of heat, cooling of electronic equipment, production of ice, etc. [18].

Phase-change problems involving melting or solidification of pure substances or of eutectic alloys are characterized by the existence of a sharp interface between the solid and liquid phases. The temperature of this interface remains constant and equal to the melting temperature (T_m) of the material. On the other hand, when the phase-change phenomena takes place in mixtures, non-eutectic alloys or impure materials, there exists a two-phase (mushy) region between the solid and liquid phases, and the phase-change takes place over an extended temperature range. In such cases, it is considered to exist an interface between the solid and mushy phases at the constant *solidus temperature* T_s and an interface between the mushy and liquid phases at the constant *liquidus temperature* T_l.

The one-dimensional solidification of pure and impure materials is illustrated in figure 3.8.1. The material, initially at a temperature T_i larger than the melting temperature (figure 3.8.1.a) or liquidus temperature (figure 3.8.1.b), is put into contact with the boundary surface at $x = 0$, which is maintained at a temperature T_0 below the melting temperature (figure 3.8.1.a) or solidus temperature (figure 3.8.1.b). As a result, solidification takes place and the solid-liquid interface in the case of pure materials (figure 3.8.1.a), as well as both the solid-mushy and mushy-liquid interfaces in the case of impure materials (figure 3.8.1.b), move towards the $x > 0$ direction. The location of such interfaces is not known *a priori* and, hence, phase-change problems are non-linear. Therefore, analytical solutions for phase-change problems are available only for simple geometries, such as one-dimensional semi-infinite medium, and for cases involving simple boundary conditions, such as the prescribed temperature at the boundary surface [18]. For general cases, phase-change problems need to be solved numerically.

Different numerical techniques for the solution of phase-change problems were developed in the past, including single-region and multiple-region methods [18,26-36]. *Single-region* methods involve one single general formulation that is valid for the solid, liquid and mushy phases, as well as for the interfaces between phases [26-31,35,36]. On the other hand, in *multiple-region* methods each phase is modeled by a different governing equation and interface conditions are used to couple the formulations of adjacent regions [18,32-34]. Single-region methods are advantageous because of their simplicity; but they cannot be extended to take into account the coupling between microscopic and macroscopic phenomena, as multiple-region methods can [32-35].

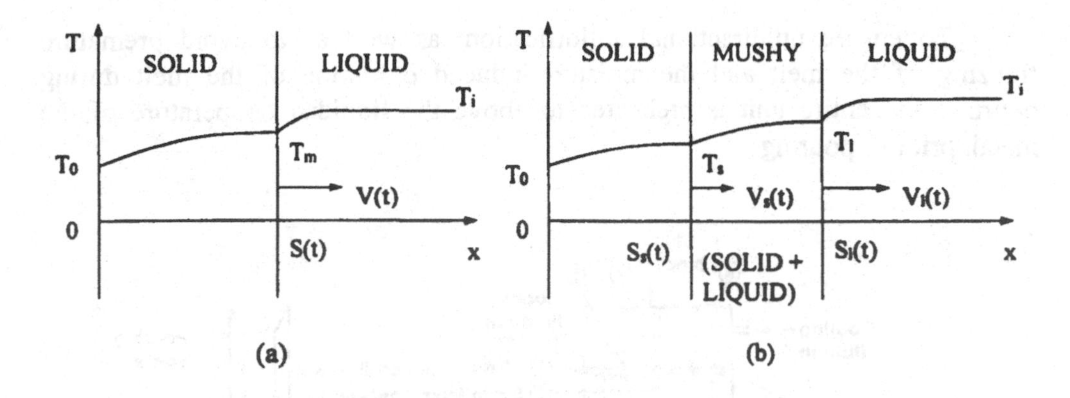

Figure 3.8.1 - One dimensional solidification of pure substances (a) and of impure substances or alloys (b).

If the effects of fluid flow in the liquid phase are negligible, convection heat transfer can be neglected and the phase-change problem can be formulated only in terms of heat conduction.

In this section we present the solution of an inverse phase-change problem, involving the estimation of the contact conductance between a solidifying metal and a metal mold [26,27]. Convective effects on the liquid phase are neglected. The transient contact conductance is estimated by using **Technique IV**, the conjugate gradient method of function estimation. The direct problem involving the solidification of the metal is solved with the Implicit Enthalpy Method, which is a very straightforward single-region method [26-31,36]. Both simulated and actual experimental data were used in the analysis. Such experimental data were obtained with an apparatus based on unidirectional solidification.

The details of the experimental apparatus and of Technique IV, as applied to the estimation of the unknown transient contact conductance, are described next.

Direct Problem

A bottom filling solidification apparatus based on the unidirectional solidification principle, as illustrated in figure 3.8.2, was constructed to study experimentally the air gap conductance between the mold and the casting. It consists of a pouring sprue, a runner and a slightly tapered rectangular mold. Molten metal poured into the sprue gradually fills the space and eventually comes into contact with the chill-plate, through which heat is extracted from the solidifying ingot. The determination of the air gap conductance between the chill-plate and the solidifying ingot is the subject of this investigation. The chill-plate, which can be made of copper, steel or any other material, is clamped to the open face of the apparatus and cooled uniformly by a water jet.

To ensure unidirectional solidification, as well as to avoid premature freezing of the melt and the moisture-induced oxidation of the melt during pouring, the entire unit is preheated to above the liquidus temperature of the metal, prior to pouring.

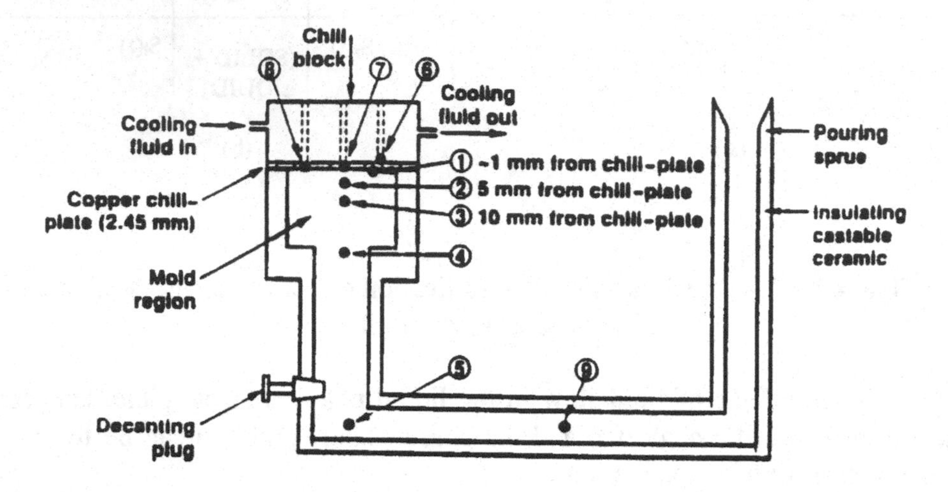

Figure 3.8.2 - Experimental solidification apparatus.

Transient temperature recordings were taken with 28 gauge (0.012") chromel-alumel thermocouples placed at pertinent locations in the chill-plate and casting region and the temperature-time data were obtained with computer controlled data acquisition system at a rate of 6 readings per second.

The transient temperature data taken as a function of time in the casting and in the metal mold are used in the inverse analysis in order to estimate the timewise variation of the unknown interface conductance.

For the present study, the direct problem is the mathematical formulation of the following solidification problem: Suppose a molten metal suddenly comes into contact with a cold chill-plate and the timewise variation of the contact conductance $h_c(t)$ between the chill-plate and the solidifying casting is known. Assuming constant properties, the mathematical formulation of such a problem is given by:

The Chill-Plate ($0 \le x \le b$):

$$k_p \frac{\partial^2 T_p(x,t)}{\partial x^2} = C_p \frac{\partial T_p(x,t)}{\partial t} \qquad \text{in } 0 < x < b \text{, for } t > 0 \qquad (3.8.1.a)$$

$$T_p = f(t) \qquad \text{at } x = 0, \qquad \text{for } t > 0 \qquad (3.8.1.b)$$

$$k_p \frac{\partial T_p}{\partial x} = h_c(t)\,(T_c - T_p) \qquad \text{at } x = b, \qquad \text{for } t > 0 \qquad (3.8.1.c)$$

$$T_p = T_{p0} \qquad \text{in } 0 < x < b, \qquad \text{for } t = 0 \qquad (3.8.1.d)$$

where $C_p = \rho_p c_p$ is the heat capacity per unit volume, while ρ_p and c_p are the density and the specific heat of the plate, respectively. Boundary condition (3.8.1.b) implies that the temperature of the outer surface of the chill-plate is available from temperature measurements by a thermocouple. This eliminates the necessity to know the convection heat transfer coefficient between the chill-plate and the cooling fluid.

The Casting Region ($b \le x \le a$):

To alleviate the tracking of the moving interface, the enthalpy form of the energy equation is used for the casting region.

$$k_c \frac{\partial^2 T_c(x,t)}{\partial x^2} = \rho_c \frac{\partial H_c(x,t)}{\partial t} \qquad \text{in } b < x < a, \text{ for } t > 0 \qquad (3.8.2.a)$$

$$k_c \frac{\partial T_c}{\partial x} = h_c(t)\,(T_c - T_p) \qquad \text{at } x = b, \qquad \text{for } t > 0 \qquad (3.8.2.b)$$

$$\frac{\partial T_c}{\partial x} = 0 \qquad \text{at } x = a, \qquad \text{for } t > 0 \qquad (3.8.2.c)$$

$$T_c = T_{c0} \qquad \text{in } b < x < a, \qquad \text{for } t = 0 \qquad (3.8.2.d)$$

where $dH_c = c_c\, dT$ is the enthalpy change of the casting material; c_c and ρ_c are respectively the specific heat and density of the casting, which initial temperature is T_{c0}.

The objective of the Direct Problem is the determination of the temperature field inside the chill-plate and the casting region.

Inverse Problem

For the inverse problem considered here, the contact conductance $h_c(t)$ is regarded as unknown and is to be estimated by using the temperature measurements of N_1 sensors located inside the chill-plate and of N_2 sensors located inside the casting region. The function $h_c(t)$ is estimated through the minimization of the following functional

$$S[h_c(t)] = \int_{t=0}^{t_f} \left\{ \sum_{i=1}^{N_1} (T_{1i} - Y_{1i})^2 + \sum_{j=1}^{N_2} (T_{2j} - Y_{2j})^2 \right\} dt \tag{3.8.3}$$

where $T_{1i} \equiv T_p(x_{1i},t)$ and Y_{1i} are the estimated and measured temperatures, respectively, at a location x_{1i} in the chill-plate. Similarly, $T_{2j} \equiv T_c(x_{2j},t)$ and Y_{2j} are the estimated and measured temperatures, respectively, at a location x_{2j} in the casting region. If an estimate is available for $h_c(t)$, the temperatures T_{1i} and T_{2j} can be computed from the solution of the direct problem defined by equations (3.8.1) and (3.8.2).

The minimization of the functional (3.8.3) with Technique IV requires the solution of the sensitivity and adjoint problems. The development of such auxiliary problems is described next.

Sensitivity Problem

Suppose $h_c(t)$ undergoes a variation $\Delta h_c(t)$. Then let ΔT_p , ΔT_c and ΔH_c be the corresponding variations of the plate temperature, casting temperature and casting enthalpy, respectively. To construct the sensitivity problem we replace T_p by $[T_p + \Delta T_p]$, T_c by $[T_c + \Delta T_c]$, H_c by $[H_c + \Delta H_c]$ and h_c by $[h_c + \Delta h_c]$ in the direct problem given by equations (3.8.1) and (3.8.2) and then subtract equations (3.8.1) and (3.8.2) from the resulting equations. The following sensitivity problem is obtained for the determination of the functions ΔT_p and ΔT_c in the chill-plate and casting, respectively.

The Chill-Plate ($0 \leq x \leq b$):

$$k_p \frac{\partial^2 \Delta T_p(x,t)}{\partial x^2} = C_p \frac{\partial \Delta T_p(x,t)}{\partial t} \qquad \text{in } 0 < x < b \text{ , for } t > 0 \tag{3.8.4.a}$$

$$\Delta T_p = 0 \qquad \text{at } x = 0, \quad \text{for } t > 0 \tag{3.8.4.b}$$

$$k_p \frac{\partial \Delta T_p}{\partial x} = h_c(t)\,(\Delta T_c - \Delta T_p) + \Delta h_c(t)\,(T_c - T_p) \quad \text{at } x = b, \quad \text{for } t > 0 \tag{3.8.4.c}$$

$$\Delta T_p = 0 \qquad \text{in } 0 < x < b, \quad \text{for } t = 0 \tag{3.8.4.d}$$

The Casting Region ($b \le x \le a$):

$$k_c \frac{\partial^2 \Delta T_c(x,t)}{\partial x^2} = C_c \frac{\partial \Delta T_c(x,t)}{\partial t} \quad \text{in } b<x<a\,, \text{ for } t>0 \tag{3.8.5.a}$$

$$k_c \frac{\partial \Delta T_c}{\partial x} = h_c(t)\,(\Delta T_c - \Delta T_p) + \Delta h_c(t)\,(T_c - T_p) \quad \text{at } x=b, \quad \text{for } t>0 \tag{3.8.5.b}$$

$$\frac{\partial \Delta T_c}{\partial x} = 0 \quad \text{at } x=a, \quad \text{for } t>0 \tag{3.8.5.c}$$

$$\Delta T_c = 0 \quad \text{in } b<x<a, \quad \text{for } t=0 \tag{3.8.5.d}$$

In equation (3.8.5.a) we replaced ΔH_c by its equivalent $c_c \Delta T_c$, since this is not a phase change problem. Therefore it can be solved with standard finite difference techniques.

Adjoint Problem

To derive the adjoint problem, we multiply equations (3.8.1.a) and (3.8.2.a) by the Lagrange Multipliers $\lambda_p(x,t)$ and $\lambda_c(x,t)$, respectively; integrate the resulting expressions over the time and correspondent space domains; and then add the resultant equation to the functional given by equation (3.8.3). The following expression results:

$$\begin{aligned} S[h_c(t)] = & \int_{t=0}^{t_f} \left\{ \sum_{i=1}^{N_1} (T_{1i} - Y_{1i})^2 + \sum_{j=1}^{N_2} (T_{2j} - Y_{2j})^2 \right\} dt \, + \\ & \int_{t=0}^{t_f} \int_{x=0}^{b} \lambda_p(x,t) \left[k_p \frac{\partial^2 T_p}{\partial x^2} - C_p \frac{\partial T_p}{\partial t} \right] dx\, dt \, + \\ & \int_{t=0}^{t_f} \int_{x=b}^{a} \lambda_c(x,t) \left[k_c \frac{\partial^2 T_c}{\partial x^2} - \rho_c \frac{\partial H_c}{\partial t} \right] dx\, dt \end{aligned} \tag{3.8.6}$$

The variation of the extended functional (3.8.6) is obtained and allowed to go to zero. After some manipulations, as outlined in section 2-4, we obtain the following adjoint problem for the Lagrange Multipliers $\lambda_p(x,t)$ and $\lambda_c(x,t)$.

The Chill-Plate ($0 \le x \le b$):

$$k_p \frac{\partial^2 \lambda_p(x,t)}{\partial x^2} + C_p \frac{\partial \lambda_p(x,t)}{\partial t} + \sum_{i=1}^{N_1} 2(T_{1i} - Y_{1i})\delta(x - x_i) = 0$$

in $0 < x < b$, for $0 < t < t_f$ (3.8.7.a)

$$\lambda_p = 0 \qquad \text{at } x = 0, \text{ for } 0 < t < t_f \qquad (3.8.7.b)$$

$$k_p \frac{\partial \lambda_p}{\partial x} = h_c(t)\,(\lambda_c - \lambda_p) \qquad \text{at } x = b, \text{ for } 0 < t < t_f \qquad (3.8.7.c)$$

$$\lambda_p = 0 \qquad \text{in } 0 < x < b, \text{ for } t = t_f \qquad (3.8.7.d)$$

The Casting Region ($b \le x \le a$):

$$k_c \frac{\partial^2 \lambda_c(x,t)}{\partial x^2} + C_c \frac{\partial \lambda_c(x,t)}{\partial t} + \sum_{j=1}^{N_2} 2(T_{2j} - Y_{2j})\delta(x - x_{2j}) = 0$$

in $b < x < a$, for $0 < t < t_f$ (3.8.8.a)

$$k_c \frac{\partial \lambda_c}{\partial x} = h_c(t)(\lambda_c - \lambda_p) \qquad \text{at } x = b, \text{ for } 0 < t < t_f \qquad (3.8.8.b)$$

$$\frac{\partial \lambda_c}{\partial x} = 0 \qquad \text{at } x = a, \text{ for } 0 < t < t_f \qquad (3.8.8.c)$$

$$\lambda_c = 0 \qquad \text{in } b < x < a, \text{ for } t = t_f \qquad (3.8.8.d)$$

Note that the adjoint problem defined by equations (3.8.7) and (3.8.8) is not a phase change problem and can be solved with standard finite difference techniques.

Gradient Equation

In the process of obtaining the adjoint problem, the following integral term is left:

$$\Delta S[h_c(t)] = \int_{t=0}^{t_f} \{[\lambda_c(b,t) - \lambda_p(b,t)][T_p(b,t) - T_c(b,t)]\}\,\Delta h_c(t)\,dt \qquad (3.8.9.a)$$

By assuming that $h_c(t)$ belongs to the space of square integrable functions in the time domain $0 < t < t_f$, we can write

$$\Delta S[h_c(t)] = \int_{t=0}^{t_f} \nabla S[h_c(t)] \Delta h_c(t)\, dt \tag{3.8.9.b}$$

Therefore, by comparing equations (3.8.9.a) and (3.8.9.b), we can obtain the gradient equation for the functional as

$$\nabla S[h_c(t)] = [\lambda_c(b,t) - \lambda_p(b,t)][T_p(b,t) - T_c(b,t)] \tag{3.8.10}$$

Iterative Procedure

The following iterative procedure of Technique IV is used for the estimation of the contact conductance $h_c(t)$:

$$h_c^{k+1}(t) = h_c^k(t) - \beta^k d^k(t) \tag{3.8.11.a}$$

where the superscript k refers to the number of iterations and the direction of descent is given by

$$d^k(t) = \nabla S[h_c^k(t)] + \gamma^k d^{k-1}(t) \tag{3.8.11.b}$$

The conjugation coefficient is obtained from the Fletcher-Reeves expression as

$$\gamma^k = \frac{\int_{t=0}^{t_f} \left\{ \nabla S\left[h_c^k(t) \right] \right\}^2 dt}{\int_{t=0}^{t_f} \left\{ \nabla S\left[h_c^{k-1}(t) \right] \right\}^2 dt} \qquad \text{for } k=1,2,3,\ldots \text{ with } \gamma^0 = 0 \tag{3.8.11.c}$$

An expression for the search step size is obtained by minimizing the functional given by equation (3.8.3) with respect to β^k. We obtain (see Note 7 in Chapter 2):

$$\beta^k = \frac{\int_{t=0}^{t_f} \left\{ \sum_{i=1}^{N_1} \Delta T_{1i}(d^k)[T_{1i} - Y_{1i}] + \sum_{j=1}^{N_2} \Delta T_{2j}(d^k)[T_{2j} - Y_{2j}] \right\} dt}{\int_{t=0}^{t_f} \left\{ \sum_{i=1}^{N_1} \left[\Delta T_{1i}(d^k) \right]^2 + \sum_{j=1}^{N_2} \left[\Delta T_{2j}(d^k) \right]^2 \right\} dt} \tag{3.8.11.d}$$

where $\Delta T_{1i}(d^k) \equiv \Delta T_{1i}(x_{1i},t), i = 1,...,N_1$ and $\Delta T_{2j}(d^k) \equiv \Delta T_{2j}(x_{2j},t), j = 1,...,N_2$, are the solutions of the sensitivity problem given by equations (3.8.4) and (3.8.5), for the chill-plate and casting regions, respectively, obtained by setting $\Delta h_c(t) = d^k(t)$.

Results

In order to examine the accuracy of the conjugate gradient method, as applied to the analysis of the inverse solidification problem previously described, we studied test cases by considering a fictitious interface conductance and using simulated temperatures as the input data for the inverse analysis. The simulated temperature data were generated by solving the direct solidification problem for aluminum, on a geometry similar to that of the experimental rig and for a specified functional form of the interface conductance.

A proper choice for the locations of the temperature sensors is important for the success of the inverse analysis. The temperature readings taken with one thermocouple located near the cooled side of the plate served as the boundary condition for the solidification problem. The temperature readings from another thermocouple located in the chill-plate near the casting side and from one more thermocouple placed in the casting region near the plate, were used for the inverse analysis. Inverse calculations were performed by using simulated temperature data with and without measurement errors. The effects of errors in the thermocouple locations on $h_c(t)$ were also examined. These matters are discussed below.

Figure 3.8.3.a shows the estimated values of the contact conductance h_c plotted as a function of time, obtained by using simulated temperature data containing no measurement errors, for the case of a 6mm thick steel plate with 2 thermocouples embedded into the plate at a distance 4mm apart, in the direction normal to the surface of the plate. A third thermocouple was located in the casting at a distance 1 mm from the surface of the chill-plate. A comparison of the exact and the estimated values of the contact conductance h_c shown in figure 3.8.3.a reveals that they are in excellent agreement. In the solution of inverse heat conduction problems with the conjugate gradient method, the value of h_c at the final time is the same as the initial guess for h_c used to start the calculation. This difficulty was alleviated by repeating the calculations with an initial guess chosen as the value of h_c at several time steps before the final time. Figure 3.8.3.b shows the estimated and the exact temperatures at the plate thermocouple located at the casting side of the plate.

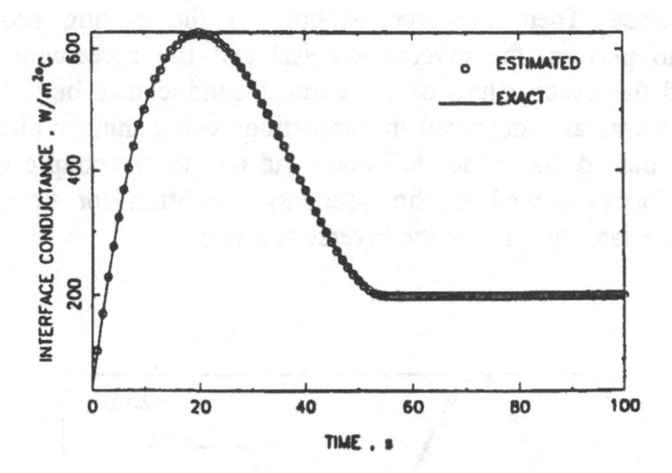

Figure 3.8.3.a - Estimated contact conductance obtained from simulated experimental data with no measurement error (6 mm thick steel chill-plate).

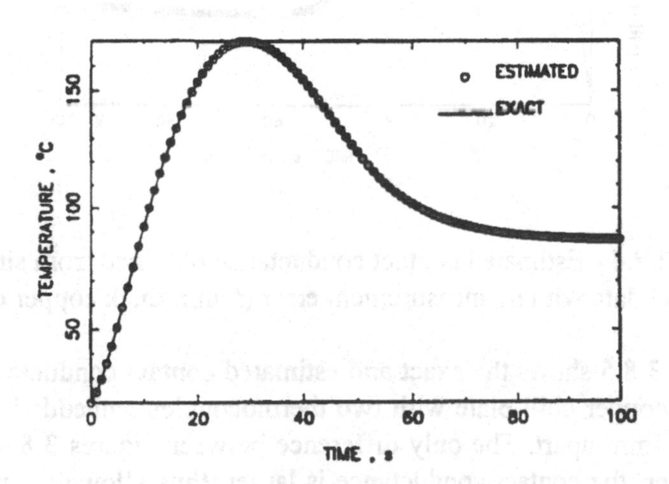

Figure 3.8.3.b - Estimated and exact simulated temperatures for the thermocouple in the chill-plate (6 mm thick steel chill-plate).

Figure 3.8.4 shows the estimated and the exact values of the contact conductance h_c for the same thermocouple configuration used before, but for a 6mm copper chill-plate. We note that during the first 5 or 6 seconds, the estimated values of the contact conductance h_c are not accurate. This is because of the small value of the temperature difference between the readings of the two thermocouples embedded in the plate, resulting from the high thermal conductivity of the copper chill-plate. Since the readings of one of the plate thermocouples are used as the boundary condition for the solidification problem and the readings of the other are used for the inverse problem, the accuracy of the estimation is impaired if the temperature difference between them becomes small. After a few seconds from the initiation of the solidification, the solid-liquid interface reaches the thermocouple located in the casting at a distance 1mm from the plate surface. Then, the thermocouple in the casting provides useful information to perform the inverse analysis and the agreement between the estimated and the exact values of the contact conductance begins to improve. Such behavior was also observed in simulations using thinner plates, with one thermocouple placed inside the chill-plate and one thermocouple in the casting region. The former served as the boundary condition for the solidification problem and the later to perform the inverse analysis.

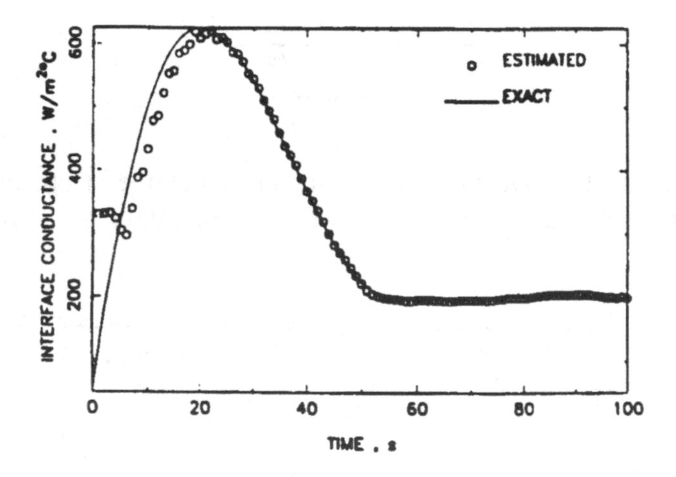

Figure 3.8.4 - Estimated contact conductance obtained from simulated experimental data with no measurement error (6 mm thick copper chill-plate).

Figure 3.8.5 shows the exact and estimated contact conductances by using a 6mm thick copper chill-plate with two thermocouples embedded into the plate at a distance 4mm apart. The only difference between figures 3.8.4 and 3.8.5 is that in the later, the contact conductance is larger, thus allowing larger heat flux across the plate. As a result, the temperature difference between the two thermocouples is larger and the accuracy of the inverse analysis is significantly improved.

Figure 3.8.6 is prepared in order to illustrate the effects of the input data containing measurement errors on the estimation of h_c with the present inverse analysis. For this case it is assumed that the standard deviation of the measurement errors is $\sigma = 0.5$ for the mold region and $\sigma = 1$ for the casting region. These values are based on the actual calibration data for the thermocouples. Since the measured data contains measurement errors, the discrepancy principle was used to terminate the iterations. The results shown in figure 3.8.6 are for a 6mm thick steel chill-plate with two thermocouples embedded into the

plate at a distance 4mm apart. The agreement between the exact and the estimated contact conductance is good.

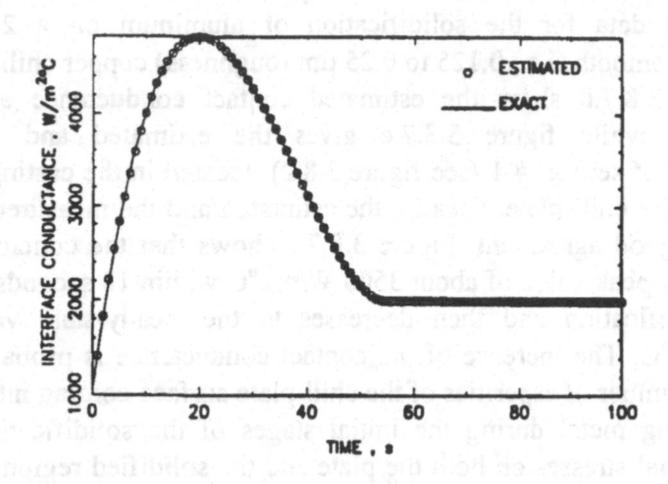

Figure 3.8.5 - Estimated contact conductance obtained from simulated experimental data with no measurement error (6 mm thick copper chill-plate).

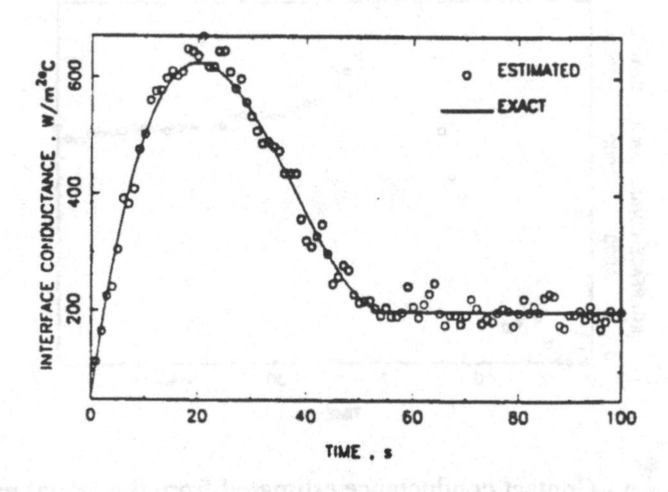

Figure 3.8.6 - Estimated contact conductance obtained from simulated experimental data with measurement errors (6 mm thick steel chill-plate).

To examine the errors associated with the misplacement of the thermocouple located in the casting region, 1mm error was assumed in the exact location of this thermocouple. The contact conductance was insensitive to such an error in the thermocouple location. For example, the contact conductance was overestimated (or underestimated) by less than two percent when the thermocouple junction in the casting was shifted towards (or away from) the chill-plate.

Finally, in figures 3.8.7.a-c we present the correlation of the actual experimental data for the solidification of aluminum on a 2.5 mm thick, substantially smooth (i.e., 0.125 to 0.25 μm roughness) copper chill-plate. Figures 3.8.7.a and 3.8.7.b show the estimated contact conductance and heat flux, respectively, while figure 3.8.7.c gives the estimated and the measured temperatures of sensor # 1 (see figure 3.8.2) located in the casting at a distance 1 mm from the chill-plate. Clearly, the estimated and the measured temperatures are in very good agreement. Figure 3.8.7.a shows that the contact conductance increases to a peak value of about 3500 W/m^2 °C within 18 seconds after the start of the solidification and then decreases to the steady-state value of about 2200 W/m^2 °C. The increase of the contact conductance is probably due to the increasing number of asperities of the chill-plate surface coming into contact with the solidifying metal during the initial stages of the solidification. After this period, thermal stresses on both the plate and the solidified region of the casting tend to break the intimate contact at the interface, resulting in the decrease of the contact conductance until a steady-state value is approached. The interface heat flux shown in figure 3.8.7.b exhibits almost the same functional behavior of the contact conductance.

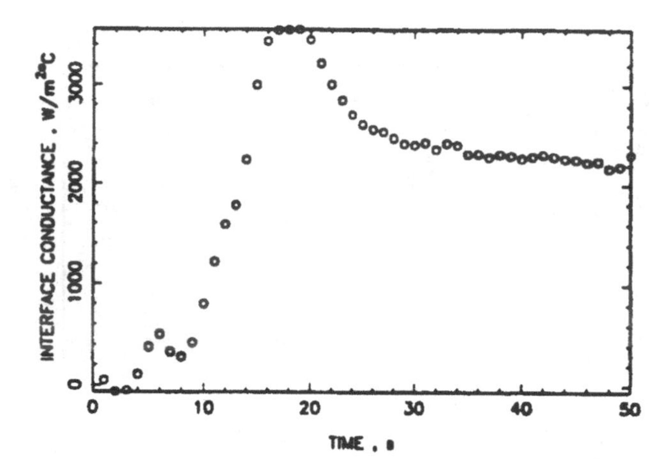

Figure 3.8.7.a - Contact conductance estimated from the actual experimental data.

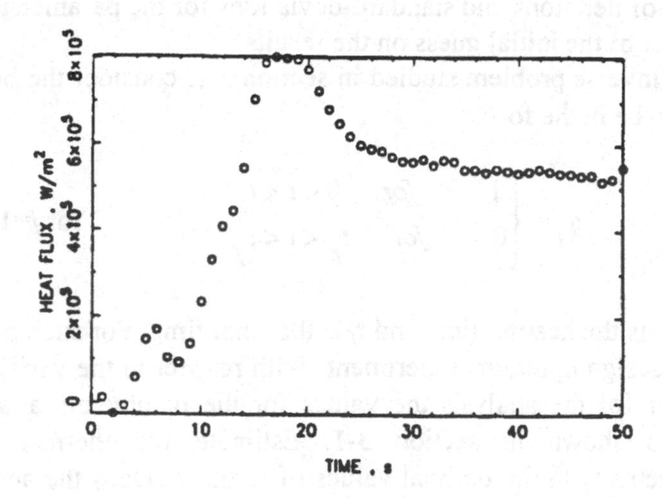

Figure 3.8.7.b - Estimated interface heat flux obtained from the actual experimental data.

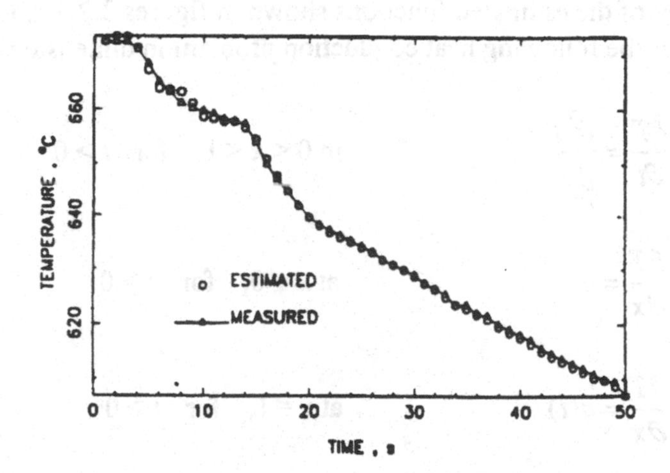

Figure 3.8.7.c - Estimated and actual experimental temperature for sensor number 1.

PROBLEMS

3-1 Consider the inverse problem studied in section 3.1, involving the estimation of the thermal conductivity components of a 3D orthotropic solid. By using the transient measurements of three sensors optimally located as described in section 3.1, solve this inverse problem by utilizing Technique II. Compare the results obtained with Technique II to those obtained with Technique I, for the estimated parameters, CPU time, number of iterations and standard-deviations for the parameters. Study also the effect of the initial guess on the results.

3-2 For the inverse problem studied in section 3.1, consider the boundary heat fluxes to be in the form

$$q_j = \begin{cases} 1 & for \quad 0 < t < t_h \\ 0 & for \quad t_h < t < t_f \end{cases} \qquad \text{for } j=1,2,3$$

where t_h is the heating time and t_f is the final time. For such boundary heat fluxes, design optimum experiments with respect to the variables t_h and t_f. Consider for the analysis the values for the parameters and the sensors locations shown in section 3-1. Estimate the thermal conductivity components with the optimal values of t_h and t_f. Does the accuracy of the estimated parameters improve, as compared to the values estimated in section 3-1? Why?

3-3 Solve the inverse problem of estimating the initial condition of section 3-2 by using Technique IV and measurements of sensors located inside the region, in addition to the measurements taken at the boundary. Does the accuracy of the estimated functions shown in figures 3.2.2.a,b improve?

3-4 Consider the following heat conduction problem in dimensionless form:

$$\frac{\partial T}{\partial t} = \frac{\partial^2 T}{\partial x^2} \qquad \text{in } 0 < x < 1, \quad \text{for } t > 0$$

$$\frac{\partial T}{\partial x} = 0 \qquad \text{at } x = 0, \quad \text{for } t > 0$$

$$\frac{\partial T}{\partial x} = q(t) \qquad \text{at } x = 1, \quad \text{for } t > 0$$

$$T = 0 \qquad \text{for } t = 0, \quad \text{in } 0 < x < 1$$

Assume $q(t)$ varies linearly with time, i.e.,

$$q(t) = a + bt$$

Then, set $a = b = 1$ in the direct problem to generate 100 equally spaced transient simulated measurements in the time interval $0 < t \leq 1$, for a sensor located at $x_{meas} = 0$. Use such simulated measured data to estimate the parameters a and b with Techniques I, II and III. Examine the effects of initial guess and random measurement errors on the final estimates.

3-5 Solve the inverse problem of estimating the boundary heat flux $q(t)$ described in problem 3-4, by using Technique IV of function estimation. Assume that no information is available on the functional form of $q(t)$, except that it belongs to the space of square integrable functions in the domain $0 < t < 1$. Use for the inverse analysis 100 equally spaced transient measurements in $0 < t \leq 1$, of a sensor located at $x_{meas} = 0$.
In order to generate the simulated measurements, utilize the following functional forms:

(i) $q(t) = 1 + t$

(ii) $q(t) = 1 + t + t^2$

(iii) $$q(t) = \begin{cases} 1 , & t \leq 0.3 \text{ and } t \geq 0.7 \\ 2 , & 0.3 < t < 0.7 \end{cases}$$

(iv) $$q(t) = \begin{cases} 1 , & t \leq 0.3 \text{ and } t \geq 0.7 \\ 5t - 0.5 , & 0.3 < t \leq 0.5 \\ -5t + 4.5 , & 0.5 < t < 0.7 \end{cases}$$

Use as initial guess a constant function $q(t) = 0$ and examine the effects of random measurement errors on the solution.

3-6 Try to improve the estimated functions of problem 3-5 in the time interval $0 < t \leq 1$, by using the following approaches:

(i) Use $q(t) = 0$ as initial guess, but consider a time interval larger than that of interest. For example, use for the final time $t_f = 1.1$, 1.25, 1.5, etc. Does the quality of the estimated functions in the time domain $0 < t \leq 1$ improve? Remember to increase the number of measurements accordingly, so that 100 measurements appear in the interval $0 < t \leq 1$.

(ii) Repeat the calculations with $t_f = 1$ and with an initial guess for $q(t)$ equal to the value estimated in problem 3-5, for a time in the neighborhood of t_f. Let's say, use now as initial guess the value estimated in problem 3-5 for $q(0.9)$. Repeat this procedure until sufficiently accurate estimates are obtained in the interval $0 < t \leq 1$.

3-7 Repeat problems 3-4 and 3-5 by using fewer transient measurements in the inverse analysis. Take, as an example, 20 measurements of the sensor located at $x_{meas} = 0$ in the time interval $0 < t \leq 1$. Are the final solutions sensitive to the number of measurements?

3-8 By examining the sensitivity coefficients for problem 3-4 and the sensitivity problem for problem 3-5, show that more accurate estimates can be obtained by locating the sensor closer to the boundary $x = 1$. Do the estimates actually improve if you locate the sensor at $x_{meas} = 0.5$, instead of $x_{meas} = 0$?

3-9 Repeat problems 3-4 and 3-5, by using in the inverse analysis the transient readings of two sensors located at $x = 0$ and $x = 0.5$. Compare the results obtained with two sensors to the results obtained in problems 3-4 and 3-5 with a single sensor.

3-10 Repeat problem 3-5 by using the *Steepest Descent Method* (the conjugation coefficient γ^k is null for all iterations), instead of Technique IV. How do the two methods compare with respect to the number of iterations required for convergence?

3-11 Repeat problem 3-5 by using a very small number for the tolerance ε in the stopping criterion, instead of using the discrepancy principle, for cases involving measurements with random errors. What happens to the stability of the estimated functions? Why?

3-12 Repeat problem 3-5 by using the additional measurement approach for the stopping criterion, for cases involving measurements with random errors. Compare the results obtained with this stopping criterion approach to those obtained by using the discrepancy principle.

3-13 Derive equations (3.5.29.a,b).

3-14 Use the approach developed in section 3-5 in order to estimate simultaneously the temperature dependencies of thermal conductivity and volumetric heat capacity.

REFERENCES

1. Sawaf, B. and Özisik, M. N., "Determining the Constant Thermal Conductivities of Orthotropic Materials by Inverse Analysis", *Int. Com Heat and Mass Transfer*, **22**, 201-211, 1995.
2. Mejias, M. M., Orlande, H. R. B. and Özisik, M. N., "Design of Optimum Experiments for the Estimation of Thermal Conductivity Components of Orthotropic Solids", *Hybrid Methods in Engineering*, **1**, 37-53,1999.
3. Mejias, M. M., Orlande, H. R. B. and Mikhailov, M. D., "Estimation of the Thermal Conductivity Components of Orthotropic Solids by using Mixed Computations", *Bull. of the Braz. Soc. Appl. Comput. Math.*, (to appear), 1999.
4. IMSL Library Edition 10.0, *User's Manual, IMSL*, Houston, 1987.
5. Silva Neto, A. J. and Özisik, M. N., "An Inverse Heat Conduction Problem of Unknown Initial Condition", *10th Int. Heat Transfer Conference*, Brighton, UK, August 14-18, 1994.
6. Silva Neto, A. J. and Özisik, M. N., "Two-Dimensional Inverse Heat Conduction Problem of Estimating the Time-Varying Strength of a Line Heat Source", *J. Applied Physics*, **71**, 5357-5362, 1992.
7. Silva Neto, A. J. and Özisik, M. N., "The Estimation of Space and Time Dependent Strength of a Volumetric Heat Source in a One-Dimensional Plate", *Int. J. Heat and Mass Transfer*, **37**, 909-915, 1994.
8. Orlande, H. R. B. and Özisik, M. N., "Determination of the Reaction Function in a Reaction-Diffusion Parabolic Problem", *ASME J. Heat Transfer*, **116**, 1041-1044, 1994.

9. Dantas, L. B. and Orlande, H. R. B., "A Function Estimation Approach for Determining Temperature-Dependent Thermophysical Properties", *Inverse Problems in Engineering*, **3**, 261-279, 1996.
10. Artyukhin, E., Inanov, G. and Nenarokonov, A., "Determining the Set of Thermophysical Properties of Materials from Unsteady-State Temperature Measurements", *High Temperature*, **31**, 199-202, 1993.
11. Artyukhin, E., "Iterative Algorithms for Estimating Temperature-Dependent Thermophysical Characteristics", *Inverse Problem in Engineering - Proceedings*, Palm Coast, FL, 101-108, 1993.
12. Artyukhin, E., "Experimental Design of Measurements for the Solution of Coefficient-Type Inverse Heat Conduction Problems", *J. Eng. Phys.*, **48**, 372-376, 1985.
13. Orlande, H. R. B. and Özisik, M. N., "Simultaneous Estimation of Thermal Diffusivity and Relaxation Time with a Hyperbolic Heat Conduction Model", Paper 15 - CI - 20, *10th Int. Heat Transfer Conference*, Brighton, UK, 403-408, 1994.
14. Cattaneo, C., "A Form of Heat Conduction Equation which Eliminates the Paradox of Instantaneous Propagation", *Compte Rendus*, **247**, 431-433, 1958.
15. Vick, B. and Özisik, M. N., "Growth and Decay of a Thermal Pulse Predicted by the Hyperbolic Heat Conduction Equation", *J. Heat Transfer*, **105**, 902-907, 1983.
16. Özisik, M. N. and Tzou, D. Y., "On the Wave Theory in Heat Conduction", *Fundamental Issues in Small Scale Heat Transfer HTD-227*, Y. Bayazitoglu and G. P. Peterson (eds.), ASME, 13-27, 1992.
17. Brown, J. B., Chung, D. Y. and Mathews, P.W., "Heat Pulses at Low Temperatures", *Phys. Lett.*, **21**, 241-243, 1966.
18. Özisik, M. N., *Heat Conduction, 2nd edition*, Wiley, New York, 1993.
19. Glass, D. E., Özisik, M. N. and Vick, B., "Non-Fourier Effects on Transient Temperature Resulting from Periodic On-Off Heat Flux", *Int. J. Heat Mass Transfer*, **30**, 1623-1632, 1987.
20. Taylor, R. E. and Maglic, K. D., "Pulse Method for Thermal Diffusivity Measurement", *Compendium of Thermophysical Property Measurement Methods 1*, K. D. Maglic (ed.), Plenum, New York, 305-336, 1984.
21. Orlande, H. R. B. and Özisik, M. N., "Inverse Problem of Estimating Interface Conductance Between Periodically Contacting Surfaces", *AIAA J. Thermophysics and Heat Transfer*, **7**, 319-325, 1993.
22. Moses , W. M. and Johnson, R. R., "Experimental Study of the Transient Heat Transfer Across Periodically Contacting Surfaces", *AIAA J. Thermophysics and Heat Transfer*, **2**, 37-42, Jan. 1988.
23. Flach, G. P. and Özisik, M. N., "Inverse Heat Conduction Problem of Periodically Contacting Surfaces", *ASME J. Heat Transfer*, **110**, 821-829, Nov. 1988.
24. Bardon, J. P., "Heat Transfer at Solid-Solid Interface: Basic Phenomenons-Recent Works", *Seminar EUROTHERM 4*, Nancy, France, June 1988.

25. Vick, B. and Özisik, M. N., "Quasi-Steady-State Temperature Distribution in Periodically Contacting Finite Regions", *ASME J. Heat Transfer*, **103**, 1991.
26. Huang, C. H., Özisik, M. N. and Sawaf, B., "Conjugate Gradient Method for Determining Unknown Contact Conductance During Metal Casting", *Int. J. Heat Mass Transfer*, **35**, 1779-1789, 1992.
27. Özisik, M. N., Orlande, H. R. B., Hector Jr., L. G. and Anyalebechi, P. N., "Inverse Problem of Estimating Interface Conductance During Solidification via Conjugate Gradient Method", *J. Mat. Proc. & Manuf. Sci.*, **1**, 213-225, 1992.
28. Voller, V. and Cross, M., "Accurate Solution of Moving Boundary Problems Using the Enthalpy Method", *Int. J. Heat Mass Transfer*, **24**, 545-556, 1981.
29. Shamsundar, N. and Rozz, E., "Numerical Methods for Moving Boundary Problems", in *Handbook of Numerical Heat Transfer*, W.J. Minkowycz, E.M. Sparrow, G.E. Schneider and R.H. Pletcher (eds.), Wiley, New York, 1988.
30. Shamsundar, N. and Sparrow, E. M., "Analysis of Multidimensional Conduction Phase Change via the Enthalpy Model", *J. Heat Transfer*, 333-340, 1975.
31. Shamsundar, N. and Sparrow, E. M., "Effect of Density Change on Multidimensional Conduction Phase Change", *J. Heat Transfer*, 550-557, 1976.
32. Ni, J. and Incropera, F. P., "Extension of the Continuum Model for Transport Phenomena Occurring During Metal Alloy Solidification - Part I", *Int. J. Heat Mass Transfer*, **38**, 1271-1284, 1995.
33. Ni, J. and Incropera, F. P., "Extension of the Continuum Model for Transport Phenomena Occurring During Metal Alloy Solidification - Part II", *Int. J. Heat Mass Transfer*, **38**, 1285-1296, 1995
34. Feller, R. J. and Beckermann, C., "Modeling of the Globulitic & Solidification of a Binary Metal Alloy", *Int. Comm. Heat Mass Transfer*, **20**, 311-322, 1993.
35. Rappaz, M., "Modeling of Microstructure Formation in Solidification Process", *Int. Mat. Rev.*, **34**, 93-123, 1989.
36. Özisik, M. N., *Finite Difference Methods in Heat Transfer*, CRC Press, Boca Raton, 1994.

Chapter 4

INVERSE CONVECTION

So far in this book we have considered problems involving conduction, which is the heat transfer mode that received most of the attention of the community dealing with inverse problems. More recently, inverse problems in which convection is the dominant heat transfer mode started to appear in the literature [1-13].

In this chapter we present the solution of inverse problems involving the estimation of inlet and boundary conditions in forced convection in parallel plate channels. For all the problems considered here we assumed the flow to be hydrodynamically developed, that is, the region of interest is sufficiently far from the inlet of the channel, so that the velocity profile does not vary with the axial direction. Analytic expressions for hydrodynamically developed velocity profiles for newtonian and non-newtonian fluids in laminar and turbulent flows can be found in standard textbooks [14-17]. Hence, such velocity profiles are considered to be known for the analysis, so that the solutions of the inverse problems to be considered here only involve the energy equation.

The solutions of the following inverse problems are considered in this chapter:

- **Estimation of the Inlet Temperature Profile in Laminar Flow [1];**
- **Estimation of the Transient Inlet Temperature in Laminar Flow [2];**
- **Estimation of the Axial Variation of the Wall Heat Flux in Laminar Flow [3];**
- **Estimation of the Transient Wall Heat Flux in Turbulent Flow [4];**
- **Simultaneous Estimation of the Spacewise and Timewise Variations of the Wall Heat Flux in Laminar Flow [5].**

We use **Technique IV** for the solution of these inverse problems, where the unknowns are estimated as a function estimation approach.

The basic steps of Technique IV include: direct problem, inverse problem, sensitivity problem, adjoint problem, gradient equation, iterative procedure, stopping criterion and computational algorithm. We present below the details of such steps of Technique IV, as applied to the solution of the inverse convection problems considered in this chapter. However, we avoid here the repetition of the details for the stopping criterion and computational algorithms, since they can be readily found in section 2-4.

4-1 ESTIMATION OF THE INLET TEMPERATURE PROFILE IN LAMINAR FLOW [1]

Direct Problem

The physical problem considered here involves laminar steady-state forced convection between parallel plates, located at a distance h from each other. The plates are subjected to a constant heat flux q, while the inlet temperature profile is given by $f(y)$, as shown in figure 4.1.1. The physical properties are assumed constant and viscous dissipation, free convection, and axial conduction effects are neglected. The mathematical formulation for this steady-state forced convection problem is given as follows:

$$k\frac{\partial^2 T}{\partial y^2} = u(y)\,\rho C_p \frac{\partial T}{\partial x} \qquad \text{in } 0<y<h, \qquad 0<x<b \tag{4.1.1.a}$$

$$k\frac{\partial T}{\partial y} = q \qquad \text{at } y=h, \qquad 0<x<b \tag{4.1.1.b}$$

$$-k\frac{\partial T}{\partial y} = q \qquad \text{at } y=0, \qquad 0<x<b \tag{4.1.1.c}$$

$$T_0 = f(y) \qquad \text{at } x=0, \qquad 0<y<h \tag{4.1.1.d}$$

where $u(y) = 6u_m \frac{y}{h}\left(1-\frac{y}{h}\right)$ is the velocity profile in the channel.

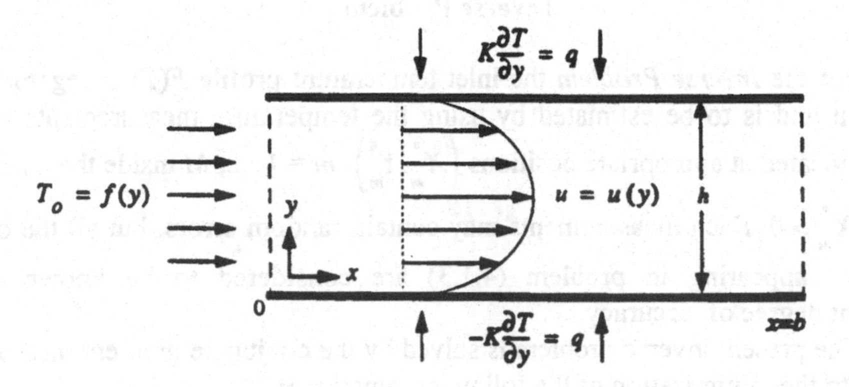

Figure 4.1.1 - Geometry and coordinates

By introducing the following dimensionless quantities

$$\Theta = \frac{T}{T_{ref}} \qquad Y = \frac{y}{h} \qquad X = \frac{x}{PeD_e} \qquad Re = \frac{u_m D_e}{\nu}$$

$$D_e = 2h \qquad Pe = RePr \qquad L = \frac{b}{PeD_e}$$

$$Q = \frac{hq}{kT_{ref}} \qquad U(Y) = \frac{u(y)}{u_m} = 6Y(1-Y) \tag{4.1.2.a - i}$$

where T_{ref} is a reference temperature value, the governing equations (4.1.1) can be expressed in dimensionless form as

$$\frac{\partial^2 \Theta}{\partial Y^2} = \frac{U(Y)}{4}\frac{\partial \Theta}{\partial X} \qquad \text{in} \quad 0 < Y < 1, \quad 0 < X < L \tag{4.1.3.a}$$

$$\frac{\partial \Theta}{\partial Y} = Q \qquad \text{at} \quad Y = 1, \quad 0 < X < L \tag{4.1.3.b}$$

$$\frac{\partial \Theta}{\partial Y} = -Q \qquad \text{at} \quad Y = 0, \quad 0 < X < L \tag{4.1.3.c}$$

$$\Theta = F(Y) \qquad \text{at} \quad X = 0, \quad 0 < Y < 1 \tag{4.1.3.d}$$

The problem given by equations (4.1.3) is denoted as a *Direct Problem*, when the inlet temperature profile $F(Y)$ and other quantities appearing in equations (4.1.3) are known. The objective of the Direct Problem is to determine the temperature field $\Theta(X,Y)$ of the fluid inside the channel.

Inverse Problem

For the *Inverse Problem* the inlet temperature profile $F(Y)$ is regarded as unknown and is to be estimated by using the temperature measurements of M sensors located at appropriate positions $\left(X_m^*, Y_m^*\right)$, $m = 1, ..., M$ inside the channel, where $X_m^* > 0$. Such measurements may contain random errors, but all the other quantities appearing in problem (4.1.3) are considered to be known with sufficient degree of accuracy.

The present inverse problem is solved by the conjugate gradient method as applied to the minimization of the following functional:

$$S[F(Y)] = \sum_{m=1}^{M} \{\Theta_m[F(Y)] - Z_m\}^2 \tag{4.1.4}$$

where Z_m and $\Theta_m[F(Y)]$ are the measured and estimated temperatures at the measurement locations, respectively, while M is the number of sensors. The estimated temperatures $\Theta_m[F(Y)]$ are obtained from the solution of the direct problem (4.1.3) by using an estimate for the inlet temperature profile $F(Y)$. Note that we assume in equation (4.1.4) the sensors to be discretely distributed in space, instead of the usual approach of considering continuous measurements for Technique IV.

Two auxiliary problems are required for the implementation of the conjugate gradient method: the *Sensitivity Problem* and the *Adjoint Problem*. The development of such problems is described next.

Sensitivity Problem

To obtain the sensitivity problem, it is assumed in the direct problem that $F(Y)$ undergoes an increment $\Delta F(Y)$. Then the temperature $\Theta(X,Y)$ changes by $\Delta\Theta(X,Y)$. By replacing in the direct problem $F(Y)$ by $[F(Y) + \Delta F(Y)]$ and $\Theta(X,Y)$ by $[\Theta(X,Y) + \Delta\Theta(X,Y)]$, subtracting from the resulting expressions the original direct problem and neglecting the second-order terms, the following sensitivity problem is obtained:

$$\frac{\partial^2 \Delta\Theta(x,y)}{\partial Y^2} = \frac{U(Y)}{4}\frac{\partial \Delta\Theta(x,y)}{\partial X} \quad \text{in} \quad 0 < Y < 1, \quad 0 < X < L \tag{4.1.5.a}$$

$$\frac{\partial \Delta\Theta}{\partial Y} = 0 \quad \text{at} \quad Y = 1, \quad 0 < X < L \tag{4.1.5.b}$$

$$\frac{\partial \Delta\Theta}{\partial Y} = 0 \qquad \text{at} \quad Y = 0, \qquad 0 < X < L \tag{4.1.5.c}$$

$$\Delta\Theta = \Delta F(Y) \qquad \text{at} \quad X = 0, \qquad 0 < Y < 1 \tag{4.1.5.d}$$

Adjoint Problem

To obtain the adjoint problem, equation (4.1.3.a) is multiplied by the Lagrange multiplier $\lambda(X,Y)$. The resulting expression is integrated over the space domain, and then added to the right-hand side of equation (4.1.4) to yield

$$S[F(Y)] = \sum_{m=1}^{M} \left(\Theta_m - Z_m\right)^2 + \int_{X=0}^{L} \int_{Y=0}^{1} \lambda \left[\frac{\partial^2\Theta}{\partial Y^2} - \frac{U}{4}\frac{\partial \Theta}{\partial X}\right] dY dX \tag{4.1.6}$$

The variation $\Delta S[F(Y)]$ of the extended functional given by equation (4.1.6) is obtained and after some algebraic manipulations, the resulting expression is allowed to go to zero. Such manipulations were described in detail in Chapter 2 and are not repeated here. The following *adjoint problem* is then obtained for the Lagrange Multiplier $\lambda(X,Y)$:

$$\frac{\partial^2 \lambda(X,Y)}{\partial Y^2} + \frac{U(Y)}{4}\frac{\partial \lambda(X,Y)}{\partial X} + 2\sum_{m=1}^{M}(\Theta - Z)\delta(X - X_m^*)\delta(Y - Y_m^*) = 0$$

$$\text{in} \quad 0 < Y < 1, \qquad 0 < X < L \tag{4.1.7.a}$$

$$\frac{\partial \lambda}{\partial Y} = 0 \qquad \text{at} \quad Y = 1, \qquad 0 < X < L \tag{4.1.7.b}$$

$$\frac{\partial \lambda}{\partial Y} = 0 \qquad \text{at} \quad Y = 0, \qquad 0 < X < L \tag{4.1.7.c}$$

$$\lambda = 0 \qquad \text{at} \quad X = L, \qquad 0 < Y < 1 \tag{4.1.7.d}$$

where $\left(X_m^*, Y_m^*\right)$ gives the measurement location of sensor m, $m = 1, \ldots, M$.

Gradient Equation

In the process of obtaining the adjoint problem, the following integral term is left:

$$\Delta S[F(Y)] = \int_{Y=0}^{1} \frac{U(Y)}{4} \lambda(0,Y) \Delta F(Y) dY \tag{4.1.8.a}$$

By assuming that $F(Y)$ belongs to the space of square integrable functions in the domain $0 < Y < 1$, we can write

$$\Delta S[F(Y)] = \int_{Y=0}^{1} \nabla S[F(Y)] \Delta F(Y) dY \tag{4.1.8.b}$$

By comparing equations (4.1.8.a) and (4.1.8.b), we can obtain the gradient equation for the functional as

$$\nabla S[F(Y)] = \frac{U(Y)}{4} \lambda(0,Y) \tag{4.1.9}$$

We note that the adjoint problem (4.1.7) involves a condition at the outlet of the channel at $X = L$, equation (4.1.7.d), instead of the inlet condition at $X = 0$ of the regular direct problem, equation (4.1.3.d). This is similar to the final condition encountered in adjoint problems of other transient inverse problems, that appeared above in the text.

Iterative Procedure

The iterative procedure of Technique IV, as applied to the estimation of the inlet temperature profile is given by:

$$F^{k+1}(Y) = F^k(Y) - \beta^k d^k(Y) \tag{4.1.10.a}$$

where the superscript k refers to the number of iterations and the direction of descent $d^k(Y)$ is given by:

$$d^k(Y) = \nabla S\left[F^k(Y)\right] + \gamma^k d^{k-1}(Y) \tag{4.1.10.b}$$

The conjugation coefficient γ^k is obtained from the Fletcher-Reeves expression as:

$$\gamma^k = \frac{\int_{Y=0}^{1}\left\{\nabla S\left[F^k(Y)\right]\right\}^2 dY}{\int_{Y=0}^{1}\left\{\nabla S\left[F^{k-1}(Y)\right]\right\}^2 dY} \quad \text{for} \quad k=1,2,... \quad \text{with } \gamma^0 = 0 \tag{4.1.10.c}$$

The search step size β^k is obtained by minimizing the functional given by equation (4.1.4) with respect to β^k. The following expression results:

$$\beta^k = \frac{\sum_{m=1}^{M}\left\{\Theta_m\left[F^k(Y)\right]-Z_m\right\}\Delta\Theta_m\left[d^k(Y)\right]}{\sum_{m=1}^{M}\left\{\Delta\Theta_m\left[d^k(Y)\right]\right\}^2} \tag{4.1.10.d}$$

where $\Delta\Theta_m\left[d^k(Y)\right]$ is the solution of the sensitivity problem (4.1.5), obtained by setting $\Delta F(Y) = d^k(Y)$.

The iterative procedure of Technique IV given by equations (4.1.10) is applied until a stopping criterion based on the *discrepancy principle* is satisfied, as described in section 2-4. Such iterative procedure can be suitably arranged in a computational algorithm, which can also be found in the same section.

Results

In order to examine the accuracy of the inverse analysis for estimating the unknown transversal variation of inlet temperature by the conjugate gradient method, we examined several strict test conditions including a function with sharp corners, a step function and a smooth function. The effects of the number of measurements, M, distribution of the measurements in the transversal direction, axial locations of the sensors, magnitude of measurement errors, and functional form of the unknown inlet temperature on the accuracy of estimations are investigated.

The problem is solved in dimensionless form for a duct with geometry illustrated in figure 4.1.1. In order to give some idea on the physical significance of the various dimensionless variables, we consider air flow at a mean velocity $u_m = 2.5$ cm/s (Reynolds number = 400) through a duct of height $h = 12.8$ cm and length $b = 63.5$ cm. The wall heat flux is taken as $q = 500$ W/m^2. Our objective is to estimate the unknown distribution of inlet temperature from the knowledge of temperature measurements taken at the downstream locations. To identify the locations of the temperature sensors, we consider dimensionless step sizes $\Delta X = 0.00025$ and $\Delta Y = 0.025$ in the axial and transversal directions, respectively.

In terms of dimensional quantities, they correspond to Δx = 2.54 cm and Δy = 0.32 cm. For M equally spaced sensors in the transversal direction, the dimensionless distance, DY, between the sensors becomes $DY = 1/(M+1)$. Simulated measurements obtained in the form given by equation (2.5.2) were used in the analysis.

Figure 4.1.2 illustrates the effects of the number of measurements on the accuracy of the estimation, for sensors located at $X^* = 20\Delta X$. The number of transversal measurements considered here includes M = 5, 9, 19 and 39, corresponding to dimensionless spacings of DY = 0.167, 0.1, 0.05 and 0.025, respectively. The results show that the accuracy of the estimations improves by increasing the number of sensors; but, even for a small number of sensors such as $M = 5$, a quite good estimation is obtained, with the exception for the point $Y = 0.667$ where exists a discontinuity in the slope of $F(Y)$.

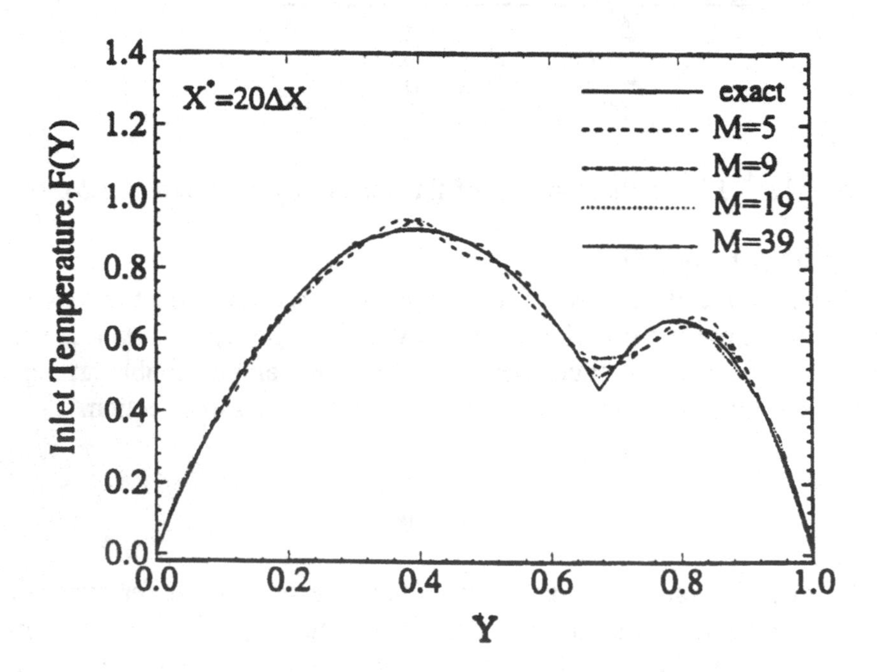

Figure 4.1.2 - Effects of number of measurements on the accuracy of inverse analysis.

Figure 4.1.3 is intended to show the effect of the axial location of the measurements on the accuracy of the estimation. In this case, five measurements are taken at each of the different axial locations $5\Delta X$, $10\Delta X$, $15\Delta X$ and $20\Delta X$ from the inlet, in order to perform the computations. It should be noted that these locations are all in the thermally developing region. As the measurements are taken at locations away from the inlet, more heat flux penetrates from both boundaries into the flow. As a result, the inlet temperature becomes difficult to recover by the inverse analysis. We also tested several cases in which the measurements were taken in the thermally fully developed region; but the measurements taken in such a region provide no information for determining the inlet temperature.

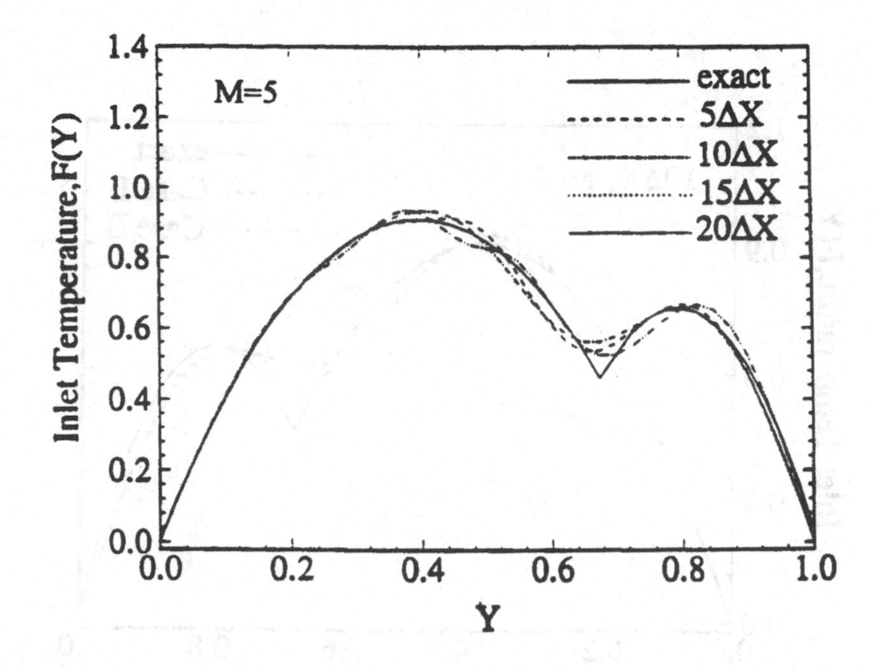

Figure 4.1.3 - Effects of axial location of the measurements on the accuracy of inverse analysis.

In the case of an inlet temperature distribution containing a slope discontinuity, the accuracy of the estimation improves if the location of the discontinuity is known *a priori* and a temperature measurement is taken at such a location. Figure 4.1.4 illustrates the effects of distribution of measurements in the transversal location on the accuracy of the estimation, for $M = 5$. In Case I, the five measurements were taken at the transversal locations of $Y = 0.167$, 0.33, 0.5, 0.667 and 0.83, where $Y = 0.667$ corresponds to the location of the slope discontinuity. The measurement locations for Case II were $Y = 0.167$, 0.33, 0.5, 0.75 and 0.83, which did not include the discontinuity location. The results show that the accuracy for Case I is better than that for Case II.

Figure 4.1.5 is intended to show the effects of magnitude of the measurement errors on the accuracy of estimation, for the cases involving measurement errors of 2.5%, 5% and 10%. As expected, the accuracy of estimation decreases with increasing the measurement error, especially at the slope discontinuity.

Figures 4.1.6 and 4.1.7 show the effects of the unknown functional form of the inlet temperature on the accuracy of the estimation. The step functional form of inlet temperature shown in figure 4.1.6 presents a very difficult case for estimation because two discontinuities are involved. Even with 39 measurements, the exact inlet temperature is not fully recovered by the inverse analysis with errorless measurements ($\sigma = 0$). The smooth function shown in figure 4.1.7 poses no difficulty for estimation by inverse analysis. The estimations using only 5 measurements with 5% and 10% error are still in good agreement with the exact solution.

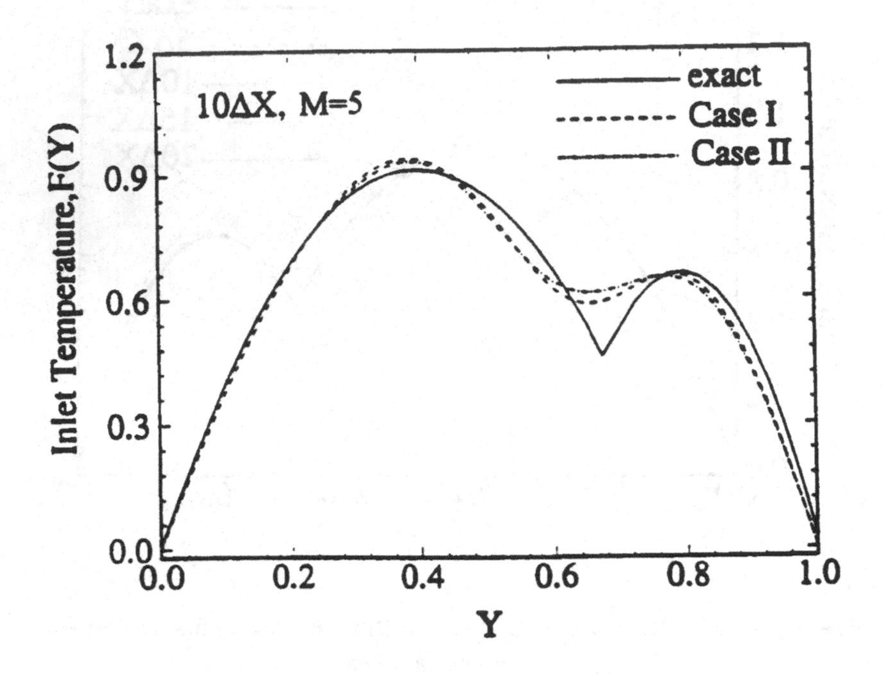

Figure 4.1.4 - Effects of transversal distribution of measurement location on the accuracy of inverse analysis.

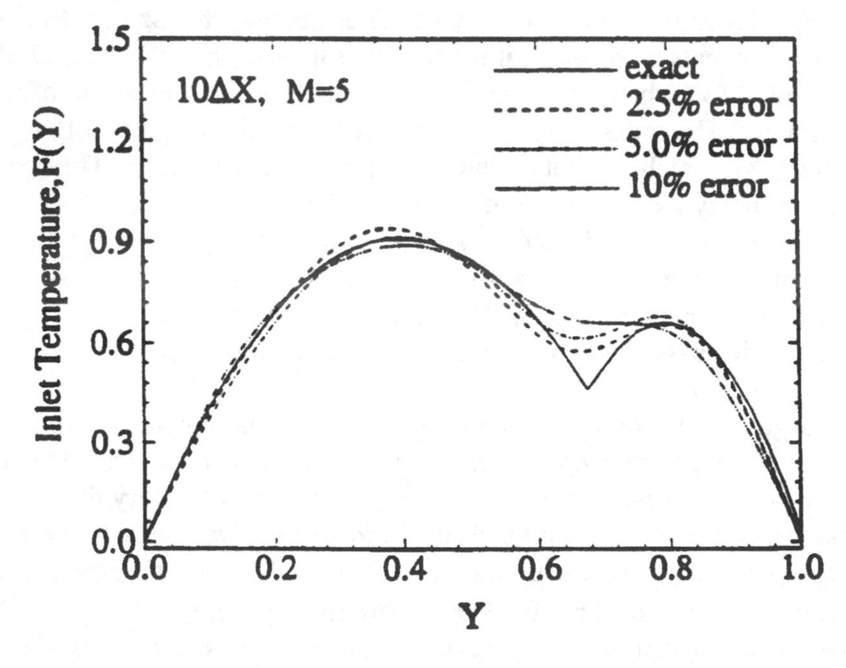

Figure 4.1.5 - Effects of the measurement error on the accuracy of inverse analysis.

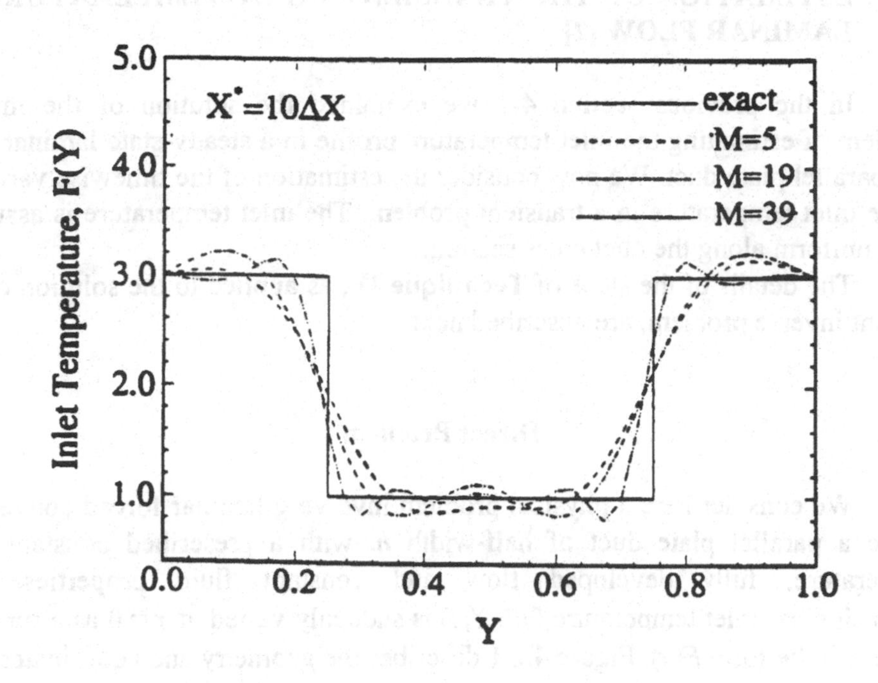

Figure 4.1.6 - Effects of the step functional form of the inlet temperature on the accuracy of inverse analysis.

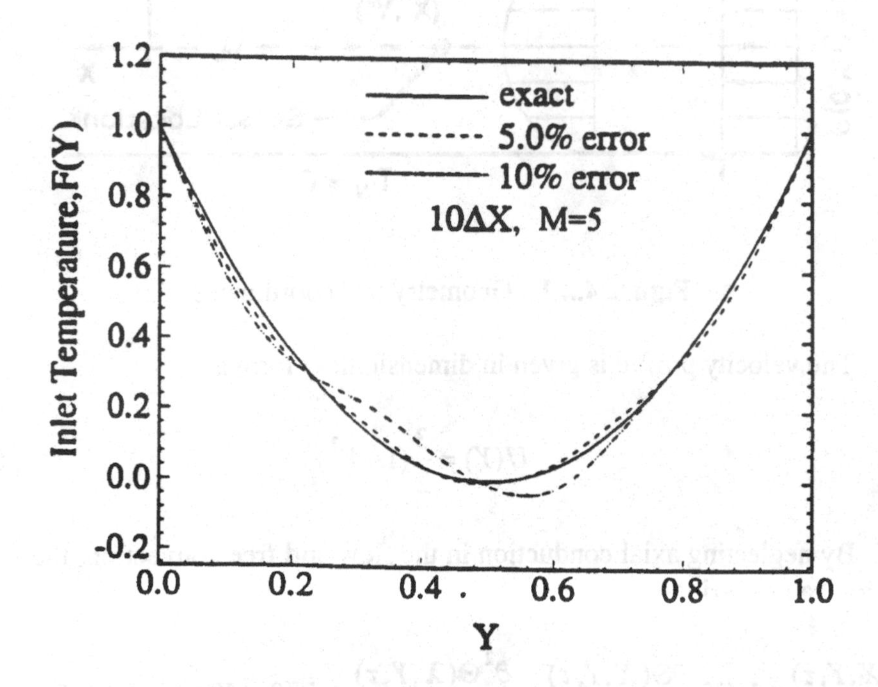

Figure 4.1.7 - Effects of the smooth functional form of the inlet temperature on the accuracy of inverse analysis.

4-2 ESTIMATION OF THE TRANSIENT INLET TEMPERATURE IN LAMINAR FLOW [2]

In the previous section 4-1 we examined the solution of the inverse problem of estimating the inlet temperature profile in a steady-state laminar flow in a parallel plate duct. We now consider the estimation of the timewise variation of the inlet temperature in a transient problem. The inlet temperature is assumed to be uniform along the duct cross section.

The details of the steps of **Technique IV**, as applied to the solution of the present inverse problem, are described next.

Direct Problem

We consider here a physical problem involving laminar forced convection inside a parallel plate duct of half-width h, with a prescribed constant wall temperature, fully developed flow and constant fluid properties. The dimensionless inlet temperature $\Theta(0,Y,\tau)$ is suddenly varied at $\tau = 0$ as a function of time in the form $F(\tau)$. Figure 4.2.1 describes the geometry and coordinates.

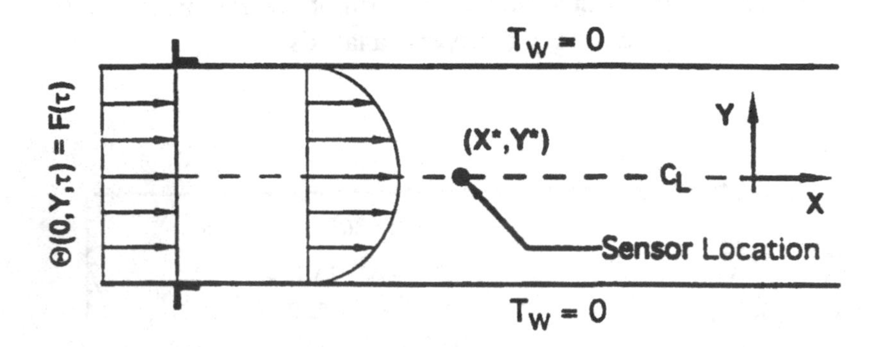

Figure 4.2.1 - Geometry and coordinates

The velocity profile is given in dimensionless form as

$$U(Y) = \frac{3}{2}(1 - Y^2) \tag{4.2.1}$$

By neglecting axial conduction in the flow and free convection, the energy equation can be written as:

$$\frac{\partial \Theta(X,Y,\tau)}{\partial \tau} + U(Y)\frac{\partial \Theta(X,Y,\tau)}{\partial X} = \frac{\partial^2 \Theta(X,Y,\tau)}{\partial Y^2} \quad \text{in } 0 < Y < 1 \;,\; 0 < X < L,$$

$$\text{for } \tau > 0 \tag{4.2.2.a}$$

with boundary conditions taken as

$$\frac{\partial\Theta(X,0,\tau)}{\partial Y}=0 \quad \text{at} \quad Y=0, \qquad 0<X<L,\ \tau>0 \tag{4.2.2.b}$$

$$\Theta(X,1,\tau)=0 \quad \text{at} \quad Y=1, \qquad 0<X<L\ ,\tau>0 \tag{4.2.2.c}$$

$$\Theta(0,Y,\tau)=F(\tau) \quad \text{at} \quad X=0, \qquad 0<Y<1\ ,\ \tau>0 \tag{4.2.2.d}$$

and the initial condition as

$$\Theta(X,Y,0)=0 \qquad \text{for}\ \tau=0, \ \text{in}\ \ 0<Y<1,\ \ 0<X<L \tag{4.2.2.e}$$

where various dimensionless terms are defined as

$$\Theta=\frac{T(x,y,t)-T_i}{T_i} \qquad \tau=\frac{\alpha t}{h^2} \qquad Y=\frac{y}{h}$$

$$X=\frac{\alpha x}{u_m h^2} \qquad U(Y)=\frac{u(y)}{u_m} \qquad F(\tau)=\frac{f(t)-T_i}{T_i} \tag{4.2.3.a-f}$$

Here α is the thermal diffusivity of the fluid, u_m is the mean velocity and T_i is the initial fluid temperature, which is assumed to be uniform and equal to the prescribed wall temperature.

We note that equation (4.2.2.b) gives the symmetry condition for the problem.

The problem (4.2.2) is a *Direct Problem* when the variation of the inlet temperature $F(\tau)$ is known. The solution of the direct problem provides the temperature field $\Theta(X,Y,\tau)$ of the fluid inside the channel.

Inverse Problem

Consider now the inlet temperature $F(\tau)$ as unknown. Such function is then to be estimated by using the transient temperature measurements of a single sensor located at an appropriate position (X^*,Y^*) inside the channel and by minimizing the following functional:

$$S[F(\tau)]=\int_{\tau=0}^{\tau_f}\{\Theta[X^*,Y^*,\tau;F(\tau)]-Z(\tau)\}^2 d\tau \tag{4.2.4}$$

where $\Theta[X^*,Y^*,\tau;F(\tau)]$ and $Z(\tau)$ are the estimated and measured temperatures, respectively, at the measurement location (X^*,Y^*). The estimated temperature is obtained from the solution of the direct problem (4.2.2) by using an estimate for $F(\tau)$.

We use Technique IV, the conjugate gradient method of function estimation, for the minimization of the functional given by equation (4.2.4). The sensitivity and adjoint problems, required for the implementation of the iterative procedure of Technique IV, are developed next.

Sensitivity Problem

In order to develop the sensitivity problem, we assume that the temperature $\Theta(X,Y,\tau)$ undergoes a variation $\Delta\Theta(X,Y,\tau)$, when the inlet temperature undergoes a variation $\Delta F(\tau)$. By substituting into the direct problem (4.2.2) $\Theta(X,Y,\tau)$ by $[\Theta(X,Y,\tau)+\Delta\Theta(X,Y,\tau)]$ and $F(\tau)$ by $[F(\tau)+\Delta F(\tau)]$, and then subtracting from the resulting equations the original direct problem, we obtain the following sensitivity problem for the sensitivity function $\Delta\Theta(X,Y,\tau)$:

$$\frac{\partial\Delta\Theta(X,Y,\tau)}{\partial\tau}+U(Y)\frac{\partial\Delta\Theta(X,Y,\tau)}{\partial X}=\frac{\partial^2\Delta\Theta(X,Y,\tau)}{\partial Y^2} \quad \text{in } 0<Y<1,\ 0<X<L, \quad \text{for } \tau>0 \qquad (4.2.5.a)$$

$$\frac{\partial\Delta\Theta(X,0,\tau)}{\partial Y}=0 \quad \text{at} \quad Y=0, \quad 0<X<L,\ \tau>0 \qquad (4.2.5.b)$$

$$\Delta\Theta(X,1,\tau)=0 \quad \text{at} \quad Y=1, \quad 0<X<L,\ \tau>0 \qquad (4.2.5.c)$$

$$\Delta\Theta(0,Y,\tau)=\Delta F(\tau) \quad \text{at} \quad X=0, \quad 0<Y<1,\ \tau>0 \qquad (4.2.5.d)$$

$$\Delta\Theta(X,Y,0)=0 \quad \text{for} \quad \tau=0, \quad 0<Y<1, \quad 0<X<L \qquad (4.2.5.e)$$

Adjoint Problem

The adjoint problem is obtained by multiplying equation (4.2.2.a) by the Lagrange Multiplier $\lambda(X,Y,\tau)$, integrating the resulting expression over the time and space domains and adding the result to the functional given by equation (4.2.4). We obtain:

$$S[F(\tau)]=\int_{\tau=0}^{\tau_f}\{\Theta[X^*,Y^*,\tau;F(\tau)]-Z(\tau)\}^2\,d\tau+$$
$$\int_{\tau=0}^{\tau_f}\int_{X=0}^{L}\int_{Y=0}^{1}\lambda(X,Y,\tau)\left[\frac{\partial\Theta}{\partial\tau}+U(Y)\frac{\partial\Theta}{\partial X}-\frac{\partial^2\Theta}{\partial Y^2}\right]dY\,dX\,d\tau \tag{4.2.6}$$

We now perturb $F(\tau)$ by $\Delta F(\tau)$ and $\Theta(X,Y,\tau)$ by $\Delta\Theta(X,Y,\tau)$ in equation (4.2.6) and subtract equation (4.2.6) from the resulting expression to get the variation $\Delta S[F(\tau)]$ of the functional $S[F(\tau)]$. By employing integration by parts, utilizing the initial and boundary conditions of the sensitivity problem and also requiring that the coefficients of $\Delta\Theta(X,Y,\tau)$ in the resulting equation should vanish, the following adjoint problem is obtained:

$$2[\Theta(X,Y,\tau)-Z(\tau)]\delta(X-X^*)\delta(Y-Y^*)-\frac{\partial\lambda(X,Y,\tau)}{\partial\tau}-U(Y)\frac{\partial\lambda(X,Y,\tau)}{\partial X}-$$
$$-\frac{\partial^2\lambda(X,Y,\tau)}{\partial Y^2}=0 \qquad \text{in } 0<Y<1,\ 0<X<L \text{ and for } 0<\tau<\tau_f \tag{4.2.7.a}$$

where $\delta(\cdot)$ is the Dirac delta function, and the boundary conditions become

$$\frac{\partial\lambda(X,0,\tau)}{\partial Y}=0 \qquad \text{at } Y=0,\ 0<X<L \text{ and } 0<\tau<\tau_f \tag{4.2.7.b}$$

$$\lambda(X,1,\tau)=0 \qquad \text{at } Y=1,\ 0<X<L \text{ and } 0<\tau<\tau_f \tag{4.2.7.c}$$

$$\lambda(L,Y,\tau)=0 \qquad \text{at } X=L,\ 0<Y<1 \text{ and } 0<\tau<\tau_f \tag{4.2.7.d}$$

$$\lambda(X,Y,\tau_f)=0 \qquad \text{for } \tau=\tau_f,\ \text{in } 0<Y<1 \text{ and } 0<X<L \tag{4.2.7.e}$$

Gradient Equation

In the process of obtaining the adjoint problem, the variation of the functional reduces to

$$\Delta S[F(\tau)]=-\int_{\tau=0}^{\tau_f}\int_{Y=0}^{1}\lambda(0,Y,\tau)\,U(Y)\,dY\,\Delta F(\tau)\,d\tau \tag{4.2.8.a}$$

By assuming that the function $F(\tau)$ belongs to the space of square integrable functions in $0 < \tau < \tau_f$, we can write

$$\Delta S[F(\tau)] = \int_{\tau=0}^{\tau_f} \nabla S[F(\tau)]\, \Delta F(\tau) d\tau \tag{4.2.8.b}$$

Hence, by comparing equations (4.2.8.a) and (4.2.8.b), we can obtain the gradient equation for the functional as:

$$\nabla S[F(\tau)] = -\int_{Y=0}^{1} \lambda(0,Y,\tau) U(Y) dY \tag{4.2.9}$$

Iterative Procedure

The iterative procedure of Technique IV, as applied to the estimation of the function $F(\tau)$ is given as

$$F^{k+1}(\tau) = F^{k}(\tau) - \beta^{k} d^{k}(\tau) \tag{4.2.10.a}$$

where k is the number of iterations. The direction of descent is obtained from

$$d^{k}(\tau) = \nabla S\left[F^{k}(\tau)\right] + \gamma^{k} d^{k-1}(\tau) \tag{4.2.10.b}$$

The conjugation coefficient is obtained from the Fletcher-Reeves expression as

$$\gamma^{k} = \frac{\int_{\tau=0}^{\tau_f} \left\{\nabla S\left[F^{k}(\tau)\right]\right\}^2 d\tau}{\int_{\tau=0}^{\tau_f} \left\{\nabla S\left[F^{k-1}(\tau)\right]\right\}^2 d\tau} \quad \text{for} \quad k=1,2,\ldots \quad \text{with } \gamma^{0} = 0 \tag{4.2.10.c}$$

The search step size is obtained by minimizing $S[F^{k+1}(\tau)]$ with respect to β^{k}, as described in Note 7 of Chapter 2. The following expression results

$$\beta^k = \frac{\int_{\tau=0}^{\tau_f} \left\{ \Theta\left[F^k(\tau)\right] - Y(\tau) \right\} \Delta\Theta\left[d^k(\tau)\right] d\tau}{\int_{\tau=0}^{\tau_f} \left\{ \Delta\Theta\left[d^k(\tau)\right] \right\}^2 d\tau} \tag{4.2.10.d}$$

where $\Delta\Theta[d^k(\tau)]$ is the solution of the sensitivity problem (4.2.5), obtained by setting $\Delta F(\tau) = d^k(\tau)$.

Results

To illustrate the accuracy of Technique IV in predicting $F(\tau)$, we examined three functional test cases; a triangular ramp, a double step and a sine curve, as illustrated in figure 4.2.2. The first two represent very difficult functions to predict, due to the discontinuities in the first derivative and in the function. As the sine curve is smooth and continuous, its estimation should not pose difficulty. Over the total experimental time of $3.6\text{x}10^{-3}$ in dimensionless terms, 200 equal time steps were considered, corresponding to a sampling frequency of $1.8\text{x}10^{-5}$. The total dimensionless length of the duct taken as $8.2\text{x}10^{-3}$, with 60 equal divisions corresponding to $\Delta X = 1.367\text{x}10^{-4}$, was long enough for all test locations to lay in the thermally developing region. Representative values for the total time and total length in dimensional terms are 30 seconds and 1.64 meters, respectively, for air with a mean velocity of 2.4 cm/s, in a duct with half width of 0.5 meters. The sensor was considered to be located at the centerline ($Y^* = 0$) and two different axial positions were studied here: $X^* = 5\Delta X$ and $X^* = 20\Delta X$. The centerline was chosen for all measurements in order to minimize the effects of the wall temperature on the readings of the sensor at the measurement location.

We use here simulated measurements in the form given by equation (2.5.2).

The time dependent inlet condition for a triangular ramp function, illustrated in figure 4.2.2.a, was assumed to vary in the form

$$F(\tau) = \begin{cases} 1111.11\tau & \text{for} \quad 0 < \tau \leq 9\text{x}10^{-4} \\ -833.33(\tau - 9\text{x}10^{-4}) + 1 & \text{for} \quad 9\text{x}10^{-4} < \tau \leq 1.5\text{x}10^{-3} \\ 0.5 & \text{for} \quad 1.5\text{x}10^{-3} < \tau \leq \tau_f \end{cases} \tag{4.2.11}$$

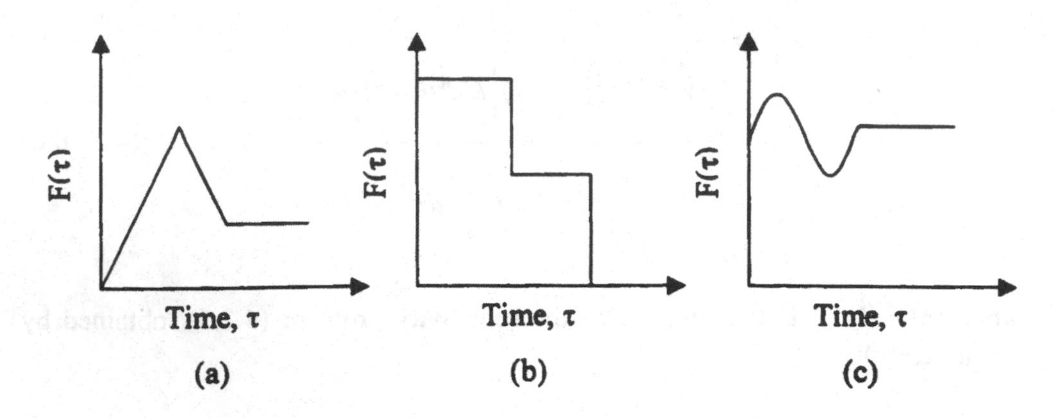

Figure 4.2.2 - Three test cases considered for the inlet temperature: (a) triangular ramp, (b) double step and (c) sine curve.

Figure 4.2.3 shows typical measured temperatures at two different downstream locations for a standard-deviation of σ =0.01. These curves show that the steady measured value was achieved after a certain time period. The inverse problem was based on all data taken before the steady temperature has been reached. Also, we chose the steady value of the measured temperature as the initial guess for the computational algorithm. This choice alleviates one of the difficulties associated with the conjugate-gradient method, that is, the final time value of the estimation is the same as the initial guess.

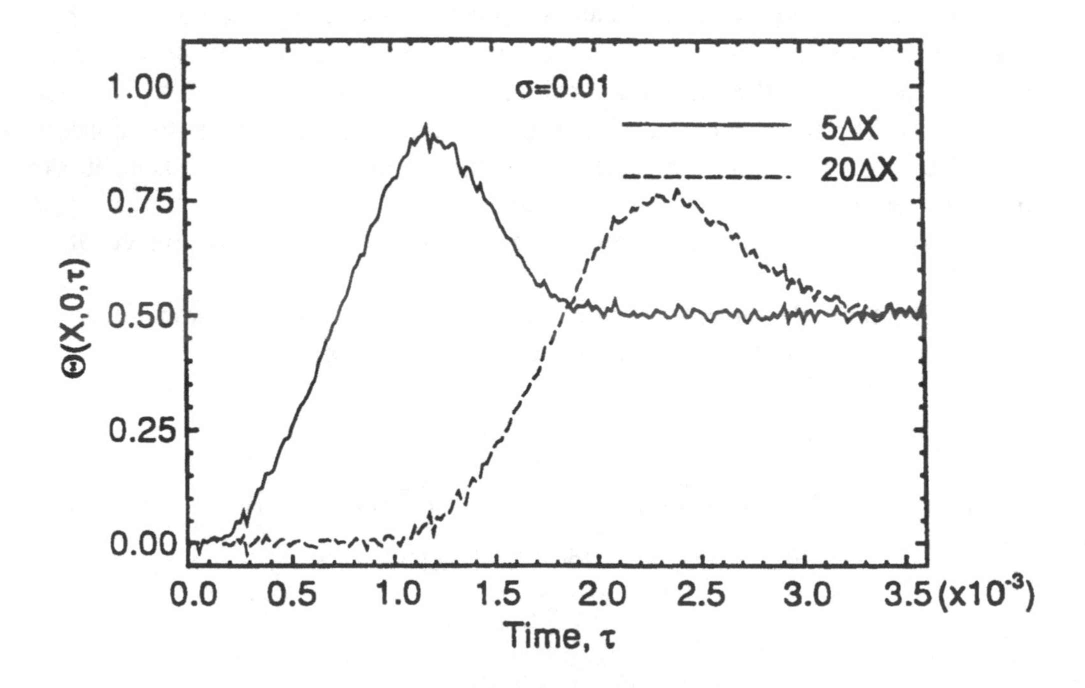

Figure 4.2.3 - Simulated measured temperatures at downstream locations $5\Delta X$ and $20\Delta X$ for triangular ramp pulse, with $\sigma = 0.01$.

Figure 4.2.4 illustrates the effects of the standard deviations $\sigma=0.01$ and $\sigma=0.03$, on the accuracy of the estimates. Here the solid lines represent the exact solution. These standard deviations represent 3% and 10% measurement error based on the maximum temperature. It is clear that, as the error increases, the accuracy of the prediction decreases. However, even for $\sigma=0.03$ the estimate is quite good. Figure 4.2.5 shows the effects of measurement location on the accuracy of the estimation. The $5\Delta X$ location, which is close to the entrance, produces more accurate results as expected. The $20\Delta X$ location shows a marked decrease in accuracy, particularly near the discontinuity in slope, with the estimate oscillating around the exact function elsewhere.

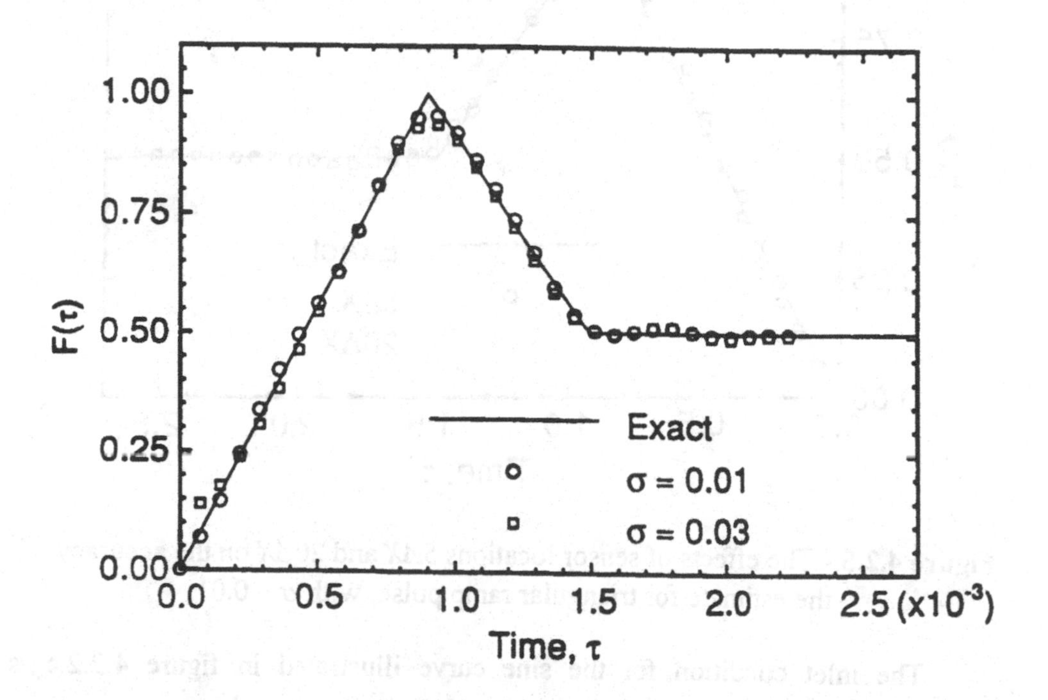

Figure 4.2.4 - The effects of standard deviation for $\sigma= 0.01$ and $\sigma= 0.03$ on the accuracy of the estimate for triangular ramp pulse. Measurements taken at downstream location $5\Delta X$.

The inlet condition for a double step function illustrated in figure 4.2.2.b was assumed in the form

$$F(\tau)=\begin{cases} 1 & \text{for} \quad 0<\tau\leq 8.2\text{x}10^{-4} \\ 0.5 & \text{for} \quad 8.2\text{x}10^{-4}<\tau\leq 1.22\text{x}10^{-3} \\ 0 & \text{for} \quad 1.22\text{x}10^{-3}<\tau\leq \tau_f \end{cases} \tag{4.2.12}$$

which represents a very strict test for the inverse analysis. Figure 4.2.6 compares the results of the inverse solutions at $X^*=5\Delta X$ and $20\Delta X$ downstream locations. For the $5\Delta X$ location the inverse solution tends to follow the discontinuities, including the second step; however, the solution oscillates after the first jump. The results for the $20\Delta X$ location follow the pulse but cannot predict the sharp corners at all.

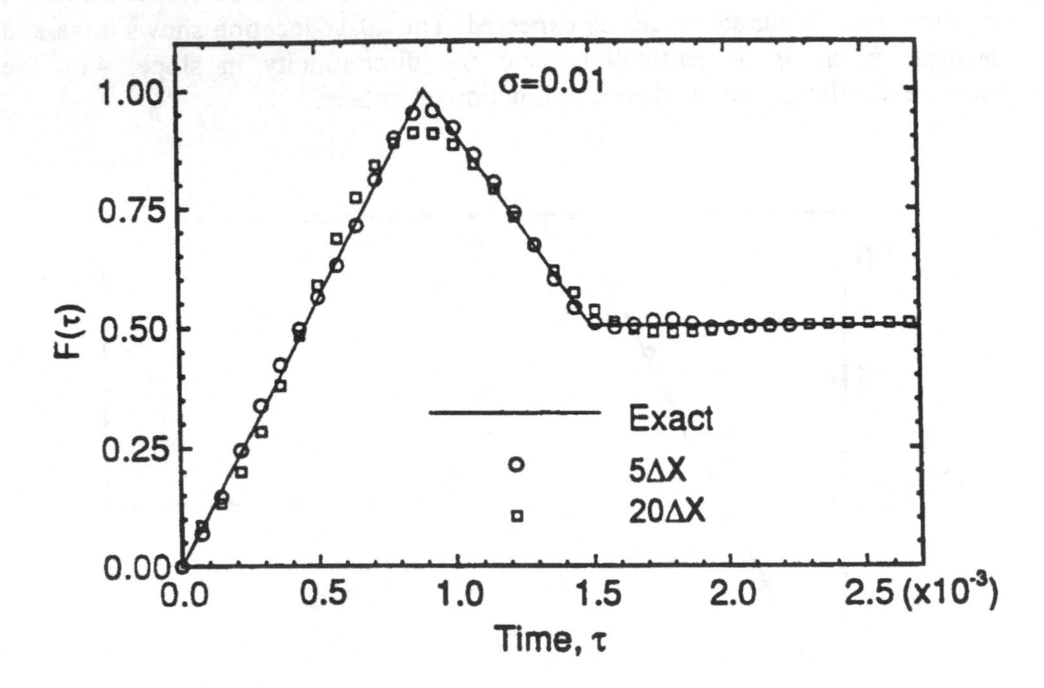

Figure 4.2.5 - The effects of sensor locations $5\Delta X$ and $20\Delta X$ on the accuracy of the estimate for triangular ramp pulse, with $\sigma = 0.01$.

The inlet condition for the sine curve illustrated in figure 4.2.2.c is assumed in the form

$$F(\tau) = \begin{cases} \sin(1111.11\pi\tau) & \text{for} \quad 0 < \tau \le 1.8\text{x}10^{-3} \\ 0 & \text{for} \quad 1.8\text{x}10^{-3} < \tau \le \tau_f \end{cases} \tag{4.2.13}$$

Since the function is smooth over the whole time domain, the inverse analysis is quite accurate for both locations $5\Delta X$ and $20\Delta X$, as apparent from figure 4.2.7.

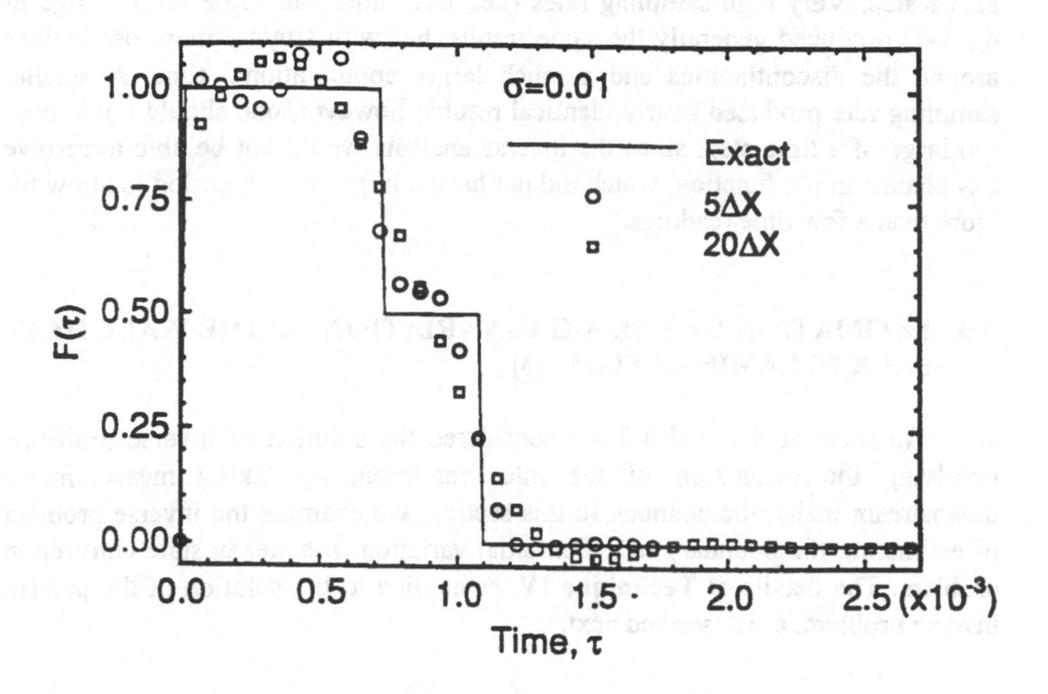

Figure 4.2.6 - The effects of sensor locations $5\Delta X$ and $20\Delta X$ on the accuracy of the estimate for double step pulse, with $\sigma = 0.01$.

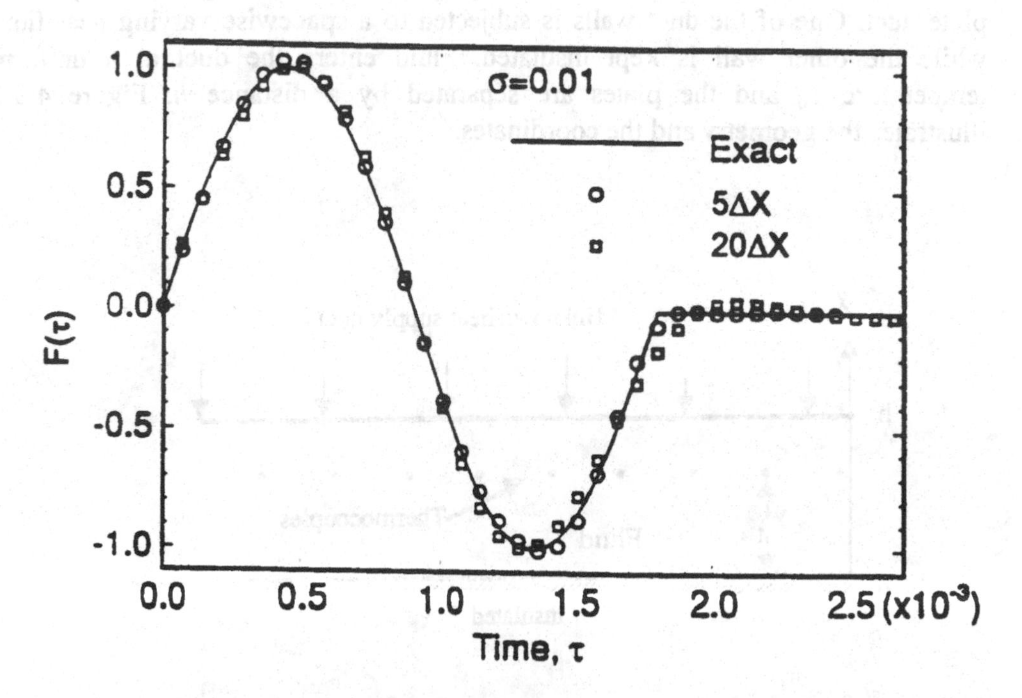

Figure 4.2.7 - The effects of sensor locations $5\Delta X$ and $20\Delta X$ on the accuracy of the estimate for a sine curve pulse, with $\sigma = 0.01$.

The effect of the sampling frequency on the accuracy of estimations was also tested. Very high sampling rates (i.e., five times the value used in figures 4.2.4-7) produced generally the same results, but with slightly more oscillations around the discontinuities and a much larger computational time. A smaller sampling rate produced nearly identical results; however, one should not choose too large of a time step, since the inverse analysis would not be able to resolve any change in the function, which did not have a large enough period to allow for more than a few time readings.

4-3 ESTIMATION OF THE AXIAL VARIATION OF THE WALL HEAT FLUX IN LAMINAR FLOW [3]

In sections 4-1 and 4-2 we considered the solution of inverse problems involving the estimation of the inlet condition, by taking measurements downstream inside the channel. In this section, we examine the inverse problem of estimating the boundary heat flux axial variation in a steady state convection problem. The details of **Technique IV**, as applied to the solution of the present inverse problem, are described next.

Direct Problem

We consider here hydrodynamically developed, thermally developing laminar forced convection of a constant property fluid flowing inside a parallel plate duct. One of the duct walls is subjected to a spacewise varying heat flux, while the other wall is kept insulated. Fluid enters the duct at a uniform temperature T_0 and the plates are separated by a distance h. Figure 4.3.1 illustrates the geometry and the coordinates.

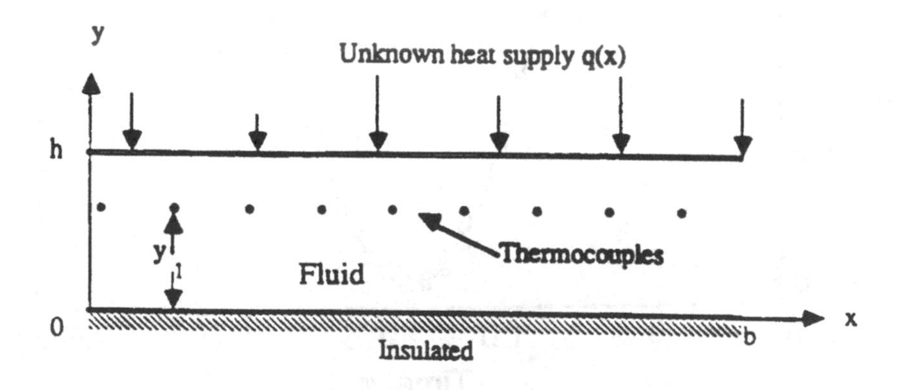

Figure 4.3.1 - Geometry, coordinates and sensors locations.

The mathematical formulation for this problem is given by

$$k\frac{\partial^2 T(x,y)}{\partial y^2} = u(y)\,\rho C_p \frac{\partial T(x,y)}{\partial x} \qquad \text{in } 0<y<h, \quad 0<x<b \tag{4.3.1.a}$$

$$\frac{\partial T(x,0)}{\partial y} = 0 \qquad \text{at } y=0, \quad 0<x<b \tag{4.3.1.b}$$

$$k\frac{\partial T(x,h)}{\partial y} = q(x) \qquad \text{at } y=h, \quad 0<x<b \tag{4.3.1.c}$$

$$T(0,y) = T_0 \qquad \text{at } x=0, \quad 0<y<h \tag{4.3.1.d}$$

where $q(x)$ is the wall heat flux. The fully developed velocity $u(y)$ is given by

$$u(y) = 6u_m \frac{y}{h}\left(1-\frac{y}{h}\right) \tag{4.3.2}$$

where u_m is the mean velocity.

The problem given by equations (4.3.1) is a *direct problem* when the wall heat flux $q(x)$, as well as the other quantities appearing in equations (4.3.1) are known. The objective of the direct problem is to determine the temperature field $T(x,y)$ of the fluid inside the channel.

Inverse Problem

For the inverse problem considered here, the wall heat flux $q(x)$ is regarded as unknown. Such a function is to be estimated by using the readings of M sensors located inside the channel at a transversal position y_1, as illustrated in figure 4.3.1. The following functional is then minimized in order to estimate $q(x)$.

$$S[q(x)] = \int_{x=0}^{b} \left\{T[x,y_1;q(x)] - Y(x)\right\}^2 dx \tag{4.3.3}$$

where $Y(x)$ are the measured temperatures at the transversal position $y = y_1$. A sufficiently large number of measurements is considered available in the axial direction, so that they can be treated as a continuous function. $T[x,y_1;q(x)]$ are the estimated temperatures at $y = y_1$, obtained from the solution of the direct problem (4.3.1) by using an estimate for $q(x)$, while b is the length of the channel containing measurements, where the wall heat flux is to be estimated.

Technique IV, the conjugate gradient method of function estimation, is applied for the minimization of the functional (4.3.3). The sensitivity and adjoint problems, required to obtain expressions for the search step size and gradient direction, are developed next.

Sensitivity Problem

It is assumed that when $q(x)$ undergoes an increment $\Delta q(x)$, the temperature $T(x,y)$ changes by an amount $\Delta T(x,y)$. Then, to construct the sensitivity problem defining the function $\Delta T(x,y)$, we replace $T(x,y)$ by $T(x,y) + \Delta T(x,y)$, and $q(x)$ by $q(x) + \Delta q(x)$ in the direct problem (4.3.1) and subtract from it the problem (4.3.1) to yield

$$k\frac{\partial^2 \Delta T(x,y)}{\partial y^2} = u(y)\,\rho C_p \frac{\partial \Delta T(x,y)}{\partial x} \qquad \text{in } 0<y<h, \;\; 0<x<b \tag{4.3.4.a}$$

$$\frac{\partial \Delta T(x,0)}{\partial y} = 0 \qquad \text{at} \quad y=0, \quad 0<x<b \tag{4.3.4.b}$$

$$k\frac{\partial \Delta T(x,h)}{\partial y} = \Delta q(x) \qquad \text{at} \quad y=h, \quad 0<x<b \tag{4.3.4.c}$$

$$\Delta T(0,y) = 0 \qquad \text{at} \quad x=0, \quad 0<y<h \tag{4.3.4.d}$$

Adjoint Problem

To derive the adjoint problem we multiply equation (4.3.1.a) by the Lagrange multiplier $\lambda(x,y)$, integrate the resulting expression over the space domain and then add this result to the functional given by equation (4.3.3). The following expression results:

$$S[q(x)] = \int_{x=0}^{b} \{T[x,y_1;q(x)] - Y(x)\}^2 dx + \int_{x=0}^{b}\int_{y=0}^{h} \lambda(x,y)\left[k\frac{\partial^2 T}{\partial y^2} - u(y)\rho C_p \frac{\partial T}{\partial x}\right] dy dx \tag{4.3.5}$$

The variation of the extended functional (4.3.5) is developed and allowed to go to zero. After some straightforward manipulations, which are left as an exercise to the reader, the following adjoint problem results:

$$u(y)\,\rho C_p \frac{\partial \lambda}{\partial x} + k\frac{\partial^2 \lambda}{\partial y^2} + 2\left[T(x,y_1) - Y(x)\right] = 0 \quad \text{in } 0<x<b,\, 0<y<h \tag{4.3.6.a}$$

$$\frac{\partial \lambda(x,0)}{\partial y} = 0 \quad \text{at} \quad y=0, \quad 0<x<b \tag{4.3.6.b}$$

$$\frac{\partial \lambda(x,h)}{\partial y} = 0 \quad \text{at} \quad y=h, \quad 0<x<b \tag{4.3.6.c}$$

$$\lambda(b,y) = 0 \quad \text{at} \quad x=b, \quad 0<y<h \tag{4.3.6.d}$$

Gradient Equation

In the process of obtaining the adjoint problem, the following integral term is left:

$$\Delta S[q(x)] = \int_{x=0}^{b} \lambda(x,h)\,\Delta q(x)\,dx \tag{4.3.7.a}$$

By assuming that $q(x)$ belongs to the space of square integrable functions in $0<x<b$, we can write

$$\Delta S[q(x)] = \int_{x=0}^{b} \nabla S[q(x)]\,\Delta q(x)\,dx \tag{4.3.7.b}$$

Therefore, by comparing equations (4.3.7.a) and (4.3.7.b), we can obtain the gradient equation as

$$\nabla S[q(x)] = \lambda(x,h) \tag{4.3.8}$$

Iterative Procedure

The following iterative procedure of Technique IV is applied to the estimation of the wall heat flux $q(x)$:

$$q^{k+1}(x) = q^k(x) - \beta^k d^k(x) \tag{4.3.9.a}$$

where the superscript k refers to the number of iterations and the direction of descent is given by

$$d^k(x) = \nabla S\left[q^k(x)\right] + \gamma^k d^{k-1}(x) \tag{4.3.9.b}$$

We use here the Fletcher-Reeves expression for the conjugation coefficient, given in the form

$$\gamma^k = \frac{\int_{x=0}^{b}\left\{\nabla S\left[q^k(x)\right]\right\}^2 dx}{\int_{x=0}^{b}\left\{\nabla S\left[q^{k-1}(x)\right]\right\}^2 dx} \quad \text{for} \quad k=1,2,... \quad \text{with } \gamma^0 = 0 \tag{4.3.9.c}$$

The step size is obtained by minimizing the functional (4.3.3) with respect to β^k. The following expression results (see Note 7 in Chapter 2):

$$\beta^k = \frac{\int_{x=0}^{b}\left[T(x,y_1) - Y(x)\right]\Delta T(d^k)\, dx}{\int_{x=0}^{b}\left[\Delta T(d^k)\right]^2 dx} \tag{4.3.9.d}$$

where $\Delta T(d^k)$ is the solution of the sensitivity problem (4.3.4), obtained by setting $\Delta q(x) = d^k(x)$.

Results

In order to illustrate the use of Technique IV, as applied to the estimation of the wall heat flux, we considered different functional forms to generate the *simulated measurements*, such as a triangular and a sinusoidal variation.

As a test-case, consider that a fluid at a temperature T_0 = 20°C enters a parallel plate duct of length b = 1.6m, with walls separated by a distance h = 0.01 m. The sensors are located at the position y_1 = 0.009 m from the lower wall and separated by 0.10 m intervals along the x direction. The fluid properties are taken as: ρ = 840 kg/m^3, k = 0.137 W/(m°C) and C_p = 2200 J/(kg°C), which refer to an engine oil. The mean velocity is taken as u_m = 0.04 m/s.

The direct, sensitivity and adjoint problems were solved by finite differences with the following grid spacing: Δx = 0.01m and Δy = 0.0002 m.

Let us consider now a triangular variation for the heat flux in the form

$$q(x) = \begin{cases} 3000 + 8750x & \text{for} \quad 0 \le x \le 0.8m \\ 10000 - 8750(x - 0.8) & \text{for} \quad 0.8 < x \le 1.6m \end{cases} \tag{4.3.10}$$

Figure 4.3.2 shows the results of the inverse analysis for the case with no measurement error (i.e., $\sigma = 0$) while figure 4.3.3 gives the results of the same calculation with a measurement error of $\sigma = 0.5$. These figures show that increasing the measurement errors decreases the accuracy of the inverse solution; but the results are still quite good.

The second example involves a sinusoidal variation for the wall heat flux in the form

$$q(x) = 7000 + 3000\sin\left(\frac{2.5x}{1.6}\pi\right) \tag{4.3.11}$$

The results obtained with measurements with no error ($\sigma = 0$), as well as measurements with random error ($\sigma = 0.5$), are very good, as can be seen in figures 4.3.4 and 4.3.5, respectively. As expected, it is easier to recover a continuous function, such as the one given by equation (4.3.11), than a function containing discontinuities in its first derivative, like the triangular variation given by equation (4.3.10).

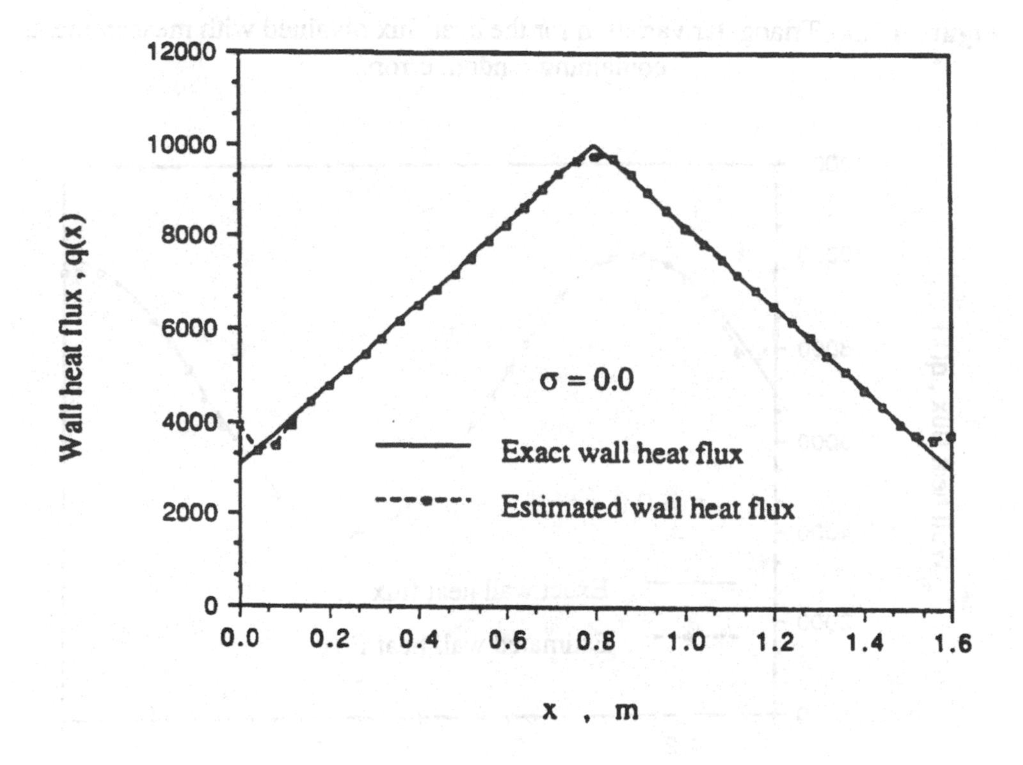

Figure 4.3.2 - Triangular variation for the heat flux obtained with errorless measurements.

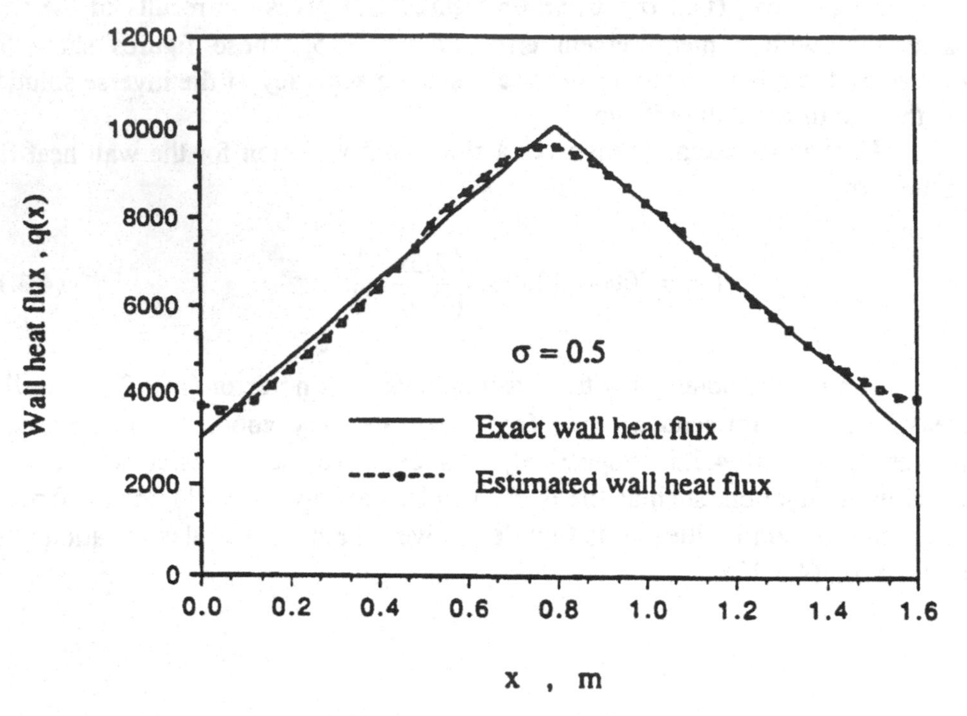

Figure 4.3.3 - Triangular variation for the heat flux obtained with measurements containing random errors.

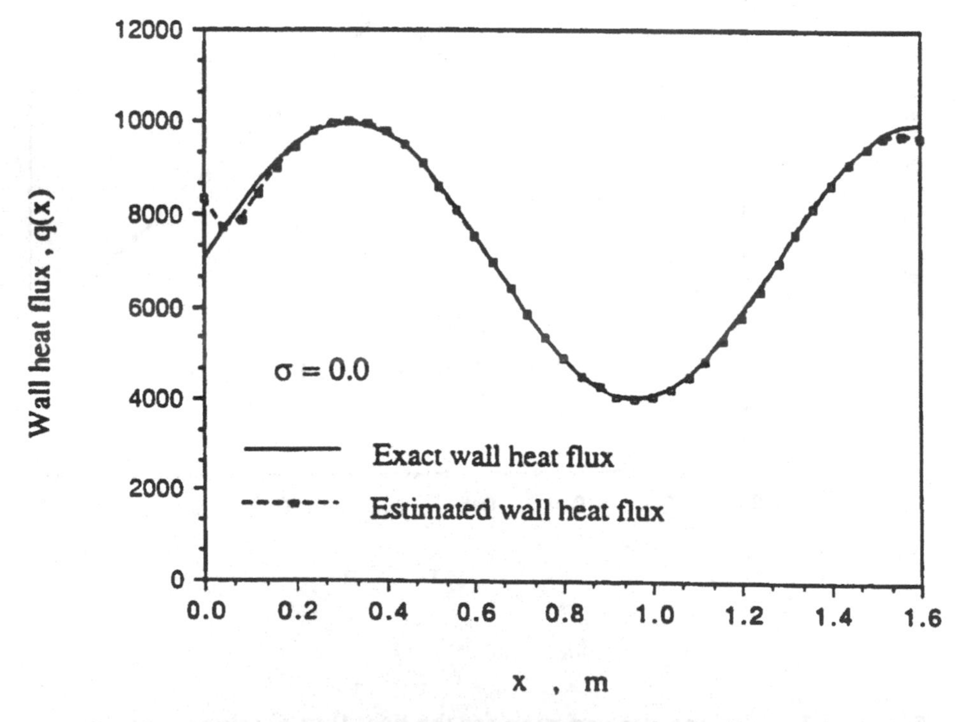

Figure 4.3.4 - Sinusoidal variation for the heat flux obtained with errorless measurements.

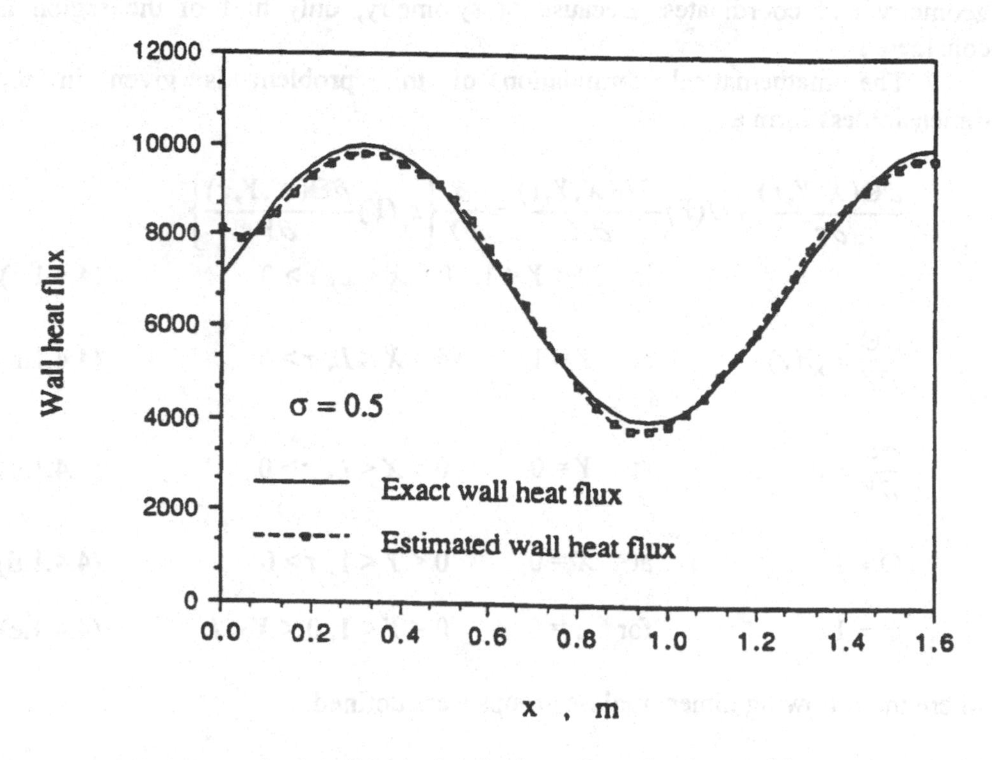

Figure 4.3.5 - Sinusoidal variation for the heat flux obtained with measurements containing random errors.

4-4 ESTIMATION OF THE TRANSIENT WALL HEAT FLUX IN TURBULENT FLOW [4]

After examining the solution of the inverse problem of estimating the axial variation of the wall heat flux in a steady-state problem in laminar flow, we now present the estimation of the transient wall heat flux in a channel with turbulent flow. A similar problem, involving the estimation of the transient wall heat flux in a channel with laminar flow, was solved in reference [7], where the effects of non-newtonian behavior of the fluid were also addressed.

The basic steps of **Technique IV**, as applied to the solution of the present inverse problem, are discussed next.

Direct Problem

We consider hydrodynamically developed, thermally developing transient heat transfer for an incompressible turbulent flow inside a parallel-plate duct, subjected to timewise varying wall heat flux at both boundaries. Axial conduction, viscous dissipation, free convection and wall conjugation effects are neglected. Flow properties are assumed constant. Figure 4.4.1 illustrates the geometry and coordinates. Because of symmetry, only half of the region is considered.

The mathematical formulation of this problem is given in the dimensionless form as

$$\frac{\partial\Theta(X,Y,\tau)}{\partial\tau}+U(Y)\frac{\partial\Theta(X,Y,\tau)}{\partial X}=\frac{\partial}{\partial Y}\left\{\varepsilon_t(Y)\frac{\partial\Theta(X,Y,\tau)}{\partial Y}\right\}$$

$$\text{in} \quad 0<Y<1,\ 0<X<L,\ \tau>0 \qquad (4.4.1.a)$$

$$\frac{\partial\Theta}{\partial Y}=Q(\tau) \quad \text{at} \quad Y=1, \quad 0<X<L,\ \tau>0 \qquad (4.4.1.b)$$

$$\frac{\partial\Theta}{\partial Y}=0 \quad \text{at} \quad Y=0, \quad 0<X<L,\ \tau>0 \qquad (4.4.1.c)$$

$$\Theta=1 \quad \text{at} \quad X=0, \quad 0<Y<1,\ \tau>0 \qquad (4.4.1.d)$$

$$\Theta=1 \quad \text{for} \quad \tau=0, \quad 0<Y<1, 0<X<L \qquad (4.4.1.e)$$

where the following dimensionless groups were defined:

$$\Theta=\frac{T}{T_0} \qquad Y=\frac{y}{h} \qquad X=\frac{16x/D_e}{RePr}$$

$$D_e=4h \qquad Pr=\frac{\nu}{\alpha} \qquad Re=\frac{u_m D_e}{\nu} \qquad (4.4.2.a\text{-}j)$$

$$U(Y)=\frac{u}{u_m} \qquad \tau=\frac{\alpha t}{h^2} \qquad Q=\frac{qh}{T_0 k} \qquad \varepsilon_t=1+\frac{\varepsilon_h}{\alpha}$$

Here, T_0 is the initial and inlet temperature of the fluid, h is the half distance between the plates and u_m is the mean fluid velocity, while ν and α are the fluid kinematic viscosity and thermal diffusivity, respectively. The fully developed turbulent velocity profile, $U(Y)$, and the total diffusivity, ε_t, were determined with the same turbulence model used in reference [18].

The problem given by equations (4.4.1) is denoted as a *direct problem* if the heat flux $Q(\tau)$ is known. The objective of the direct problem is to determine the temperature field $\Theta(X,Y,\tau)$ of the fluid inside the channel.

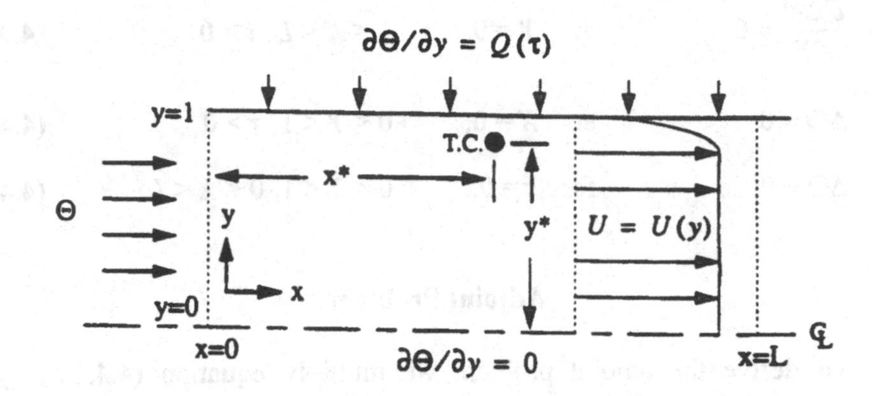

Inverse Problem

For the inverse problem considered here, the wall heat flux $Q(\tau)$ is regarded as unknown, and is to be estimated by minimizing the following functional

$$S[Q(\tau)] = \int_{\tau=0}^{\tau_f} \{\Theta[X^*,Y^*,\tau;Q(\tau)] - Z(\tau)\}^2 d\tau \tag{4.4.3}$$

where $\Theta[X^*,Y^*,\tau;Q(\tau)]$ is the estimated temperature at the sensor position (X^*, Y^*), which can be obtained from the solution of the direct problem (4.4.1) by using an estimate for the unknown heat flux. $Z(\tau)$ is the measured temperature.

The *sensitivity and adjoint problems*, required for the implementation of the iterative procedure of Technique IV, are developed next.

Sensitivity Problem

When the wall heat flux $Q(\tau)$ undergoes an increment $\Delta Q(\tau)$, the temperature $\Theta(X,Y,\tau)$ also changes by the amount $\Delta\Theta(X,Y,\tau)$. To construct the sensitivity problem we replace $\Theta(X,Y,\tau)$ and $Q(\tau)$ in the direct problem (4.4.1) by $[\Theta(X,Y,\tau)+\Delta\Theta(X,Y,\tau)]$ and $[Q(\tau)+\Delta Q(\tau)]$, respectively, and then subtract from the resulting equations the original direct problem. We obtain

$$\frac{\partial \Delta\Theta(X,Y,\tau)}{\partial \tau} + U(Y)\frac{\partial \Delta\Theta(X,Y,\tau)}{\partial X} = \frac{\partial}{\partial Y}\left\{\varepsilon_t(Y)\frac{\partial \Delta\Theta(X,Y,\tau)}{\partial Y}\right\}$$

$$\text{in} \quad 0 < Y < 1, \quad 0 < X < L, \ \tau > 0 \tag{4.4.4.a}$$

$$\frac{\partial \Delta\Theta}{\partial Y} = \Delta Q(\tau) \qquad \text{at} \quad Y = 1, \qquad 0 < X < L,\ \tau > 0 \tag{4.4.4.b}$$

$$\frac{\partial \Delta\Theta}{\partial Y} = 0 \qquad \text{at} \quad Y = 0, \qquad 0 < X < L,\ \tau > 0 \tag{4.4.4.c}$$

$$\Delta\Theta = 0 \qquad \text{at} \quad X = 0, \qquad 0 < Y < 1,\ \tau > 0 \tag{4.4.4.d}$$

$$\Delta\Theta = 0 \qquad \text{for} \quad \tau = 0, \qquad 0 < Y < 1, 0 < X < L \tag{4.4.4.e}$$

Adjoint Problem

To derive the adjoint problem we multiply equation (4.4.1.a) by the Lagrange multiplier $\lambda(X,Y,\tau)$, integrate the resulting expression over the space and time domains and then add it to equation (4.4.3) to yield

$$S[Q(\tau)] = \int_{\tau=0}^{\tau_f} (\Theta - Z)^2 d\tau + \int_{\tau=0}^{\tau_f} \int_{X=0}^{L} \int_{Y=0}^{1} \lambda \left\{ \frac{\partial}{\partial Y}\left(\varepsilon_t \frac{\partial \Theta}{\partial Y}\right) - \frac{\partial \Theta}{\partial \tau} - U \frac{\partial \Theta}{\partial X} \right\} dY dX d\tau \tag{4.4.5}$$

The variation of the extended functional (4.4.5) is obtained and after some algebraic manipulations it is allowed to go to zero. The following adjoint problem results:

$$\frac{\partial \lambda}{\partial \tau} + U(Y)\frac{\partial \lambda}{\partial X} + \frac{\partial}{\partial Y}\left(\varepsilon_t \frac{\partial \lambda}{\partial Y}\right) + 2(\Theta - Z)\delta(X - X^*)\delta(Y - Y^*) = 0$$

$$\text{in} \quad 0 < Y < 1, \qquad 0 < X < L, 0 < \tau < \tau_f \tag{4.4.6.a}$$

$$\frac{\partial \lambda}{\partial Y} = 0 \qquad \text{at} \quad Y = 1, \qquad 0 < X < L, 0 < \tau < \tau_f \tag{4.4.6.b}$$

$$\frac{\partial \lambda}{\partial Y} = 0 \qquad \text{at} \quad Y = 0, \qquad 0 < X < L, 0 < \tau < \tau_f \tag{4.4.6.c}$$

$$\lambda = 0 \qquad \text{at} \quad X = L, \qquad 0 < Y < 1, 0 < \tau < \tau_f \tag{4.4.6.d}$$

$$\lambda = 0 \qquad \text{for} \quad \tau = \tau_f, \qquad 0 < Y < 1, 0 < X < L \tag{4.4.6.e}$$

Gradient Equation

In the process of obtaining the adjoint problem, the following integral term is left:

$$\Delta S[Q(\tau)] = -\int_{\tau=0}^{\tau_f}\int_{X=0}^{L}\lambda(X,1,\tau)\varepsilon_t(1)\Delta Q(\tau)dXd\tau \tag{4.4.7.a}$$

For $Q(\tau)$ belonging to the space of square integrable functions in the time domain $0 < \tau < \tau_f$, we can write

$$\Delta S[Q(\tau)] = \int_{\tau=0}^{\tau_f}\nabla S[Q(\tau)]\Delta Q(\tau)d\tau \tag{4.4.7.b}$$

Hence, by comparing equations (4.4.7.a) and (4.4.7.b), we can obtain the gradient equation in the form

$$\nabla S[Q(\tau)] = -\int_{X=0}^{L}\lambda(X,1,\tau)\varepsilon_t(1)dX \tag{4.4.8}$$

Iterative Procedure

The iterative procedure of Technique IV, as applied to the estimation of the unknown function $Q(\tau)$, is given by

$$Q^{k+1}(\tau) = Q^k(\tau) - \beta^k d^k(\tau) \tag{4.4.9.a}$$

The direction of descent $d^k(\tau)$, used to advance from iteration k to k+1, is obtained as

$$d^k(\tau) = \nabla S\left[Q^k(\tau)\right] + \gamma^k d^{k-1}(\tau) \tag{4.4.9.b}$$

The Fletcher-Reeves expression for the conjugation coefficient is given by:

$$\gamma^k = \frac{\int_{\tau=0}^{\tau_f}\left\{\nabla S\left[Q^k(\tau)\right]\right\}^2 d\tau}{\int_{\tau=0}^{\tau_f}\left\{\nabla S\left[Q^{k-1}(\tau)\right]\right\}^2 d\tau} \quad \text{for} \quad k=1,2,... \quad \text{with } \gamma^0 = 0 \tag{4.4.9.c}$$

An expression for the search step size β^k is obtained by minimizing the functional given by equation (4.4.3) with respect to β^k. The following expression results (see Note 7 in Chapter 2):

$$\beta^k = \frac{\int_{\tau=0}^{\tau_f} \left\{ \Theta\left[Q^k(\tau)\right] - Z(\tau) \right\} \Delta\Theta\left[d^k(\tau)\right] d\tau}{\int_{\tau=0}^{\tau_f} \left\{ \Delta\Theta\left[d^k(\tau)\right] \right\}^2 d\tau} \tag{4.4.9.d}$$

where $\Delta\Theta[d^k(\tau)]$ is the solution of the sensitivity problem (4.4.4), obtained by setting $\Delta Q(\tau) = d^k(\tau)$.

Results

We use simulated measurements in the form given by equations (2.5.2), in order to evaluate the accuracy of the inverse analysis for estimating $Q(\tau)$ with Technique IV.

In the present study, we investigated the following three different timewise variations of the wall heat flux $Q(\tau)$, with functional forms illustrated in figure 4.4.2 and specified as given below:

$$\text{Case(A)}: Q(\tau) = \begin{cases} 80 & 0 \le \tau \le 0.3 \\ 50 & 0.3 < \tau \le 0.6 \end{cases} \tag{4.4.10.a}$$

$$\text{Case(B)}: Q(\tau) = \begin{cases} 50 + (2.228\tau - 0.1175\tau^2 + 0.0016\tau^3)/2.5 & 0 \le \tau \le 0.35 \\ 60 & 0.35 < \tau \le 0.6 \end{cases} \tag{4.4.10.b}$$

$$\text{Case(C)}: Q(\tau) = \begin{cases} 50 + 100\tau & 0 \le \tau \le 0.3 \\ 50 + 100(0.6 - \tau) & 0.3 < \tau \le 0.6 \end{cases} \tag{4.4.10.c}$$

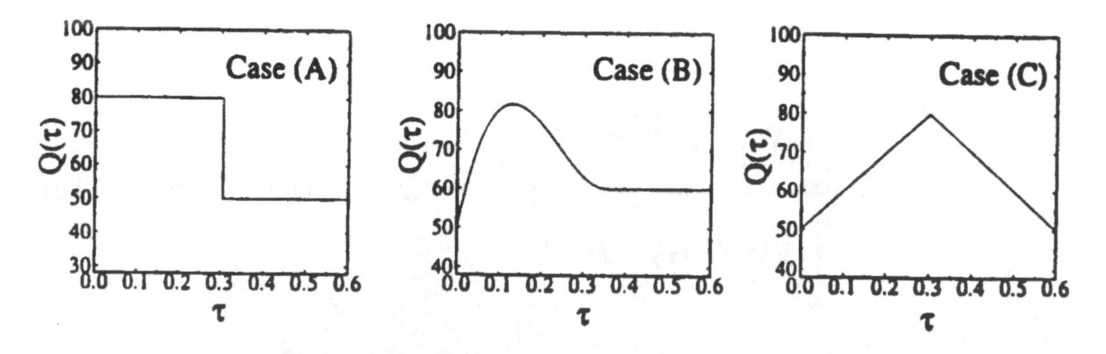

Figure 4.4.2 - Different functional forms tested for the wall heat flux $Q(\tau)$.

We consider here a turbulent flow with Re = 10^5 and Pr = 1.

Let us consider initially the functional form given by case (A). Figure 4.4.3 illustrates the effects of the transversal sensor location on the inverse problem solution, for measurements with a standard deviation of $\sigma = 0.005$ which corresponds to an error of up to 1.25%, obtained with a sensor located at $X^* = 5D_e$. Different transversal locations were examined, including $Y^* = 1, 0.9$, 0.8 and 0.7. The location $Y^* = 1$ would correspond to a sensor located at the wall. As apparent from figure 4.4.3, the accuracy of the estimation decreases as the sensor is moved away from the boundary. In fact, for $Y^* < 0.7$ the sensor is located outside the thermal boundary layer. Hence, temperature measurements taken in the region $Y^* < 0.7$ for $X^* = 5D_e$ bring no useful information for the estimation of $Q(\tau)$.

Figure 4.4.4 shows the effect of axial location of the sensor on the accuracy of the estimation. In this figure we examine three axial locations $X^* = 5D_e$, $7D_e$ and $9D_e$ with the transversal position taken as $Y^* = 0.9$ and the standard deviation $\sigma = 0.01$ (which corresponds to 2.5% measurement error). Clearly, increasing the axial location X^* of the sensor decreases the accuracy of the estimation, because the sensor location moves to the fully developed region.

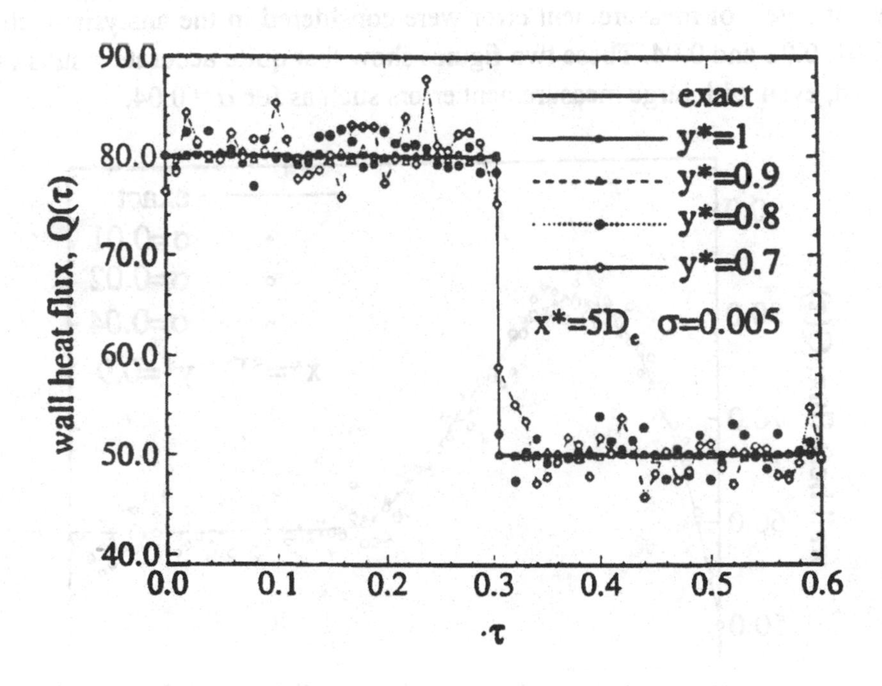

Figure 4.4.3 - Effect of the transversal location Y^* of the sensor on the accuracy of estimations for case (A).

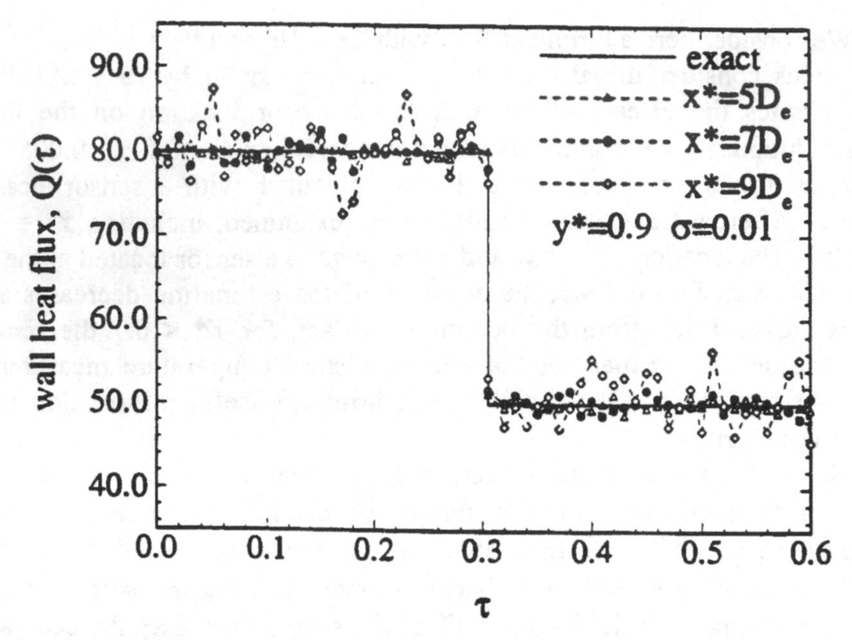

Figure 4.4.4 - Effect of the axial location X^* of the sensor on the accuracy of estimations for case (A).

Figures 4.4.5 and 4.4.6 illustrate the effects of the measurement error on the inverse problem solution for the functional forms of cases (B) and (C), respectively. The sensor is considered to be located at $X^* = 5D_e$ and $Y^* = 0.9$. Different levels of measurement error were considered in the analysis, including $\sigma = 0.01$, 0.02 and 0.04. These two figures show that quite accurate results can be obtained, even with large measurement errors such as for $\sigma = 0.04$.

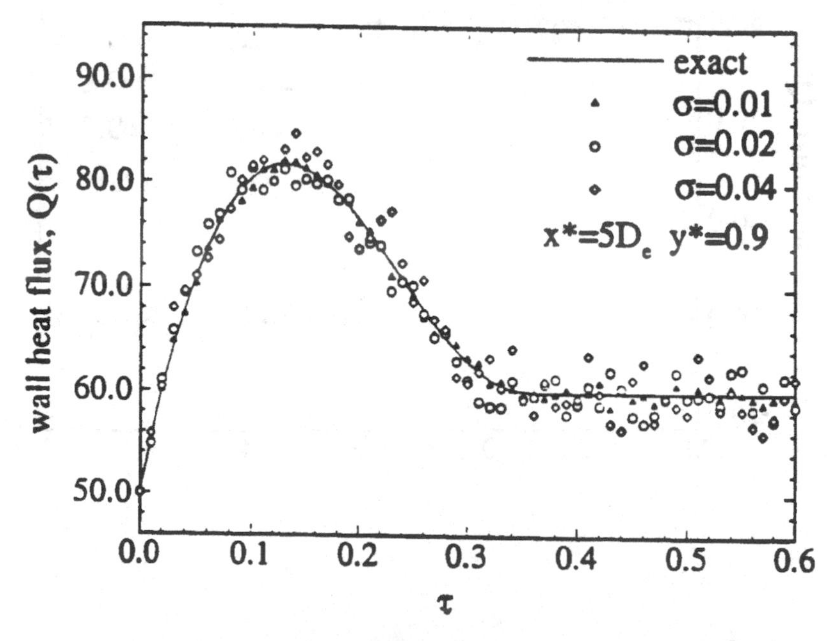

Figure 4.4.5 - Effect of the measurement errors on the accuracy of estimations for case (B).

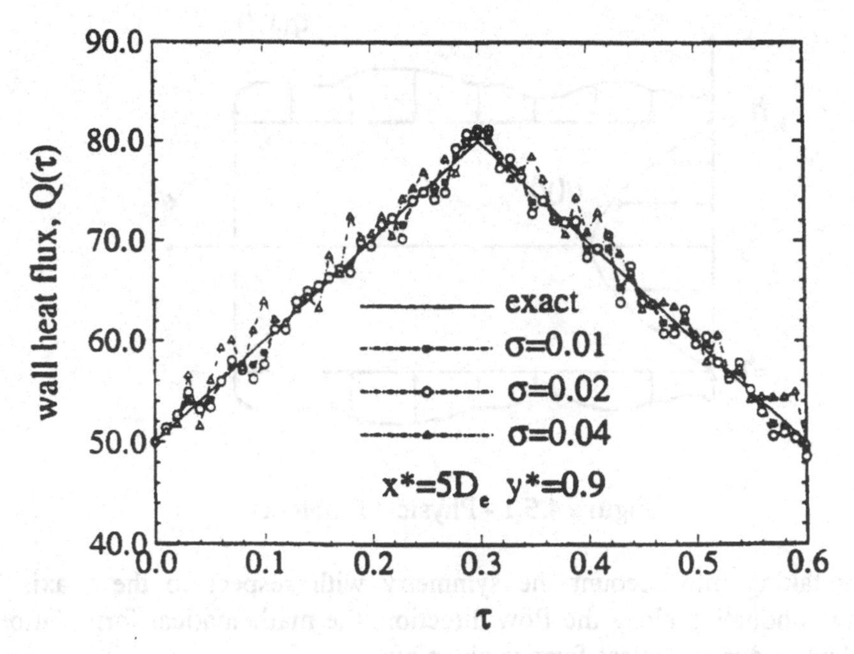

Figure 4.4.6 - Effect of the measurement errors on the accuracy of estimations for case (C).

4-5 ESTIMATION OF THE SPACEWISE AND TIMEWISE VARIATIONS OF THE WALL HEAT FLUX IN LAMINAR FLOW [5]

In this section we present the solution of the inverse problem of estimating the wall heat flux in a parallel plate channel, by using **Technique IV**, the conjugate gradient method with adjoint problem. The unknown heat flux is supposed to vary in time and along the channel flow direction. We examine the accuracy of the present function estimation approach by using transient simulated measurements of several sensors located inside the channel. The inverse problem is solved for different functional forms of the unknown wall heat flux, including those containing sharp corners and discontinuities, which are the most difficult to be recovered by an inverse analysis. The effects on the inverse problem solution of the number of sensors, as well as their locations, are also addressed.

Direct Problem

The physical problem considered here is the laminar hydrodynamically developed flow between parallel plates of a fluid with constant properties. The inlet temperature is maintained at a constant value T_0, which is also assumed to be the initial fluid temperature. For times greater than zero, the plates are subjected to a time and space-dependent heat flux, as illustrated in Figure 4.5.1.

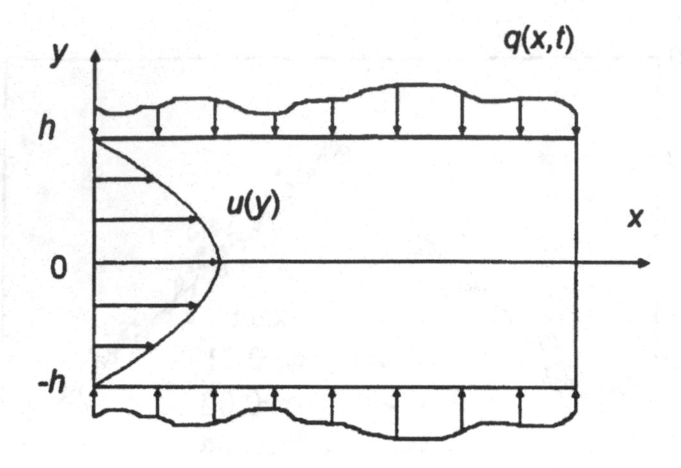

Figure 4.5.1 - Physical Problem

By taking into account the symmetry with respect to the x-axis and neglecting conduction along the flow direction, the mathematical formulation of this problem in dimensionless form is given by:

$$\frac{\partial \Theta}{\partial \tau}+U(Y)\frac{\partial \Theta}{\partial X}=\frac{\partial^2 \Theta}{\partial Y^2} \quad \text{in } 0<Y<1, \quad 0<X<L, \quad \text{for } \tau>0 \tag{4.5.1.a}$$

$$\frac{\partial \Theta}{\partial Y}=0 \quad \text{at } Y=0, \quad 0<X<L, \quad \text{for } \tau>0 \tag{4.5.1.b}$$

$$\frac{\partial \Theta}{\partial Y}=Q(X,\tau) \quad \text{at } Y=1, \quad 0<X<L, \quad \text{for } \tau>0 \tag{4.5.1.c}$$

$$\Theta=0 \quad \text{at } X=0, \quad 0<X<L, \quad \text{for } \tau>0 \tag{4.5.1.d}$$

$$\Theta=0 \quad \text{for } \tau=0, \quad \text{in } 0<Y<1, \quad 0<X<L \tag{4.5.1.e}$$

where the following dimensionless groups were introduced:

$$Y=\frac{y}{h}; \quad X=\frac{\alpha x}{u_m h^2}; \quad \Theta=\frac{T-T_0}{\frac{q_0 h}{k}}; \quad \tau=\frac{\alpha t}{h^2} \tag{4.5.2.a-d}$$

$$U(Y)=\frac{u(y)}{u_m}=\frac{3}{2}\left[1-\left(\frac{y}{h}\right)^2\right] \tag{4.5.2.e}$$

α and k are the fluid thermal diffusivity and conductivity, respectively, h is the channel half-width and u_m is the mean fluid velocity. The wall heat flux is written as

$$q(x,t) = q_0 Q(X,\tau) \tag{4.5.3}$$

where q_0 is a constant reference value with units of heat flux and $Q(X,\tau)$ is a dimensionless function of X and τ.

The direct problem given by equations (4.5.1) is concerned with the determination of the temperature field of the fluid inside the channel, when the boundary heat flux $Q(X,\tau)$ at $Y = 1$ is known.

Inverse Problem

For the inverse problem, the heat flux $Q(X,\tau)$ at $Y = 1$ is considered to be unknown and is to be estimated by using the transient readings of M temperature sensors located inside the channel. We assume that no information is available regarding the functional form of the unknown wall heat flux, except that it belongs to the space of square integrable functions in the domain $0 < \tau < \tau_f$ and $0 < X < L$, where τ_f is the duration of the experiment and L is the length of the test-section in the channel.

The solution of such inverse problem is obtained by minimizing the following functional

$$S[Q(X,\tau)] = \int_{\tau=0}^{\tau_f} \sum_{m=1}^{M} \left\{ \Theta\left[X_m^*, Y_m^*, \tau; Q(X,\tau)\right] - Z_m(\tau) \right\}^2 d\tau \tag{4.5.4}$$

where $Z_m(\tau)$ is the measured temperature at the sensor location (X_m^*, Y_m^*) inside the channel and $\Theta[X_m^*, Y_m^*, \tau; Q(X,\tau)]$ is the estimated temperature at the same location. Such estimated temperature is obtained from the solution of the direct problem given by equations (4.5.1), by using an estimate for the unknown heat flux $Q(X,\tau)$.

The development of the sensitivity and adjoint problems, required for the implementation of the iterative procedure of Technique IV, are described next.

Sensitivity Problem

The sensitivity problem is obtained by assuming that the heat flux $Q(X,\tau)$ is perturbed by an amount $\Delta Q(X,\tau)$. Such perturbation in the heat flux causes a perturbation $\Delta\Theta(X,Y,\tau)$ in the temperature $\Theta(X,Y,\tau)$. By substituting $\Theta(X,Y,\tau)$ by

$[\Theta(X,Y,\tau)+\Delta\Theta(X,Y,\tau)]$ and $Q(X,\tau)$ by $[Q(X,\tau)+\Delta Q(X,\tau)]$ in the direct problem given by equations (4.5.1), and then subtracting from the resulting expressions the original direct problem, we obtain the following sensitivity problem for the determination of the sensitivity function $\Delta\Theta(X,Y,\tau)$:

$$\frac{\partial\Delta\Theta}{\partial\tau}+U(Y)\frac{\partial\Delta\Theta}{\partial X}=\frac{\partial^2\Delta\Theta}{\partial Y^2} \qquad \text{in } 0<Y<1,\ 0<X<L, \text{ for } \tau>0 \qquad (4.5.5.a)$$

$$\frac{\partial\Delta\Theta}{\partial Y}=0 \qquad \text{at } Y=0, \quad 0<X<L, \text{ for } \tau>0 \qquad (4.5.5.b)$$

$$\frac{\partial\Delta\Theta}{\partial Y}=\Delta Q(X,\tau) \qquad \text{at } Y=1, \quad 0<X<L, \text{ for } \tau>0 \qquad (4.5.5.c)$$

$$\Delta\Theta=0 \qquad \text{at } X=0, \quad 0<Y<1, \text{ for } \tau>0 \qquad (4.5.5.d)$$

$$\Delta\Theta=0 \qquad \text{for } \tau=0, \text{ in } 0<Y<1, 0<X<L \qquad (4.5.5.e)$$

Adjoint Problem

In order to obtain the adjoint problem, we multiply the differential equation (4.5.1.a) of the direct problem by the Lagrange multiplier $\lambda(X,Y,\tau)$ and integrate over the time and space domains. The resulting expression is then added to equation (4.5.4) to obtain the following extended functional:

$$S[Q(X,\tau)]=\int_{\tau=0}^{\tau_f}\int_{X=0}^{L}\int_{Y=0}^{1}\left\{\sum_{m=1}^{M}[\Theta(X,Y,\tau)-Z(\tau)]^2\delta\left(X-X_m^*\right)\delta\left(Y-Y_m^*\right)+\left[\frac{\partial\Theta}{\partial\tau}+U(Y)\frac{\partial\Theta}{\partial X}-\frac{\partial^2\Theta}{\partial Y^2}\right]\lambda(X,Y,\tau)\right\}dY\,dX\,d\tau$$

(4.5.6)

where $\delta(\cdot)$ is the Dirac delta function.

The variation of the extended functional (4.5.6) is obtained and, after some manipulations, the resulting expression is allowed to go to zero in order to obtain the following adjoint problem for the Lagrange multiplier $\lambda(X,Y,\tau)$:

$$-\frac{\partial\lambda}{\partial\tau}-U(Y)\frac{\partial\lambda}{\partial X}-\frac{\partial^2\lambda}{\partial Y^2}+2\sum_{m=1}^{M}(\Theta-Z)\delta(X-X_m^*)\delta(Y-Y_m^*)=0$$

$$\text{in } 0<Y<1, \quad 0<X<L, \quad \text{for } \tau>0 \qquad (4.5.7.a)$$

$$\frac{\partial \lambda}{\partial Y} = 0 \qquad \text{at } Y = 0, \qquad 0 < X < L, \text{ for } \tau > 0 \qquad (4.5.7.b)$$

$$\frac{\partial \lambda}{\partial Y} = 0 \qquad \text{at } Y = 1, \qquad 0 < X < L, \text{ for } \tau > 0 \qquad (4.5.7.c)$$

$$\lambda = 0 \qquad \text{at } X = L, \qquad 0 < Y < 1, \text{ for } \tau > 0 \qquad (4.5.7.d)$$

$$\lambda = 0 \qquad \text{for } \tau = \tau_f, \text{ in } 0 < Y < 1 \quad 0 < X < L \qquad (4.5.7.e)$$

Gradient Equation

In the process of obtaining the adjoint problem, the following integral term is left:

$$\Delta S[Q(X,\tau)] = -\int_{\tau=0}^{\tau_f}\int_{X=0}^{L} \lambda(X,1,\tau)\Delta Q(X,\tau)\,dX\,d\tau \qquad (4.5.8.a)$$

From the hypothesis that $Q(X,\tau)$ belongs to the space of square integrable functions in $0 < X < L$ and $0 < \tau < \tau_f$, we can write

$$\Delta S[Q(X,\tau)] = \int_{\tau=0}^{\tau_f}\int_{X=0}^{L} \nabla S[Q(X,\tau)]\Delta Q(X,\tau)\,dX\,d\tau \qquad (4.5.8.b)$$

Therefore, by comparing equations (4.5.8.a) and (4.5.8.b), we obtain the gradient equation for the functional as

$$\nabla S[Q(X,\tau)] = -\lambda(X,1,\tau) \qquad (4.5.9)$$

Iterative Procedure

The iterative algorithm of Technique IV, as applied to the estimation of the unknown heat flux $Q(X,\tau)$, is given by

$$Q^{k+1}(X,\tau) = Q^{k}(X,\tau) - \beta^{k} d^{k}(X,\tau) \qquad (4.5.10.a)$$

where the superscript k denotes de number of iterations.

The direction of descent $d^k(X,\tau)$ is obtained as a conjugation of the gradient direction and of the previous direction of descent as

$$d^k(X,\tau) = \nabla S\left[Q^k(X,\tau)\right] + \gamma^k d^{k-1}(X,\tau) \tag{4.5.10.b}$$

where the conjugation coefficient is obtained from the Fletcher-Reeves expression as:

$$\gamma^k = \frac{\int_{\tau=0}^{\tau_f}\int_{X=0}^{L}\left\{\nabla S\left[Q^k(\tau)\right]\right\}^2 dX d\tau}{\int_{\tau=0}^{\tau_f}\int_{X=0}^{L}\left\{\nabla S\left[Q^{k-1}(\tau)\right]\right\}^2 dX d\tau} \quad \text{for } k=1,2,... \quad \text{with } \gamma^0 = 0 \tag{4.5.10.c}$$

An expression for the search step size β^k is obtained by minimizing the functional given by equation (4.5.4) with respect to β^k (see Note 7 in Chapter 2). We obtain

$$\beta^k = \frac{\int_{\tau=0}^{\tau_f}\sum_{m=1}^{M}\left(\Theta_m - Z_m\right)\Delta\Theta_m(d^k)\,d\tau}{\int_{\tau=0}^{\tau_f}\sum_{m=1}^{M}\left[\Delta\Theta_m(d^k)\right]^2 d\tau} \tag{4.5.10.d}$$

where $\Delta\Theta_m(d^k)$ is the solution of the sensitivity problem (4.5.5), obtained by setting $\Delta Q(X,\tau) = d^k(X,\tau)$.

Results

We use transient simulated measurements in order to assess the accuracy of the present approach of estimating the unknown wall heat flux $Q(X,\tau)$. The simulated temperature measurements were obtained from equation (2.5.2).

For the cases considered below, we have taken the total experiment duration τ_f as 0.08 and the channel test-length L as 0.004, while the heat flux at the boundary $Y = 1$ was assumed in the form:

$$Q(X,\tau) = Q_X(X) + Q_\tau(\tau) \tag{4.5.11}$$

The direct, sensitivity and adjoint problems were solved with finite-differences by using an upwind discretization for the convection term and an implicit discretization in time. The domain was discretized by using 101 and 81 points in the X and Y directions, respectively, while using 41 time steps. Such number of points was chosen by comparing the solution of the direct problem for the local Nusselt number obtained by finite-differences, with a known analytical solution [19].

By examining equations (4.5.7.d) and (4.5.7.e), we note that the gradient of the functional given by equation (4.5.9) is null at the final time τ_f and the final axial position L. Therefore, the initial guess used for the iterative process remains unchanged at τ_f and at L. In the examples shown below, we use as an initial guess for the final time and for the final position the exact values for $Q(X,\tau)$, which are assumed available. For other times and axial positions, we take $Q(X,\tau)$ null as the initial guess for the conjugate gradient method. We lose no generality with such an approach, since we can always choose τ_f and L sufficiently larger than the respective experimental time and test section length of interest, so that the initial guess has no influence on the solution, as illustrated in section 2-5.

Figures 4.5.2.a-c present the results obtained for a boundary heat flux containing a triangular variation in X and a step variation in time, in the form:

$$Q_X(X) = \begin{cases} 1, \text{ for } X \le 0.001 \text{ and } X \ge 0.003 \\ 1000X, \text{ for } 0.001 < X \le 0.002 \\ -1000X + 4, \text{ for } 0.002 < X < 0.003 \end{cases} \tag{4.5.12}$$

$$Q_\tau(\tau) = \begin{cases} 1, \text{ for } \tau \le 0.02 \text{ and } \tau \ge 0.06 \\ 2, \text{ for } 0.02 < \tau < 0.06 \end{cases} \tag{4.5.13}$$

For such case, we have used in the inverse analysis 21 sensors located at $Y^* = 0.95$. The first sensor was located at $X_1^* = 0.00004$ and the last one at $X_{21}^* = 0.00396$. The others were equally spaced, so that $X_m^* = 0.0002(m-1)$, for $m = 2, ..., 20$. Figures 4.5.2 show the results for errorless measurements (dashed lines), as well as for measurements with a standard deviation $\sigma = 0.01\Theta_{max}$ (symbols), where Θ_{max} is the maximum temperature measured by the sensors. In Figure 4.5.2.a, we have the results for the axial variation for 3 different times, where $Q_\tau(0.002) = Q_\tau(0.07) = 1$ and $Q_\tau(0.04) = 2$ from equation (4.5.13). The unknown heat fluxes for such times were accurately predicted, so that the results for $\tau = 0.002$ and $\tau = 0.07$ fall in the curve at the bottom, while those for $\tau = 0.04$ fall in the curve at the top of Figure 4.5.2.a. The predicted heat flux is in good agreement with the exact one for both errorless measurements and measurements with random error. Figures 4.5.2.b-c show the results obtained for the flux variation in time for different axial positions. The results for $X = 0.0004$ and $X = 0.0036$ fall on the same curve in Figure 4.5.2.b as expected, since $Q_X(0.0004) = Q_X(0.0036) = 1$ from equation (4.5.12). The results shown in Figure 4.5.2.c for $X = 0.002$, where $Q(X,\tau)$ has a peak in X, are also in good agreement with the exact functional form assumed for $Q(X,\tau)$.

The RMS error (e_{RMS}) for the results shown in Figures 4.5.2 obtained with errorless measurements, is 0.014. We define the RMS error here as:

$$e_{RMS} = \frac{1}{N}\sqrt{\sum_{i=1}^{N}\left[Q_{ex}(X_i,\tau_i)-Q_{est}(X_i,\tau_i)\right]^2} \tag{4.5.14}$$

where N is the total number of measurements used in the inverse analysis, while Q_{ex} and Q_{est} are the exact and estimated heat fluxes, respectively.

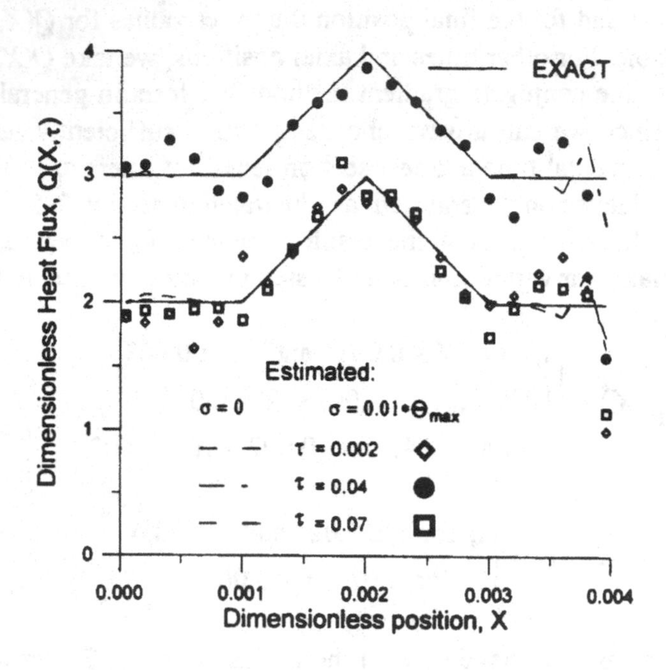

Figure 4.5.2.a - Inverse problem solution for different times obtained with 21 sensors. Triangular variation in the axial direction given by equation (4.5.12).

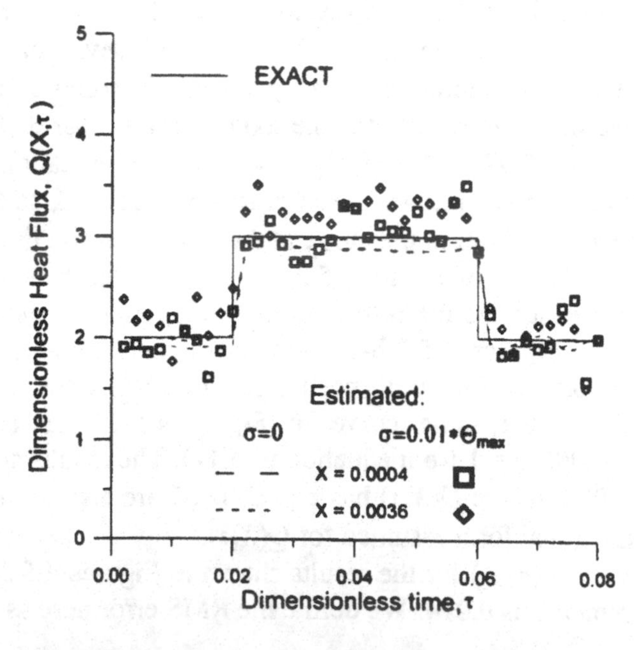

Figure 4.5.2.b - Inverse problem solution for different axial positions obtained with 21 sensors. Step variation in time given by equation (4.5.13).

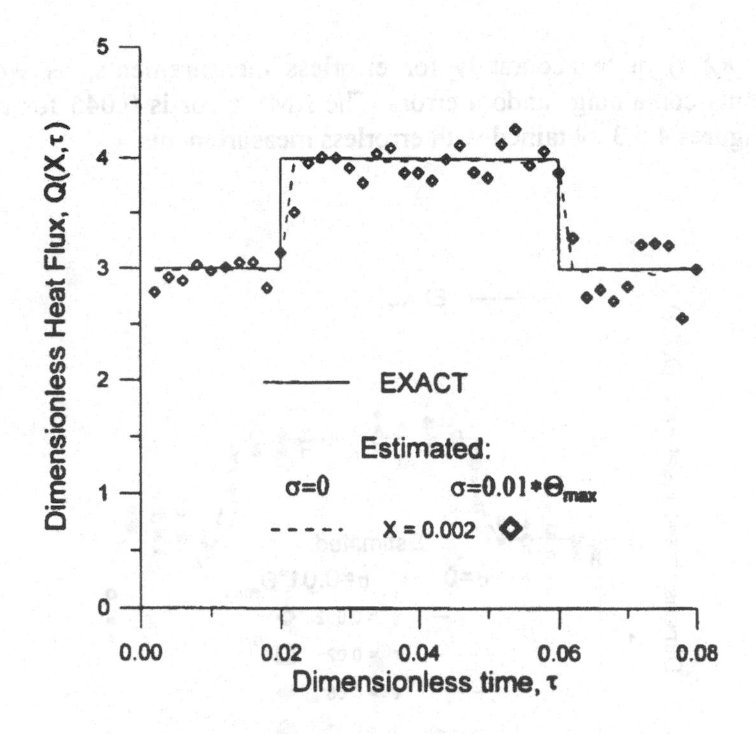

Figure 4.5.2.c - Inverse problem solution for $X = 0.002$ obtained with 21 sensors. Step variation in time given by equation (4.5.13).

Figures 4.5.3.a-c present the results obtained for a heat flux with a step variation in X and with a triangular variation in time, in the form:

$$Q_X(X) = \begin{cases} 1, & \text{for } X \leq 0.001 \text{ and } X \geq 0.003 \\ 2, & \text{for } 0.001 < X < 0.003 \end{cases} \tag{4.5.15}$$

$$Q_\tau(\tau) = \begin{cases} 1, & \text{for } \tau \leq 0.02 \text{ and } \tau \geq 0.06 \\ 50\tau, & \text{for } 0.02 < \tau \leq 0.04 \\ -50\tau + 4, & \text{for } 0.04 < \tau < 0.06 \end{cases} \tag{4.5.16}$$

where the dashed lines show the results obtained with errorless measurements and the symbols show the results obtained with measurements with a standard deviation of $\sigma = 0.01\ \Theta_{max}$. The 21 sensors used for this case are located at $Y^* = 0.95$ and at the same axial positions as for the case shown in Figures 4.5.2. Figure 4.5.3.a shows the axial variation of $Q(X,\tau)$ for different times that correspond to $Q_\tau(\tau) = 1$, as given by equations (4.5.16). Similarly, Figure 4.5.3.b shows the axial variation of $Q(X,\tau)$ for $\tau = 0.04$, when $Q_\tau(\tau)$ has a peak, i.e., $Q_\tau(\tau) = 2$ as given by equation (4.5.16). In figure 4.5.3.c, we have the results for the variation of $Q(X,\tau)$ in time for three different axial positions, so that, in accordance with equations (4.5.15) we have $Q_X(0.0004) = Q_X(0.0036) = 1$ and $Q_X(0.002) = 2$. As for the case presented in figures 4.5.2, figures 4.5.3 show that the present function estimation approach is capable of recovering the unknown heat flux $Q(X,\tau)$ quite accurately for errorless measurements, as well as for measurements containing random errors. The RMS error is 0.045 for the results shown in figures 4.5.3, obtained with errorless measurements.

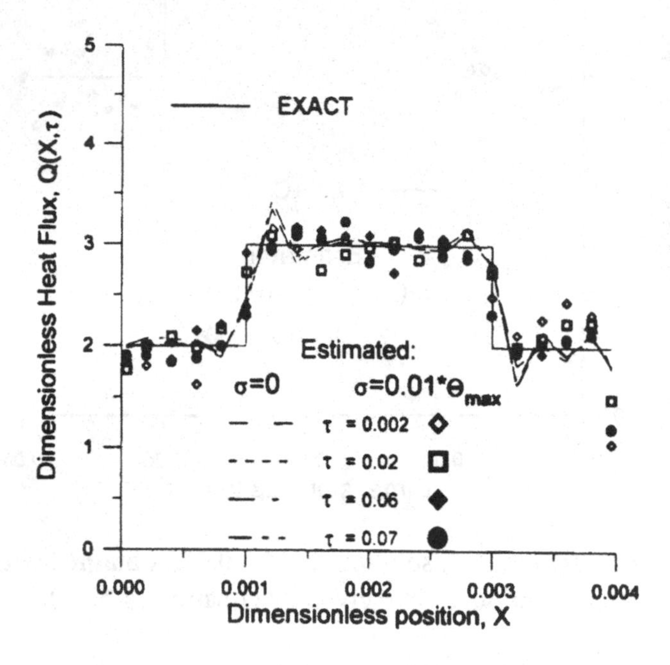

Figure 4.5.3.a - Inverse problem solution for different times obtained with 21 sensors. Step variation in the axial direction given by equation (4.5.15).

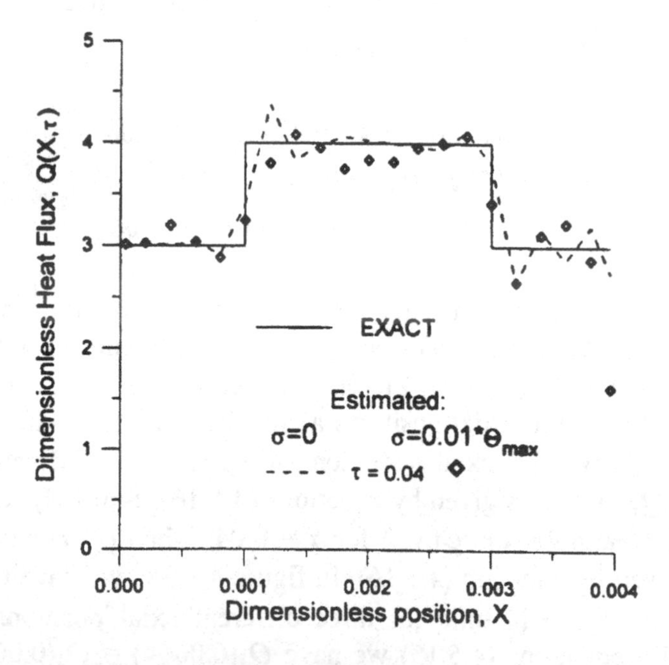

Figure 4.5.3.b - Inverse problem solution for $\tau = 0.04$ obtained with 21 sensors. Step variation in the axial direction given by equation (4.5.15).

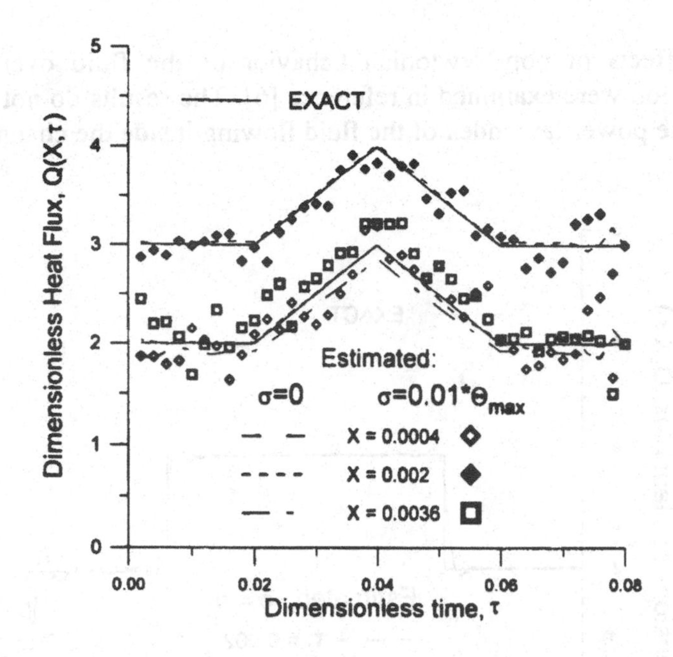

Figure 4.5.3.c - Inverse problem solution for different axial positions obtained with 21 sensors. Triangular variation in time given by equation (4.5.16).

The results shown above in figures 4.5.2 and 4.5.3 can be generally improved by using more measurements in the inverse analysis. Let's consider, for example, the estimation of the axial variation of $Q(X,\tau)$ shown in Figure 4.5.3.a. In Figure 4.5.4, we present the estimation of $Q(X,\tau)$ for the same case studied in Figure 4.5.3.a, but using the errorless measurements of 101 sensors instead of 21. The sensors were equally spaced along the channel length and at $Y = 0.95$. The time frequency of measurements was considered to be the same as for Figure 4.5.3.a. By comparing figures 4.5.3.a and 4.5.4, we can clearly notice the improvement in the estimation of $Q(X,\tau)$ by using more sensors along the channel. The RMS error obtained with 101 sensors is 0.013 as compared to 0.045 obtained by using 21 sensors.

For inverse heat conduction problems dealing with the estimation of a boundary condition, the sensors should be located as close to the boundary with the unknown condition as possible, in order to improve the estimation. Such is also the case for inverse convection problems. We have estimated $Q(X,\tau)$ for $Q_X(X)$ and $Q_\tau(\tau)$ given by equations (4.5.15) and (4.5.16), respectively, and by using the errorless measurements of 21 sensors located at the same axial positions as for Figures 4.5.3, but at $Y = 0.9$, instead of at $Y = 0.95$. The RMS error has increased to 0.238, as compared to 0.045 obtained with the sensors located at $Y = 0.95$.

We note in figures 4.5.2-4 that generally the agreement between the estimated solutions and the exact functional form assumed for $Q(X,\tau)$ tends to deteriorate near the final axial position and near the final time. This is due to the very small values of the gradient of the functional, equation (4.5.9), in such regions, as can be noticed by examining equations (4.5.7.d,e).

The effects of non-Newtonian behavior of the fluid over the inverse problem solution were examined in reference [6]. The results do not appear to be sensitive to the power-law index of the fluid flowing inside the channel.

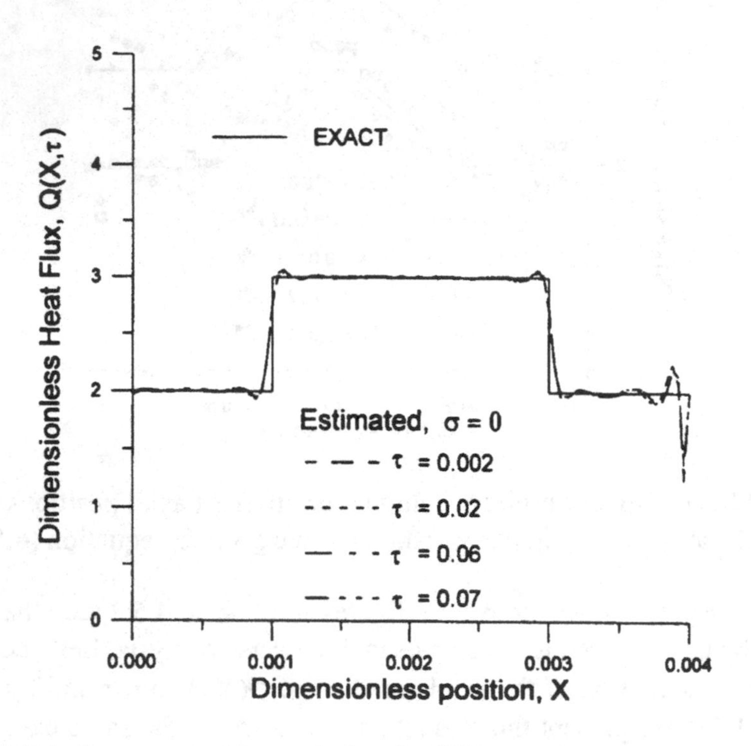

Figure 4.5.4 - Inverse problem solution for different times obtained with 101 sensors. Step variation in the axial direction given by equation (4.5.15).

PROBLEMS

4-1 Solve the inverse problem examined in section 4-1, by using a parameter estimation approach, instead of the function estimation approach. In order to do so, approximate the unknown inlet temperature profile as

$$F(Y) = \sum_{j=1}^{N} P_j C_j(Y)$$

where P_j are unknown coefficients and $C_j(Y)$ are known trial functions, given in the form of a Fourier series as

$$C_j(Y) = \cos\left[(j-1)\frac{\pi}{2}Y\right] \qquad \text{for } j=1,3,5,\ldots$$

$$C_j(Y) = \sin\left[j\frac{\pi}{2}Y\right] \qquad \text{for } j=2,4,6,\ldots$$

Try to reproduce the results shown in figure 4.1.6 by using the parameter estimation approach of Technique I. Examine the effects of number of trial functions, number of sensors, sensor locations and measurement errors on the estimated function.

4-2 Make a comparison of Techniques I and IV, as applied to the estimation of the inlet temperature profile shown in figure 4.1.6.

4-3 Consider the physical problem involving laminar hydrodynamically developed flow in a parallel plate channel. The plates, located at a distance $2h$ from each other, are maintained at a constant temperature T_i, which is also the initial temperature of the fluid inside the channel. For times $t > 0$, the inlet temperature profile $f(y,t)$ varies in time and across the channel. Formulate this forced convection problem in dimensionless form, by using the dimensionless variables given by equations (4.2.3).

4-4 Use Technique IV for the estimation of the unknown dimensionless inlet temperature profile $F(Y,\tau)$ in problem 4-3, by using measurements taken downstream. Generate the simulated measurements by assuming the unknown function to be written as the product of a function of Y by a function of τ, i.e., $F(Y,\tau)=F_Y(Y)F_\tau(\tau)$. Test the following functions for $F_Y(Y)$ and $F_\tau(\tau)$:

$$F_Y(Y)=\begin{cases} 1 & \text{for } 0<Y\le 0.3 \text{ and } 0.7\le Y<1 \\ 2 & \text{for } 0.3<Y<0.7 \end{cases}$$

$$F_Y(Y)=\begin{cases} 1 & \text{for } 0<Y\le 0.3 \text{ and } 0.7\le Y<1 \\ 5Y-0.5 & \text{for } 0.3<Y\le 0.5 \\ -5Y+4.5 & \text{for } 0.5<Y<0.7 \end{cases}$$

$$F_\tau(\tau)=\begin{cases} 1 & \text{for } 0<\tau\le 0.0006 \text{ and } 0.0014\le\tau<0.0020 \\ 2 & \text{for } 0.0006<\tau<0.0014 \end{cases}$$

$$F_\tau(\tau)=\begin{cases} 1 & \text{for } 0<\tau\le 0.0006 \text{ and } 0.0014\le\tau<0.0020 \\ 2500\tau-0.5 & \text{for } 0.0006<\tau\le 0.0010 \\ -2500\tau+4.5 & \text{for } 0.0010<\tau<0.0014 \end{cases}$$

Examine the effects of number of sensors, sensor locations and measurement errors on the estimated functions.

4-5 Solve the inverse problem examined in section 4-3, by using a parameter estimation approach, instead of the function estimation approach. Approximate the unknown boundary heat flux by a Fourier series, similarly to the estimation of the inlet temperature profile in problem 4-1 above. Make a comparison of Techniques I and IV, as applied to the functional forms shown in figures 4.3.2 and 4.3.4. Examine the effects of number of sensors, sensor locations and measurement errors on the estimated functions.

4-6 For the hydrodynamically developed flow of Non-Newtonian fluids following the power-law model for the shear stress [14], the velocity profile in a parallel plate channel of half-width h is given by:

$$u(y) = u_m \left(\frac{1+2n}{1+n} \right) \left[1 - \left(\frac{y}{h} \right)^{\frac{n+1}{n}} \right]$$

where u_m is the mean fluid velocity and n is the fluid power-law index. Solve the inverse problem of estimating the transient wall heat flux (supposed uniform along the channel) by using Technique IV. Consider available the transient readings of a single sensor for the inverse analysis. By using functional forms containing sharp corners and discontinuities to generate the simulated measured data, examine the effects of sensor location and measurement errors on the inverse problem solution. Also, verify if such solution is affected by the fluid power-law index n.

4-7 Repeat problem 4-6 by considering now the unknown boundary heat flux to be a function of time and of the axial position x, that is, $q \equiv q(x,t)$, as in section 4-5. Examine the number and position of sensors required to obtain accurate estimates for the unknown function.

4-8 Is it possible to use temperature measurements to estimate the power-law index n? Consider as the physical problem the hydrodynamically developed flow of a power-law fluid in a parallel plate channel of half-width h, subjected to constant heat flux q_0 on both walls. Assume the initial temperature of the fluid to be T_0, which is also the uniform inlet temperature.

4-9 Repeat problem 4-8 by assuming the walls to be maintained at the constant temperature T_0, instead of being supplied the heat flux q_0.

4-10 Use Techniques I and II to estimate a uniform inlet temperature T_0, by using transient temperature measurements taken downstream in a parallel plate channel. Assume hydrodynamically develop laminar flow of a Newtonian fluid inside the channel, which is subjected to a constant wall heat flux q_0 on both boundaries. Utilize the concepts of design of optimum experiments discussed in Note 2 of Chapter 2, for determining experimental variables, such as the sensor location and duration of the experiment, for the estimation of T_0.

4-11 Is the solution of forced convection inverse problems, in which the flow is not hydrodynamically developed, more involved than those inverse convection problems considered so far? Why?

4-12 Consider the heating of a fluid in the entry region of a parallel plate channel of width $2h$, subjected to the flux distribution $q(x,t)$ on both walls. The inlet and initial fluid temperature is T_0. The velocity profile at the channel inlet is uniform and parallel to the walls. Give the mathematical formulation of such physical problem.

4-13 Derive all the basic steps for the estimation of the boundary heat flux $q(x,t)$ in problem 4-12, by using Technique IV.

4-14 Why are inverse free convection problems difficult to solve?

4-15 Consider a fluid inside a square cavity of side a. The boundaries at $y = 0$ and $y = a$ are supposed insulated, while the boundary at $x = a$ is supposed to be maintained at the constant temperature T_0, which is also the initial temperature of the fluid. The heat flux distribution at $x = 0$ is given by $q(y,t)$. Give the mathematical formulation of such physical problem.

4-16 Derive all the basic steps for the estimation of the boundary heat flux $q(y,t)$ in problem 4-15, by using Technique IV.

REFERENCES

1. Liu, F. B. and Özisik, M. N., "Estimation of Inlet Temperature Profile in Laminar Duct Flow", *Inverse Problems in Engineering*, **3**, 131-141, 1996.
2. Bokar, J. C. and Özisik, M. N., "Inverse Analysis for Estimating the Time Varying Inlet Temperature in Laminar Flow Inside a Parallel Plate Duct", *Int. J. Heat Mass Transfer*, **38**, 39-45, 1995.
3. Huang, C. H. and Özisik, M. N., "Inverse Problem of Determining Unknown Wall Heat Flux in Laminar Flow Through a Parallel Plate", *Numerical Heat Transfer, Part. A*, **21**, 55-70, 1992.
4. Liu, F. B. and Özisik, M. N., "Inverse Analysis of Transient Turbulent Forced Convection Inside Parallel Plates", *Int. J. Heat Mass Transfer*, **39** 2615-2618, 1996.
5. Machado, H. A. and Orlande, H. R. B., "Inverse Analysis of Estimating the Timewise and Spacewise Variation of the Wall Heat Flux in a Parallel Plate Channel", *Int. J. Num. Meth. Heat & Fluid Flow*, **7**, 696-710, 1997.
6. Machado, H. A. and Orlande, H. R. B., "Estimation of the Timewise and Spacewise Variation of the Wall Heat Flux to a Non-Newtonian Fluid in a Parallel Plate Channel", *Proceedings of the Int. Symp. On Transient Convective Heat Transfer*, 587-596, Cesme, Turkey, August, 1996.
7. Machado, H. A. and Orlande, H. R. B., "Inverse Problem for Estimating the Heat Flux to a Non-Newtonian Fluid in a Parallel Plate Channel", *Journal of the Brazilian Society of Mechanical Sciences*, **20**, 51-61, 1998.
8. Moutsouglou, A., "An Inverse Convection Problem", *J. Heat Transfer*, **111**, 37-43, 1989.
9. Moutsouglou, A., "Solution of an Elliptic Inverse Convection Problem Using a Whole Domain Regularization Technique", *AIAA J. Thermophysics*, **4**, 341-349, 1990.
10. Raghunath, R., "Determining Entrance Conditions From Downstream Measurements", *Int. Comm. Heat Mass Transfer*, **20**, 173-183, 1993.
11. Li, M., Prud'homme, M. and Nguyen, T. "A Numerical Solution for the Inverse Natural - Convection Problem", *Numerical Heat Transfer, Part B*, 307-321, 1995.

12. Szczygiel, I., "Estimation of the Boundary Conditions in Conventional Heat Transfer Problems", *in Inverse Problems in Heat Transfer and Fluid Flow., vol. 2, HTD - Vol. 340*, 17-24, G.S. Dulikravich and K.A. Woodburry (eds.) , ASME, 1997.
13. Moaveni, S. "An Inverse Problem Involving Thermal Energy Equation", *in Inverse Problems in Heat Transfer and Fluid Flow.*, vol. 2, *HTD - Vol. 340*, 49-54, G. S. Dulikravich and K. A. Woodburry (eds.), ASME, 1997.
14. Bird, R. B. Stewart, W. E. and Lightfoot, E. N., *Transport Phenomena*, John Wiley & Sons, New York, 1960.
15. Kakac, S., Shah, R. and Aung, W. (eds.), *Handbook of Single-Phase Convection Heat Transfer*, John Wiley & Sons, New York, 1987.
16. Schlichting, H., *Boundary-Layer Theory*, McGraw Hill, New York, 1979.
17. Kays, W. M. and Crawford, M. E., *Convective Heat Transfer*, 3rd ed., McGraw Hill, New York, 1993.
18. Kim, W. S. and Özisik, M. N., "Turbulent Forced Convection Inside a Parallel Plate Channel with Periodic Variation of Inlet Temperature", *J. Heat Transfer*, **111**, 882-888, 1989.
19. Cotta, R. M. and Özisik, M. N. "Laminar Forced Convection to Non-Newtonian Fluids in Ducts with Prescribed Wall Heat Flux", *International Comm. of Heat & Mass Transfer*, **13**, 325-334, 1986.

Chapter 5

INVERSE RADIATION

In the study of radiation heat transfer, a distinction is made between radiation transfer as a *surface phenomenon* and as a *bulk phenomenon*. For opaque materials such as metals, woods, rocks, etc, the radiation emitted or absorbed by the body is said to originate from the immediate vicinity of the surface (i.e., within about 1 μm), hence the radiation transport is regarded as a surface phenomenon. In the case of semi-transparent materials such as glass, salt, crystals and gases at elevated temperatures, the emission or absorption of radiation occurs at all depths within the medium. Hence the radiation problem is considered as a bulk phenomenon.

A semi-transparent medium may *scatter radiation* in addition to absorbing and emitting it. That is, when a beam of radiation strikes a semi-transparent body, some of the incident beam is reflected from the surface, the remaining portion penetrates into the medium, where part of the radiation energy is absorbed by the body, and the remaining portion passes out through the medium, if the medium is not a strong absorber. The scattering of radiation is important in porous particulate media, like powders and foams, which are widely used in industrial high technology applications and in thermal insulation. The radiation scattering properties of such materials are characterized by a *spectral scattering coefficient* σ_λ and a *phase function* $p\left(\hat{\Omega}\cdot\hat{\Omega}'\right)$, where $\hat{\Omega}$ and $\hat{\Omega}'$ denote the directions of the incident and scattered radiation beams, respectively. Then, the dot product $\hat{\Omega}.\hat{\Omega}'$ is the cosine of the angle between the scattered and the incident rays. To characterize the radiation absorption characteristics of the medium, a *spectral radiation absorption coefficient* κ_λ is introduced. When the medium is in local thermodynamic equilibrium and Kirchoff law is valid, the spectral absorption coefficient also characterizes the spectral emission coefficient. The sum of the spectral scattering and absorption coefficients is called the *spectral extinction coefficient* β_λ, i.e.

$$\beta_\lambda = \kappa_\lambda + \sigma_\lambda$$

and the ratio of the scattering to extinction coefficient, i.e.,

$$\omega_\lambda = \frac{\sigma_\lambda}{\beta_\lambda} \quad , \quad \text{for} \quad 0 \le \omega_\lambda \le 1$$

is called the *spectral single scattering albedo.*

The limiting case of $\omega_\lambda = 0$ characterizes a medium that completely absorbs the incident radiation at the wavelength λ, whereas $\omega_\lambda = 1$ characterizes a medium which completely scatters radiation of the wavelength λ.

A fundamental quantity in the study of radiation transfer in participating media is the spectral radiation intensity, $I_\lambda(s,\hat{\Omega})$, where $\hat{\Omega}$ is the direction of propagation and s is the path of propagation. It represents the flow of radiation energy per unit area normal to the direction of propagation of the radiation beam, per unit wavelength, per unit solid angle, per unit time. If the energy per unit time is measured in Watt, the wavelength λ is measured in μm, and the solid angle in steradian, sr, then the dimension of the spectral radiation intensity, $I_\lambda(s,\hat{\Omega})$, becomes

$$W / \left(m^2 . \mu m . sr\right)$$

where the area is measured perpendicular to the direction of propagation of the radiation beam.

In the study of radiation transfer, the radiation intensity is the fundamental quantity, which is obtained from the solution of the *Equation of Radiative Transfer* (ERT). In the study of conduction or convection, the temperature T of the body is the fundamental quantity, which is obtained from the solution of the standard energy equation. The radiation intensity being a directional quantity, its determination from the solution of the equation of radiative transfer is a much more difficult matter than the determination of temperature T from the solution of the standard energy equation. Therefore, a considerable amount of effort has been devoted to the solution of ERT. During the past three decades a variety of numerical, exact analytic, approximate analytic solutions of ERT have been reported in literature. The reader should consult references [1-7] for in-depth discussion of the derivation of ERT and its solution with various techniques. In the solution of an *inverse problem of radiation in participating media*, the solution of the *direct radiation problem* is needed; it is most important that such solution be highly accurate. Here we present a brief discussion of the commonly used techniques for the solution of the Equation of Radiative Transfer.

Methods for Solving the Equation of Radiative Transfer

The Discrete Ordinates Method: This method, first proposed by Chandrasekhar [6], was adapted by Hyde and Truelove [8] and Fiveland [9] for solving radiation problems of participating media. Since then, numerous applications of this method have appeared in the literature [8-20]. The method appears to be very promising for solving complex radiation transfer problems encountered in combustors, heaters and furnaces [7,14]

The Spherical Harmonics Method (P_N approximation): This method, originally proposed by Jeans [21] in connection with radiation transfer in stars, transforms the equation of radiative transfer into a set of simultaneous partial differential equations. A detailed description of this method can be found in references [1,22-26]. A shortcoming of the method is that low order approximations are accurate only for optically thick media.

The Galerkin Method: This method is specially suitable for solving one-dimensional radiation problems of absorbing, emitting and isotropically scattering media. The method can accommodate problems of spatially varying albedo, $\omega(x)$, and anisotropic medium. The applications of the method can be found in references [27-33].

Simple Differential Approximations: The equation of radiative transfer can be transformed into a simple ordinary differential equation for the determination of the net radiation heat flux; but the accuracy of such simple solutions is generally very poor and is not recommended for use in the inverse analysis. One such approximation is the *Schuster-Schwarzchild* (or the *two-flux*) approximation and the other is the *Eddington approximation* [1,3,5]. To improve the accuracy of the differential approximation, Modest [5] proposed a *modified differential approximation*. However, when applying such approximations, care must be exercised in order to stay within the range of validity of the model.

The Zonal Method: Developed by Hottel [4] for heat transfer in furnaces, it approximates the spatial behavior by separating the medium into a finite number of isothermal sub-volumes and surface area zones. An energy balance is then performed for radiative exchange between any two zones. The procedure leads to a set of simultaneous equations for the determination of unknown temperatures or heat fluxes.

One-Dimensional Equation of Radiative Transfer

In this chapter we present inverse radiation problems for an absorbing, emitting, isotropically scattering plane-parallel medium and for a medium with spherical symmetry. Therefore, we first present here the governing equations of radiative transfer applicable for such situations involving gray media. For the plane-parallel medium it is given by:

$$\mu \frac{\partial I(\tau,\mu)}{\partial \tau} + I(\tau,\mu) = g(\tau) + \frac{\omega}{2}\int_{-1}^{1} I(\tau,\mu')d\mu' \qquad 0<\tau<\tau_0 \qquad -1<\mu<1$$

$$g(\tau) = (1-\omega)\frac{\bar{n}^2 \bar{\sigma} T^4(\tau)}{\pi}$$

and for the case of spherical symmetry we have:

$$\mu \frac{\partial I(r,\mu)}{\partial r} + \frac{1}{r}(1-\mu^2)\frac{\partial I(r,\mu)}{\partial \mu} + I(r,\mu) = g(r) + \frac{\omega}{2}\int_{-1}^{1} I(r,\mu')d\mu'$$

$$g(r) = (1-\omega)\frac{\bar{n}^2 \bar{\sigma} T^4(r)}{\pi}$$

Here, I is the radiation intensity, T is the temperature, $\bar{n}$ is the refractive index, $\bar{\sigma}$ is the Stefan-Boltzmann constant, ω is the single scattering albedo, τ and r are the optical variables and μ is the cosine of the angle between the direction of the radiation intensity and the positive τ axis.

Inverse radiation problems of participating media arise in a variety of engineering applications, including, among others, remote sensing of the atmosphere, estimation of the temperature profile in combustion systems, and the estimation of radiation properties κ, β, σ or ω of the participating medium. In most cases, it is desirable to avoid the use of detectors within the medium. In such cases, inverse analysis allows the use of exit radiation intensities for estimating the unknown radiation properties or the radiation source term in the medium. Typical applications of the estimation of radiation properties of porous materials, such as fiberglass and foam insulation, as well as the estimation of the phase function for packed sphere systems, can be found in the references [34-38].

To illustrate the applications of the inverse radiation technique, we consider in this chapter the solution of the following three distinct radiation problems:

- Estimation of unknown temperature profile in an absorbing, emitting, isotropically scattering plane-parallel medium by utilizing the measured exit intensities [39].
- Simultaneous estimation of temperature profile and surface reflectivity by utilizing the measured exit radiation intensities [40].
- Estimation of radiation source term in a solid semi transparent gray sphere, from the measured exit radiation intensities [41].

We present the solutions of such inverse problems by using a parameter estimation approach. Therefore, the sensitivity coefficients are required for the solution procedures. We illustrate here the computation of the sensitivity coefficients through the solution of ***sensitivity problems***, as outlined in section 2.1.

The solution of other inverse radiation problems of interest can be found in references [42-46].

5-1 IDENTIFICATION OF THE TEMPERATURE PROFILE IN AN ABSORBING, EMITTING AND ISOTROPICALLY SCATTERING MEDIUM [39]

The inverse radiation problem considered here is concerned with the estimation of the unknown temperature source term $g(\tau) \equiv g[T(\tau)]$ in an absorbing, emitting, isotropically scattering plane-parallel gray plate of optical thickness τ_0, from the knowledge of the measured exit radiation intensities $Y(\mu)$ at the boundary surface $\tau = 0$ and $Z(\mu)$ at the boundary surface $\tau = \tau_0$, as illustrated in figure 5.1.1.

Technique II, the conjugate gradient method of minimization, is used for solving the present inverse radiation problem. The basic steps in the solution include the followings: direct problem, inverse problem, sensitivity problem, iterative procedure, stopping criterion and computational algorithm. Details of such steps are described next.

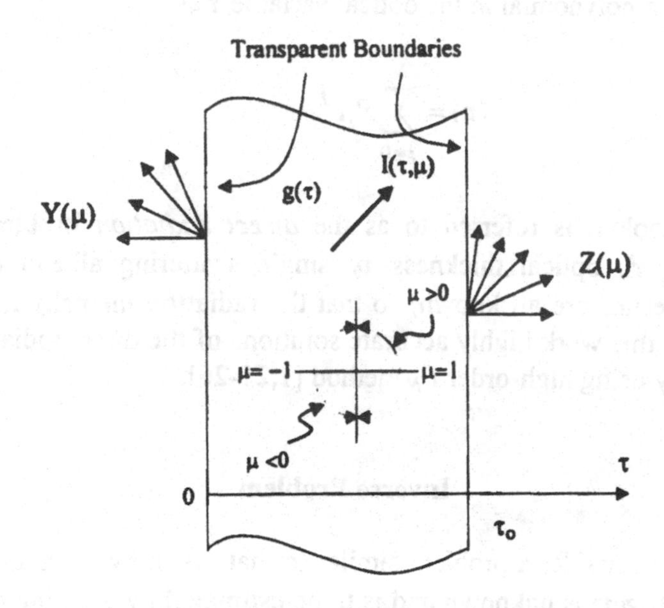

Figure 5.1.1 - Geometry and coordinates.

Direct Problem

The mathematical formulation of the direct problem associated with the inverse problem considered here is given by

$$\mu \frac{\partial I(\tau,\mu)}{\partial \tau} + I(\tau,\mu) = g(\tau) + \frac{\omega}{2} \int_{-1}^{1} I(\tau,\mu') d\mu' \qquad 0 < \tau < \tau_0 \qquad -1 < \mu < 1 \tag{5.1.1.a}$$

where
$$g(\tau) = (1-\omega) \frac{\bar{n}^2 \bar{\sigma} T^4(\tau)}{\pi} \tag{5.1.1.b}$$

with boundary conditions

$$I(0,\mu) = 0 \qquad \mu > 0 \tag{5.1.1.c}$$

and

$$I(\tau_0, -\mu) = 0 \qquad \mu > 0 \tag{5.1.1.d}$$

Such boundary conditions represent transparent boundary surfaces with no incident radiation.

The source term involving the fourth power of the temperature is represented by a polynomial in the optical variable τ as

$$g(\tau) = \sum_{j=0}^{N} P_j \tau^j \tag{5.1.2}$$

This problem is referred to as the *direct radiation* problem when the source term $g(\tau)$, optical thickness τ_0, single scattering albedo ω and other radiation properties are all known, so that the radiation intensity $I(\tau,\mu)$ is to be determined. In this work highly accurate solutions of the direct radiation problem are obtained by using high-order P_N method [1,21-26].

Inverse Problem

We now consider a problem similar to that given by equations (5.1.1) but the source term $g(\tau)$ is unknown and is to be estimated by utilizing the *measured exit radiation* intensities.

The problem defined by equations (5.1.1) with the source term or the temperature unknown, but measured exit intensities known, is an *inverse problem* which can be solved by the minimization of the following objective function:

$$S(\mathbf{P}) = \int_{-1}^{0}[I(0,\mu;\mathbf{P}) - Y(\mu)]^2\, d\mu + \int_{0}^{1}[I(\tau_0,\mu;\mathbf{P}) - Z(\mu)]^2\, d\mu \qquad (5.1.3)$$

where $Y(\mu)$ and $Z(\mu)$ are the measured exit radiation intensities at the surfaces $\tau = 0$ and $\tau = \tau_0$, respectively, while $I(0,\mu,\mathbf{P})$ and $I(\tau_0,\mu,\mathbf{P})$ are the estimated exit radiation intensities at the surfaces $\tau = 0$ and $\tau = \tau_0$, respectively, obtained from the solution of the direct problem (5.1.1), by using an estimate for the vector of unknown parameters **P**. Since the unknown source term $g(\tau)$ is approximated by equation (5.1.2), the inverse radiation problem is reduced to an estimation problem in $(N+1)$ dimensional space. We note that a sufficiently large number of measurements is considered available, so that they are assumed as continuous functions $Y(\mu)$ at $\tau = 0$ and $Z(\mu)$ at $\tau = \tau_0$.

The objective function given by equation (5.1.3) is minimized by differentiating $S(\mathbf{P})$ with respect to each of the unknown coefficients P_j, $j = 0,\ldots,N$. Then, the resulting expression for $\partial S(\mathbf{P})/\partial P_j$ contains the ***sensitivity coefficients*** $\partial I/\partial P_j$, which can be determined from the solution of the sensitivity problem developed as described below.

Sensitivity Problem

The differentiation of the direct problem given by equations (5.1.1) with respect to P_j results in the following ***sensitivity problem*** for the determination of the sensitivity coefficients $\partial I/\partial P_j$:

$$\mu\frac{\partial}{\partial\tau}\left(\frac{\partial I(\tau,\mu)}{\partial P_j}\right) + \left(\frac{\partial I(\tau,\mu)}{\partial P_j}\right) = \tau^j + \frac{\omega}{2}\int_{-1}^{1}\left(\frac{\partial I(\tau,\mu')}{\partial P_j}\right)d\mu' \quad 0<\tau<\tau_0, -1<\mu<1$$

(5.1.4.a)

$$\left(\frac{\partial I(0,\mu)}{\partial P_j}\right) = 0 \qquad \mu > 0 \qquad (5.1.4.b)$$

$$\left(\frac{\partial I(\tau_0,-\mu)}{\partial P_j}\right) = 0 \qquad \mu > 0 \qquad (5.1.4.c)$$

for $j = 0, 1, \ldots, N$. The vector

$$[\nabla I]^T = \left[\frac{\partial I}{\partial P_0}, \frac{\partial I}{\partial P_1}, \ldots, \frac{\partial I}{\partial P_N}\right] \qquad (5.1.5)$$

is the sensitivity coefficient vector which can be determined from the solution of the sensitivity problem. Clearly, the solution procedure for this problem is the same as that for the direct problem (5.1.1), with $g(\tau)$ replaced by τ^j. We note that the present parameter estimation problem is linear, since the sensitivity problem is independent of the unknown parameters P_j, $j = 0,...,N$. Therefore, it only needs to be solved once, as it will be apparent later in the computational algorithm.

Now, we develop an expression for the components of the gradient $\partial S(\mathbf{P})/\partial P_j$, by differentiating $S(\mathbf{P})$ given by equation (5.1.3) with respect to P_j. We obtain

$$\frac{\partial S(\mathbf{P})}{\partial P_j} = \int_{-1}^{0} 2[I(0,\mu;\mathbf{P}) - Y(\mu)]\frac{\partial I(0,\mu;\mathbf{P})}{\partial P_j}d\mu + \int_{1}^{0} 2[I(\tau_0,\mu;\mathbf{P}) - Z(\mu)]\frac{\partial I(\tau_0,\mu;\mathbf{P})}{\partial P_j}d\mu \tag{5.1.6.a}$$

for $j = 0,1,...,N$, where the vector

$$[\nabla S(\mathbf{P})]^T = \left[\frac{\partial S}{\partial P_0}, \frac{\partial S}{\partial P_1}, \ldots, \frac{\partial S}{\partial P_N}\right] \tag{5.1.6.b}$$

is the gradient of the objective function.

Iterative Procedure

To determine the unknown vector **P**, we consider the iterative procedure of Technique II, and write

$$\mathbf{P}^{k+1} = \mathbf{P}^k - \beta^k \mathbf{d}^k \tag{5.1.7}$$

where β^k is the *step size* and $\mathbf{d}^k$ is the *direction of descent* at the kth iteration. Here, $\mathbf{d}^k$ is determined from

$$\mathbf{d}^k = \nabla S(\mathbf{P}^k) + \gamma^k \mathbf{d}^{k-1} \tag{5.1.8}$$

where the conjugation coefficient γ^k is computed here with the Fletcher-Reeves expression

$$\gamma^k = \frac{\left[\nabla S(\mathbf{P}^k)\right]^T \left[\nabla S(\mathbf{P}^k)\right]}{\left[\nabla S(\mathbf{P}^{k-1})\right]^T \left[\nabla S(\mathbf{P}^{k-1})\right]} \qquad \text{for } k=1,2,... \quad \text{with } \gamma^0=0 \tag{5.1.9}$$

The step size β^k in going from iteration k to $k+1$ is determined from the condition $\min_{\beta^k} S(\mathbf{P}^{k+1})$ or $\min_{\beta^k} S(\mathbf{P}^k - \beta^k \mathbf{d}^k)$, that is,

$$\min_{\beta^k} S(\mathbf{P}^{k+1}) = \min_{\beta^k} \left\{ \int_{-1}^{0} [I(0,\mu;\mathbf{P}^k - \beta^k \mathbf{d}^k) - Y(\mu)]^2 \, d\mu + \right.$$

$$\left. + \int_{0}^{1} [I(\tau_0,\mu;\mathbf{P}^k - \beta^k \mathbf{d}^k) - Z(\mu)]^2 \, d\mu \right\} \tag{5.1.10}$$

By using a Taylor series expansion and performing the minimization, as described in Note 3 in Chapter 2, we obtain the following expression for the search step size

$$\beta^k = \frac{\Delta_1}{\Delta_2} \tag{5.1.11.a}$$

where

$$\Delta_1 = \int_{-1}^{0} [I(0,\mu;\mathbf{P}^k) - Y(\mu)][\nabla I(0,\mu;\mathbf{P}^k)]^T \mathbf{d}^k \, d\mu +$$
$$\int_{0}^{1} [I(\tau_0,\mu;\mathbf{P}^k) - Z(\mu)][\nabla I(\tau_0,\mu;\mathbf{P}^k)]^T \mathbf{d}^k \, d\mu \tag{5.1.11.b}$$

$$\Delta_2 = \int_{-1}^{0} \left\{ [\nabla I(0,\mu;\mathbf{P}^k)]^T \mathbf{d}^k \right\}^2 d\mu + \int_{0}^{1} \left\{ [\nabla I(\tau_0,\mu;\mathbf{P}^k)]^T \mathbf{d}^k \right\}^2 d\mu \tag{5.1.11.c}$$

Stopping Criterion

Once $\mathbf{d}^k$ is calculated from equation (5.1.8) and β^k from equation (5.1.11.a), we can use the iterative procedure given by equation (5.1.7) to obtain new estimates $\mathbf{P}^{k+1}$, until a stopping criterion based on the *discrepancy principle* is satisfied. Such stopping criterion is given by

$$S\left(\mathbf{P}^{k+1}\right) < 2\sigma^2 \tag{5.1.12}$$

where σ is the standard deviation of the measurement errors.

Computational Algorithm

The computational algorithm for Technique II, as applied to the present inverse radiation problem, can be summarized as follows: Assume $\mathbf{P}^k$ is known at the kth iteration, then

Step 1. Solve the sensitivity problem given by equations (5.1.4) and compute the sensitivity coefficient vector ∇I, given by equation (5.1.5).

Step 2. Solve the direct problem given by equations (5.1.1) and compute the exit intensities $I(0,\mu;\mathbf{P}^k)$ and $I(\tau_0,\mu;\mathbf{P}^k)$ at the surfaces $\tau = 0$ and $\tau = \tau_0$, respectively.

Step 3. Check the stopping criterion given by equation (5.1.12). Continue if not satisfied.

Step 4. Knowing ∇I, $I(0,\mu;\mathbf{P}^k)$, $I(\tau_0,\mu;\mathbf{P}^k)$ and the measured exit intensities $Y(\mu)$ and $Z(\mu)$, compute the gradient $\nabla S(\mathbf{P}^k)$ from equation (5.1.6.b).

Step 5. Knowing $\nabla S(\mathbf{P}^k)$, compute γ^k from equation (5.1.9); then compute the direction of descent $\mathbf{d}^k$ from equation (5.1.8).

Step 6. Knowing ∇I, $I(0,\mu;\mathbf{P}^k)$, $I(\tau_0,\mu;\mathbf{P}^k)$, $Y(\mu)$, $Z(\mu)$ and $\mathbf{d}^k$, compute the step size β^k from equation (5.1.11).

Step 7. Knowing β^k and $\mathbf{d}^k$, compute $\mathbf{P}^{k+1}$ from equation (5.1.7) and return to step 2.

The reader should notice that the sensitivity problem is only solved once in Step 1 of the above algorithm, since the present parameter estimation problem is linear.

Results

Numerical results are now presented in order to illustrate that the computational procedure used here works well and produces results which are exact when the simulated measured data contain no measurement errors. The results become less accurate as the standard deviation of the measured data is increased.

To illustrate the feasibility of this approach under conditions encountered in fires and furnaces, we considered temperatures ranging from 800 K to 1800 K [7] and single scattering albedo varying from $\omega = 0.2$ to $\omega = 0.35$. Also, we represented the source term $g(\tau)$ as a polynomial of degree four in the optical variable τ in the form

$$g(\tau) = 1 + 20\tau + 44\tau^2 - 128\tau^3 + 64\tau^4 \quad \mathrm{W/cm^2} \tag{5.1.13}$$

as well as in a sinusoidal variation in the form

$$g(\tau) = 5 + 3\sin(2\pi\tau) \quad \text{W/cm}^2 \tag{5.1.14}$$

Simulated measured exit intensities $Y(\mu)$ and $Z(\mu)$ containing errors were generated by adding random errors of standard deviation σ to the exact exit intensities, computed from the solution of the direct problem.

Figures 5.1.2 and 5.1.3 show the results obtained with errorless measurements ($\sigma = 0$) and measurements containing random errors ($\sigma = 0.03$), respectively, for the polynomial variation of the source term. We note in figure 5.1.2 that the exact function is perfectly recovered when errorless measurements are used in the analysis. The results obtained with measurements containing random errors are also quite accurate, as illustrated in figure 5.1.3.

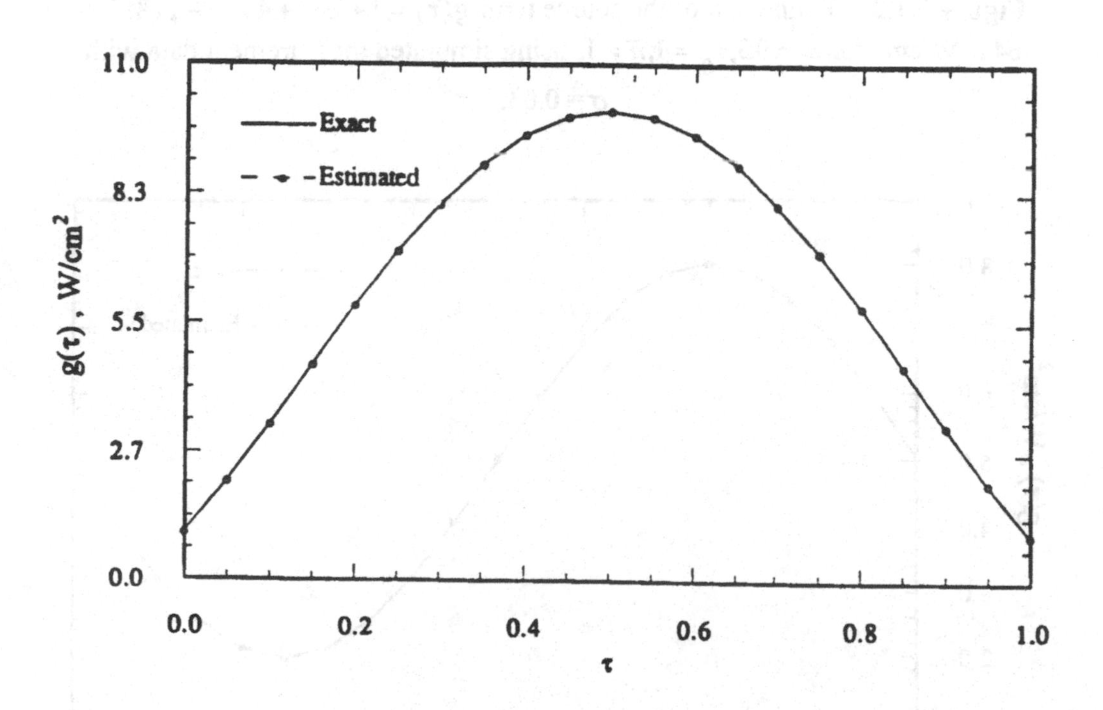

Figure 5.1.2 - Estimation of the source term $g(\tau) = 1 + 20\tau + 44\tau^2 - 128\tau^3 + 64\tau^4$ W/cm² for $\omega = 0.3, \tau_0 = 1, \bar{n} = 1$, using simulated exact measurement data with $\sigma = 0$.

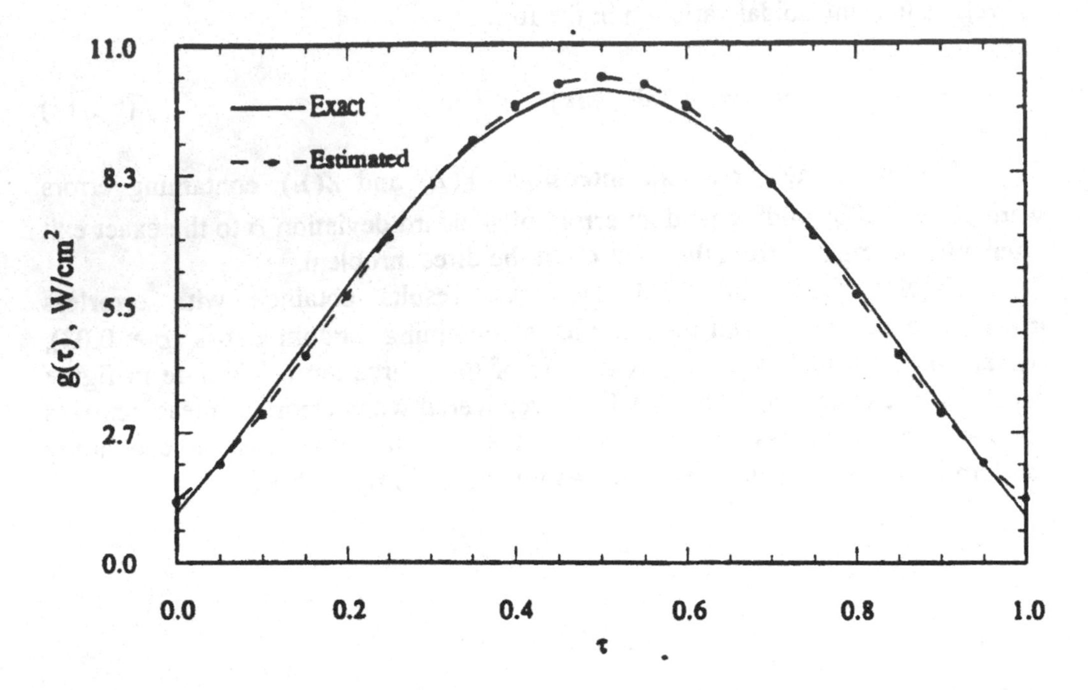

Figure 5.1.3 - Estimation of the source term $g(\tau) = 1 + 20\tau + 44\tau^2 - 128\tau^3 + 64\tau^4$ W/cm² for $\omega = 0.3, \tau_0 = 1, \bar{n} = 1$, using simulated measurement data with $\sigma = 0.03$.

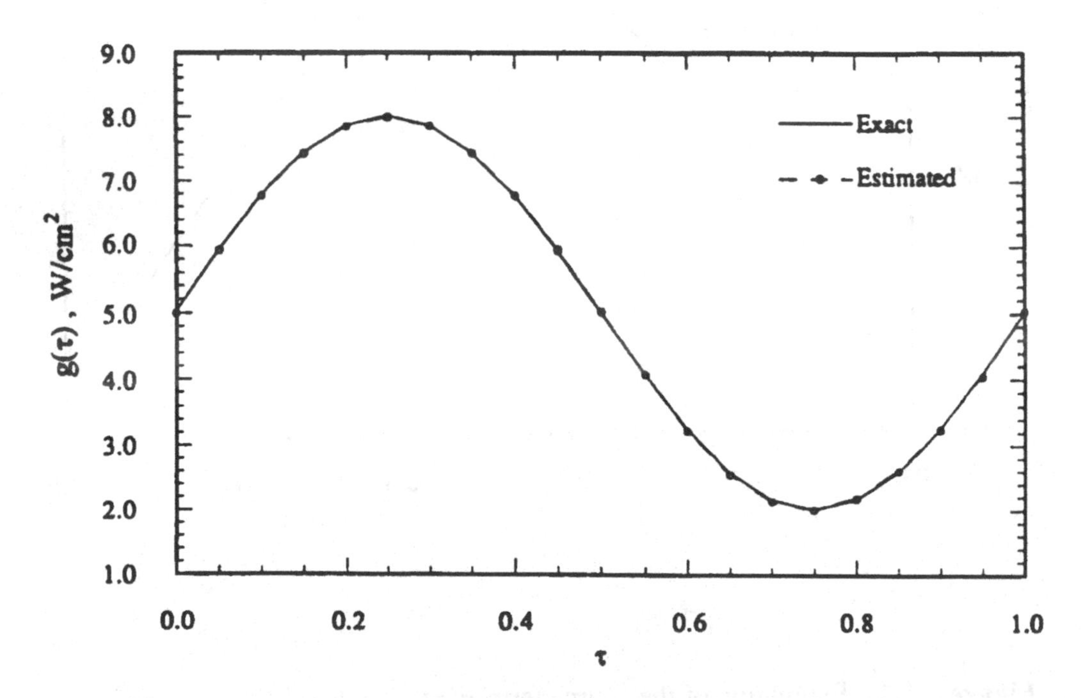

Figure 5.1.4 - Estimation of the source term $g(\tau) = 5 + 3\sin(2\pi\tau)$ W/cm² for $\omega = 0.3, \tau_0 = 1, \bar{n} = 1$, using simulated exact measurement data with $\sigma = 0$.

The results obtained for the sinusoidal variation of the source term are similar to those obtained for the polynomial variation, as presented in figures 5.1.4 and 5.1.5.

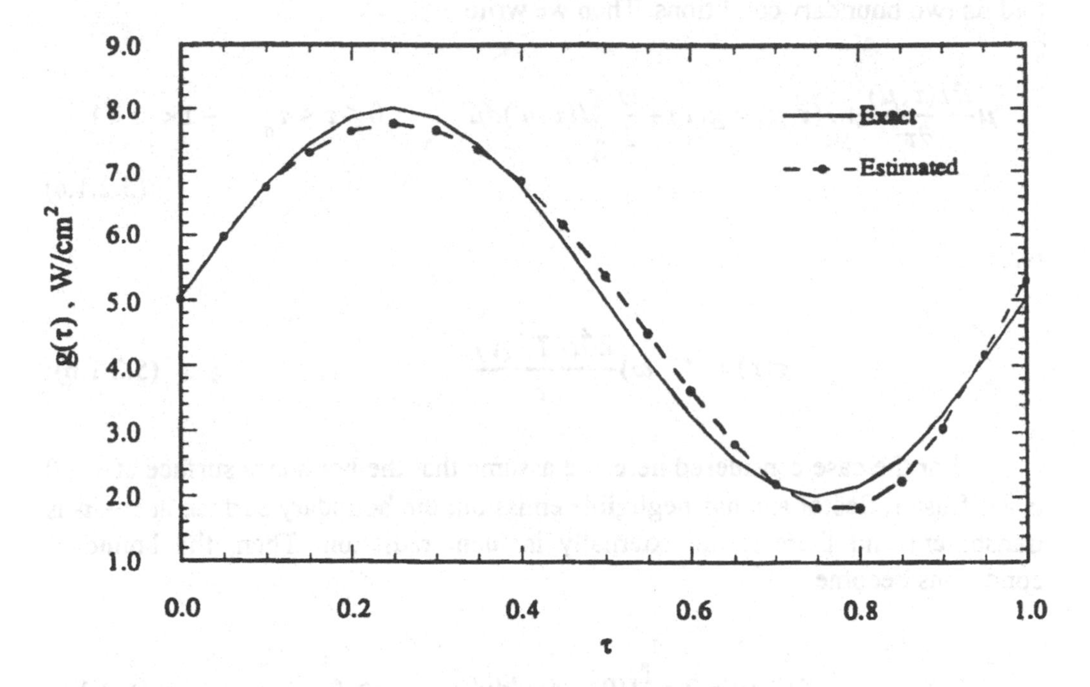

Figure 5.1.5 - Estimation of the source term $g(\tau) = 5 + 3\sin(2\pi\tau)$ W / cm^2 for $\omega = 0.3, \tau_0 = 1, \bar{n} = 1$, using simulated measurement data with $\sigma = 0.03$.

5-2 SIMULTANEOUS ESTIMATION OF TEMPERATURE PROFILE AND SURFACE REFLECTIVITY [40]

This section is concerned with simultaneous estimation of the unknown temperature distribution $T(\tau)$ and the diffuse reflectivity ρ of the boundary surface at $\tau = 0$, of an absorbing, emitting, isotropically scattering plane parallel slab of optical thickness τ_0. We assume that the exit radiation intensities at the boundary surfaces $\tau = 0$ and $\tau = \tau_0$ can be measured experimentally. Various mathematical approaches have been applied to solve the inverse radiation problems of participating media. Here we consider **Technique II**, the conjugate gradient method of minimization, to solve the inverse radiation problem and use high-order P_N method to solve the corresponding direct radiation problem [1,21-26]. The basic steps in the analysis with this approach consists of the followings: direct problem, inverse problem, sensitivity problems, iterative procedure, stopping criterion and computational algorithm. We summarize below the pertinent details of each of these basic steps.

Direct Problem

The mathematical formulation of the direct problem, associated with the inverse problem considered here, consists of the equation of radiation transfer and its two boundary conditions. Then we write

$$\mu \frac{\partial I(\tau,\mu)}{\partial \tau} + I(\tau,\mu) = g(\tau) + \frac{\omega}{2} \int_{-1}^{1} I(\tau,\mu')d\mu' \qquad 0<\tau<\tau_0 \qquad -1<\mu<1 \tag{5.2.1.a}$$

with

$$g(\tau) = (1-\omega)\frac{\bar{n}^2 \bar{\sigma} T^4(\tau)}{\pi} \tag{5.2.1.b}$$

For the case considered here, we assume that the boundary surface at $\tau = 0$ is a diffuse reflector and has negligible emission; the boundary surface at $\tau = \tau_0$ is transparent and there is no externally incident radiation. Then, the boundary conditions become

$$I(0,\mu) = 2\rho \int_{0}^{1} I(0,-\mu')\mu' d\mu' \qquad \mu > 0 \tag{5.2.1.c}$$

and

$$I(\tau_0, -\mu) = 0 \qquad \mu > 0 \tag{5.2.1.d}$$

where ρ, $0 \le \rho \le 1$, is the reflectivity of the boundary surface at $\tau = 0$.

The source term involving the fourth power of the temperature is approximated by a polynomial in the optical variable τ as

$$g(\tau) = \sum_{n=0}^{N} a_n \tau^n \tag{5.2.2}$$

Figure 5.2.1 shows the geometry and coordinates. Here N is the degree of the polynomial utilized in the approximation.

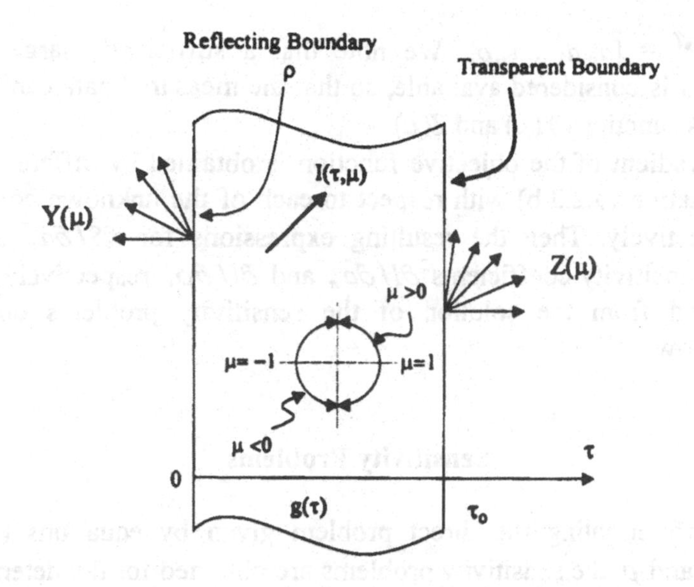

Figure 5.2.1 - Geometry and coordinates.

Inverse Problem

For the inverse problem considered here, the source term $g(\tau)$, as well as the surface reflectivity ρ, are regarded unknown. Since the source term was written in the form given by equation (5.2.2), such inverse problem involves the estimation of $(N+2)$ parameters, that is, $(N+1)$ coefficients a_n, $n = 0,\ldots,N$, of the source function and the reflectivity ρ at the surface $\tau = 0$. The vector of unknown parameters is then given by

$$\mathbf{P}^T = [a_0, a_1, \cdots, a_N, \rho] \tag{5.2.3.a}$$

Such vector is to be estimated by using measurements of the exit intensities at the surfaces $\tau = 0$ and $\tau = \tau_0$.

The estimation of the unknown source term $g(\tau)$ and the boundary surface reflectivity ρ, from the knowledge of the exit intensities measured at different directions, can be recast as a problem of minimization of the following objective function $S(\mathbf{P})$:

$$S(\mathbf{P}) = \int_{-1}^{0} [I(0,\mu;\mathbf{P}) - Y(\mu)]^2 d\mu + \int_{0}^{1} [I(\tau_0,\mu;\mathbf{P}) - Z(\mu)]^2 d\mu \tag{5.2.3.b}$$

where $Y(\mu)$ and $Z(\mu)$ are the measured exit radiation intensities at the surfaces $\tau = 0$ and $\tau = \tau_0$, respectively, while $I(0,\mu;\mathbf{P})$ and $I(\tau_0,\mu;\mathbf{P})$ are the estimated exit radiation intensities at the surfaces $\tau = 0$ and $\tau = \tau_0$, respectively, obtained from the solution of the direct problem (5.2.1) by using estimated values for the parameters $\mathbf{P}^T = [a_0, a_1, \ldots a_N, \rho]$. We note that a sufficiently large number of measurements is considered available, so that the measured data can be assumed as continuous functions $Y(\mu)$ and $Z(\mu)$.

The gradient of the objective function is obtained by differentiating $S(\mathbf{P})$ given by equation (5.2.3.b) with respect to each of the unknown coefficients a_n and ρ, respectively. Then the resulting expressions for $\partial S / \partial a_n$ and $\partial S / \partial \rho$ contain the sensitivity coefficients $\partial I / \partial a_n$ and $\partial I / \partial \rho$, respectively, which can be determined from the solution of the sensitivity problems developed as described below.

Sensitivity Problems

By differentiating the direct problem given by equations (5.2.1) with respect to a_n and ρ, the sensitivity problems are obtained for the determination of the sensitivity coefficients $\partial I / \partial a_n$ (n=0,1,...,N) and $\partial I / \partial \rho$, respectively. By differentiating (5.2.1) with respect to a_n gives

$$\mu \frac{\partial}{\partial \tau}\left(\frac{\partial I(\tau,\mu)}{\partial a_n}\right) + \left(\frac{\partial I(\tau,\mu)}{\partial a_n}\right) = \tau^n + \frac{\omega}{2}\int_{-1}^{1}\left(\frac{\partial I(\tau,\mu')}{\partial a_n}\right) d\mu' \quad 0<\tau<\tau_0 \quad -1<\mu<1 \tag{5.2.4.a}$$

with boundary conditions

$$\left(\frac{\partial I(0,\mu)}{\partial a_n}\right) = 2\rho \int_0^1 \left(\frac{\partial I(0,\mu')}{\partial a_n}\right) \mu' d\mu' \qquad \mu>0 \tag{5.2.4.b}$$

$$\left(\frac{\partial I(\tau_0,-\mu)}{\partial a_n}\right) = 0 \qquad \mu>0 \tag{5.2.4.c}$$

for $n = 0,1,\ldots,N$. The differentiation of (5.2.1) with respect to ρ gives

$$\mu \frac{\partial}{\partial \tau}\left(\frac{\partial I(\tau,\mu)}{\partial \rho}\right) + \left(\frac{\partial I(\tau,\mu)}{\partial \rho}\right) = \frac{\omega}{2}\int_{-1}^{1}\left(\frac{\partial I(\tau,\mu')}{\partial \rho}\right) d\mu' \quad 0<\tau<\tau_0 \quad -1<\mu<1 \tag{5.2.5.a}$$

with boundary conditions

$$\left(\frac{\partial I(0,\mu)}{\partial \rho}\right) = 2\int_0^1 I(0,-\mu')\mu' d\mu' + 2\rho \int_0^1 \left(\frac{\partial I(0,-\mu')}{\partial \rho}\right) \mu' d\mu' \qquad \mu>0 \tag{5.2.5.b}$$

$$\left(\frac{\partial I(\tau_0,-\mu)}{\partial \rho}\right)=0 \qquad \mu>0 \tag{5.2.5.c}$$

The row vector

$$[\nabla I]^T=\left[\frac{\partial I}{\partial a_0},\frac{\partial I}{\partial a_1},\ldots,\frac{\partial I}{\partial a_N},\frac{\partial I}{\partial \rho}\right] \tag{5.2.6}$$

is the sensitivity coefficient vector which can be determined from the solution of the above sensitivity problems. We note that the reflectivity ρ is unknown and it appears in the formulations of both sensitivity problems (5.2.4) and (5.2.5). Therefore, the present estimation problem is nonlinear. If the reflectivity were known, the estimation problem would be linear, as illustrated in section 5.1.

Next, we develop expressions for the components of the gradient, i.e., $\partial S/\partial a_n$ and $\partial S/\partial \rho$, by differentiating $S(\mathbf{P})$ given by equation (5.2.3.b) with respect to a_n and ρ to obtain, respectively,

$$\frac{\partial S}{\partial a_n}=\int_{-1}^{0} 2[I(0,\mu;\mathbf{P})-Y(\mu)]\frac{\partial I(0,\mu;\mathbf{P})}{\partial a_n}d\mu+\int_{0}^{1} 2[I(\tau_0,\mu;\mathbf{P})-Z(\mu)]\frac{\partial I(\tau_0,\mu;\mathbf{P})}{\partial a_n}d\mu \tag{5.2.7.a}$$

for $n=0,1,\ldots,N$, and

$$\frac{\partial S}{\partial \rho}=\int_{-1}^{0} 2[I(0,\mu;\mathbf{P})-Y(\mu)]\frac{\partial I(0,\mu;\mathbf{P})}{\partial \rho}d\mu+\int_{0}^{1} 2[I(\tau_0,\mu;\mathbf{P})-Z(\mu)]\frac{\partial I(\tau_0,\mu;\mathbf{P})}{\partial \rho}d\mu \tag{5.2.7.b}$$

where the row vector $[\nabla S(\mathbf{P})]^T$, defined by

$$[\nabla S(\mathbf{P})]^T=\left[\frac{\partial S}{\partial a_0},\frac{\partial S}{\partial a_1},\ldots\frac{\partial S}{\partial a_N},\frac{\partial S}{\partial \rho}\right] \tag{5.2.7.c}$$

is the gradient of the objective function.

Iterative Procedure

To determine the unknown vector $\mathbf{P}$ defined above, we consider the following iterative minimization procedure of Technique II

$$\mathbf{P}^{k+1}=\mathbf{P}^k-\beta^k\mathbf{d}^k \tag{5.2.8}$$

where β^k is the *step size* and $\mathbf{d}^k$ is the *direction of descent* at the kth iteration, determined from

$$\mathbf{d}^k = \nabla S(\mathbf{P}^k) + \gamma^k \mathbf{d}^{k-1} \tag{5.2.9}$$

The *conjugation coefficient* γ^k is computed from the Fletcher-Reeves expression

$$\gamma^k = \frac{\left[\nabla S(\mathbf{P}^k)\right]^T \left[\nabla S(\mathbf{P}^k)\right]}{\left[\nabla S(\mathbf{P}^{k-1})\right]^T \left[\nabla S(\mathbf{P}^{k-1})\right]} \qquad \text{for } k = 1,2,3\ldots \text{ with } \gamma^0 = 0 \tag{5.2.10}$$

Here, the step size β^k in going from $\mathbf{P}^k$ to $\mathbf{P}^{k+1}$ is determined from the condition

$$\min_{\beta^k} S(\mathbf{P}^{k+1}) = \min_{\beta^k} \left\{ \int_{-1}^{0} [I(0,\mu;\mathbf{P}^k - \beta^k \mathbf{d}^k) - Y(\mu)]^2 d\mu + \int_{0}^{1} [I(\tau_0,\mu;\mathbf{P}^k - \beta^k \mathbf{d}^k) - Z(\mu)]^2 d\mu \right\} \tag{5.2.11}$$

By using a Taylor series expansion and performing the minimization as described in Note 3 in Chapter 2, we obtain the following expression for the search step size

$$\beta^k = \frac{\Delta_1}{\Delta_2} \tag{5.2.12.a}$$

where

$$\Delta_1 = \int_{-1}^{0} [I(0,\mu;\mathbf{P}^k) - Y(\mu)][\nabla I(0,\mu;\mathbf{P}^k)]^T \mathbf{d}^k d\mu + \int_{0}^{1} [I(\tau_0,\mu;\mathbf{P}^k) - Z(\mu)][\nabla I(\tau_0,\mu;\mathbf{P}^k)]^T \mathbf{d}^k d\mu \tag{5.2.12.b}$$

and

$$\Delta_2 = \int_{-1}^{0} \left\{[\nabla I(0,\mu;\mathbf{P}^k)]^T \mathbf{d}^k\right\}^2 d\mu + \int_{0}^{1} \left\{[\nabla I(\tau_0,\mu;\mathbf{P}^k)]^T \mathbf{d}^k\right\}^2 d\mu \tag{5.2.12.c}$$

Stopping Criterion

Once $\mathbf{d}^k$ is calculated from equation (5.2.9) and β^k from equation (5.2.12.a), the iterative process defined by equation (5.2.8) can be used to obtain new estimates $\mathbf{P}^{k+1}$, until a stopping criterion based on the *discrepancy principle* is satisfied. Hence, the stopping criterion is given by

$$S\left(\mathbf{P}^{k+1}\right) < 2\sigma^2 \tag{5.2.13}$$

where σ is the standard deviation of the measurement errors.

Computational Algorithm

The computational algorithm of Technique II, as applied to the present parameter estimation problem, can be summarized as follows. Assume $\mathbf{P}^k$ is known at the kth iteration, then

Step 1. Solve the direct problem given by equations (5.2.1) and compute the exit radiation intensities $I(0,\mu;\mathbf{P}^k)$ and $I(\tau_0,\mu;\mathbf{P}^k)$ at the surfaces $\tau = 0$ and $\tau = \tau_0$, respectively.

Step 2. Check the stopping criterion given by equation (5.2.13). Continue if not satisfied.

Step 3. Solve the sensitivity problems given by equations (5.2.4) and (5.2.5), and compute the sensitivity coefficient vector ∇I defined by equation (5.2.6).

Step 4. Knowing ∇I, $I(0,\mu;\mathbf{P}^k)$, $I(\tau_0,\mu;\mathbf{P}^k)$ and the measured exit radiation intensities $Y(\mu)$ and $Z(\mu)$, compute the gradient $\nabla S(\mathbf{P}^k)$ from equation (5.2.7.c).

Step 5. Knowing $\nabla S(\mathbf{P}^k)$, compute the conjugation coefficient γ^k from equation (5.2.10). Then compute the direction of descent $\mathbf{d}^k$ from equation (5.2.9).

Step 6. Knowing ∇I, $I(0,\mu;\mathbf{P}^k)$, $I(\tau_0,\mu;\mathbf{P}^k)$, $Y(\mu)$, $Z(\mu)$ and $\mathbf{d}^k$, compute the step size β^k from equation (5.2.12.a).

Step 7. Knowing β^k and $\mathbf{d}^k$, compute $\mathbf{P}^{k+1}$ from equation (5.2.8) and return to step 1.

Results

Numerical results are now presented, in order to give some idea of the accuracy of the conjugate gradient method of minimization in the solution of the inverse problem of simultaneous estimation the source term $g(\tau)$ and the diffuse reflectivity ρ of the boundary surface at $\tau = 0$, for absorbing, emitting, isotropically scattering gray plate of optical thickness τ_0. The temperatures considered lie between 800 K to 1800 K, which is encountered in fires and furnaces. The single scattering albedo was chosen as $\omega = 0.3$, which is encountered in coal flames. In order to simulate the measured exit intensities $Y(\mu)$ and $Z(\mu)$ containing measurement errors, random errors of standard deviation σ were added to the exact exit intensities, computed from the solution of the direct problem. The source term was expressed as a polynomial of degree four in the optical variable τ. The two different forms of such representation considered here include

$$g(\tau) = 1 + 10\tau + 75\tau^2 - 170\tau^3 + 85\tau^4 \quad \mathrm{W/cm^2}, \quad \text{in } 0 \le \tau \le 1 \tag{5.2.14.a}$$

$$g(\tau) = 1 + 0.1\tau^2 + 0.01\tau^4 \quad \mathrm{W/cm^2}, \quad \text{in } 0 \le \tau \le 5 \tag{5.2.14.b}$$

Figure 5.2.2 shows simultaneous estimation of the source term $g(\tau)$ (or temperature distribution $T(\tau)$) and the reflectivity ρ, by choosing the source term $g(\tau)$ in the form given by equation (5.2.14.a), for $\tau_0 = 1.0$, $\omega = 0.3$, $\rho = 0.9$ and $\sigma = 0.05$. The agreement between the exact and the estimated results for both the source term $g(\tau)$ and reflectivity is good. For the case of no measurement errors (i.e., $\sigma = 0$), the estimated results agreed with the exact ones within the accuracy of the graphical representation.

Figure 5.2.3 presents results similar to those shown by figure 5.2.2, except for the presence of errors with standard deviation $\sigma = 0.1$. The agreement between the estimated and the exact results is still quite good.

Figures 5.2.4 and 5.2.5 are for a plate of optical thickness $\tau_0 = 5$ and a source term $g(\tau)$ given by equation (5.2.14.b). Figure 5.2.5, for a standard deviation $\sigma = 0.1$, does not seem to be in good agreement with the exact results. The estimation of reflectivity is not good because a very low value (i.e. $\rho = 0.1$) is to be estimated, and because the optical thickness is too large.

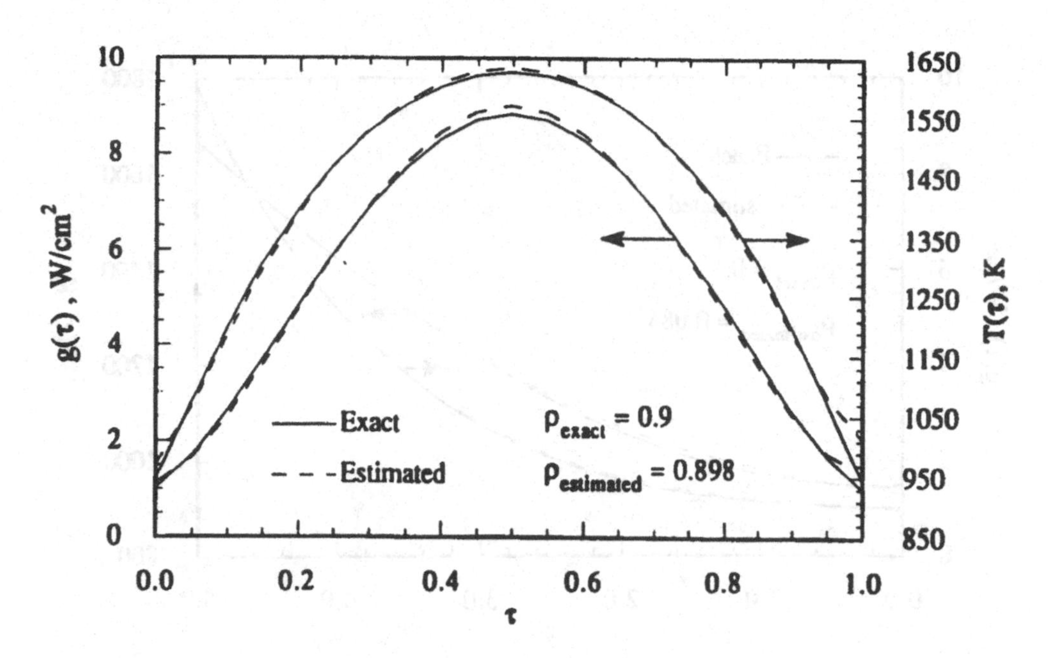

Figure 5.2.2 - Simultaneous estimation of source term and surface reflectivity with measurement error $\sigma = 0.05$, $g(\tau) = 1 + 10\tau + 75\tau^2 - 170\tau^3 + 85\tau^4$, $\omega = 0.3, \tau_0 = 1, \bar{n} = 1$.

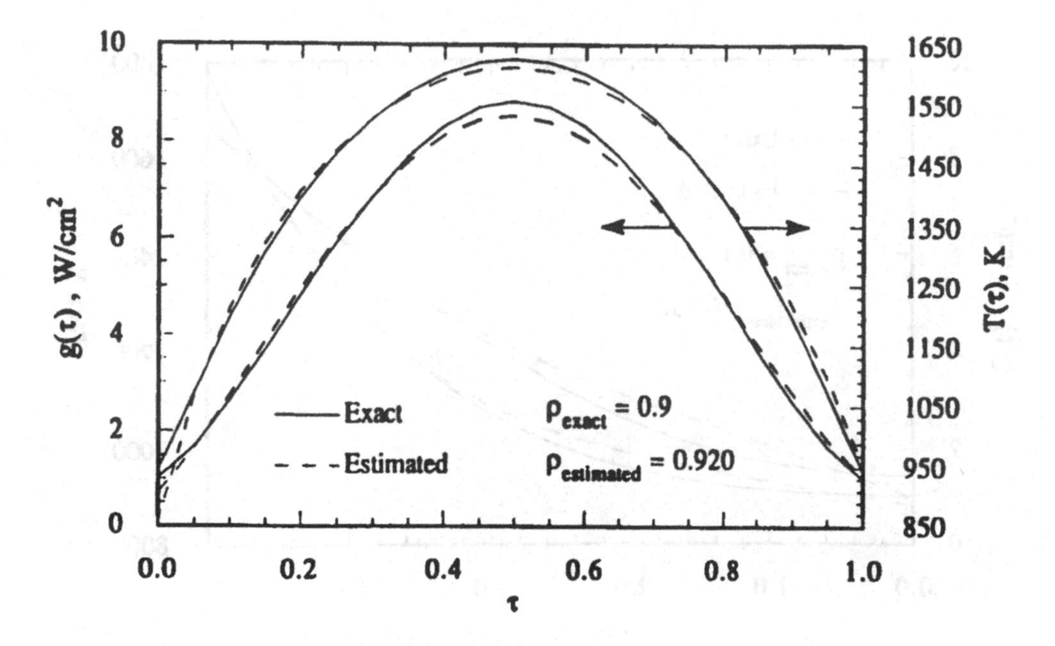

Figure 5.2.3 - Simultaneous estimation of source term and surface reflectivity with measurement error $\sigma = 0.1$, $g(\tau) = 1 + 10\tau + 75\tau^2 - 170\tau^3 + 85\tau^4$, $\omega = 0.3, \tau_0 = 1, \bar{n} = 1$.

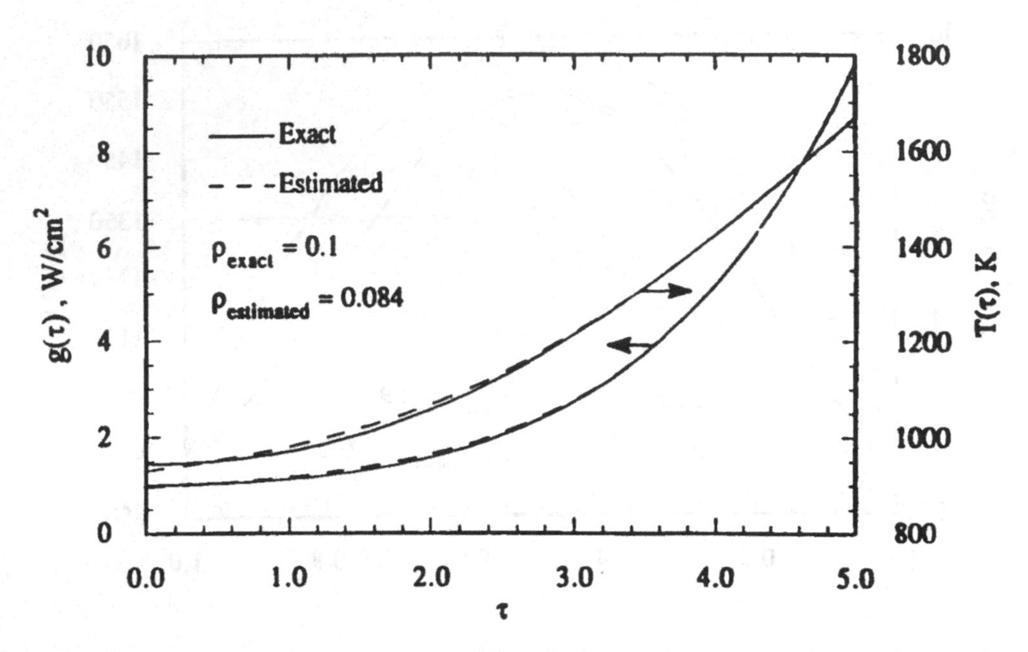

Figure 5.2.4 - Simultaneous estimation of source term and surface reflectivity with measurement error $\sigma = 0.05$, $\omega = 0.3$, $\tau_0 = 5, \bar{n} = 1$, $g(\tau) = 1 + 0.1\tau^2 + 0.01\tau^4$.

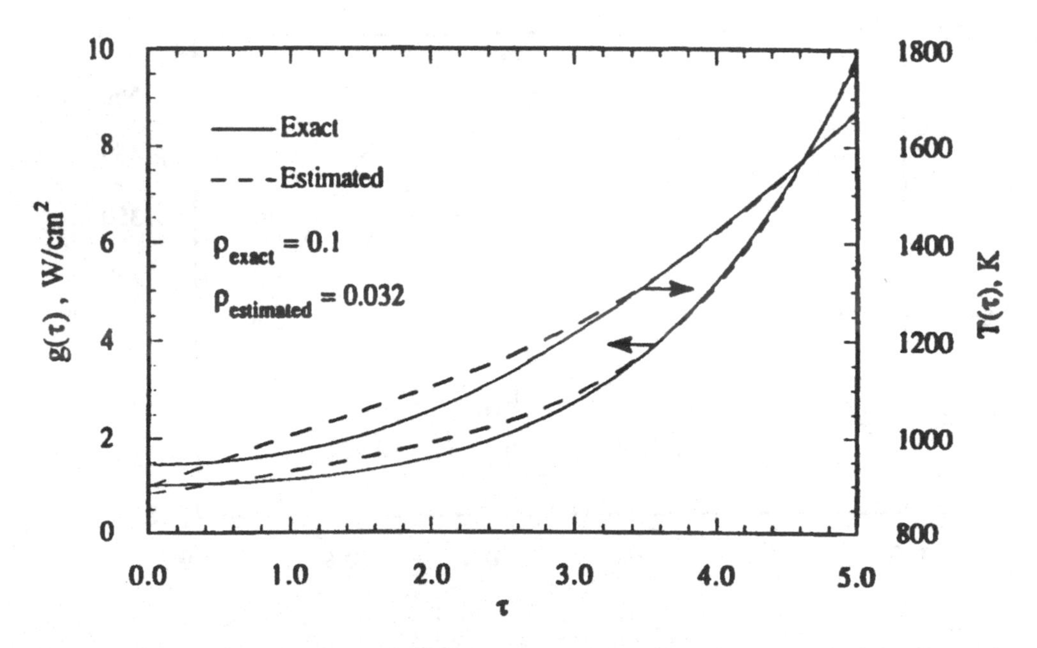

Figure 5.2.5 - Simultaneous estimation of source term and surface reflectivity with measurement error $\sigma = 0.1$, $\omega = 0.3$, $\tau_0 = 5, \bar{n} = 1$, $g(\tau) = 1 + 0.1\tau^2 + 0.01\tau^4$.

5-3 ESTIMATION OF THE RADIATION SOURCE TERM IN A SEMITRANSPARENT SOLID SPHERE [41]

The inverse radiation problem of estimating the unknown temperature distribution and radiation properties in an absorbing, emitting and scattering medium has received a good deal of attention [34-40, 42-46]. However, such works have been limited to the plane-parallel medium. In this section we examine the inverse problem of estimating the unknown temperature field in an absorbing, emitting, isotropically scattering semi-transparent solid sphere, by utilizing the measured exit radiation intensities.

The inverse analysis utilizes **Techniques I and II**, the Levenberg-Marquardt method and the conjugate gradient method, respectively. Both methods require the solution of the *direct problem* and the determination of the *sensitivity coefficients*, which are obtained here by solving sensitivity problems. The solution methodologies for the conjugate gradient method and Levenberg-Marquardt method of minimization are discussed below, after the formulation of the direct, inverse and sensitivity problems.

Direct Problem

For an absorbing, emitting, isotropically-scattering gray solid sphere of optical radius R with transparent boundary, the equation of radiative transfer can be expressed as [30]:

$$\mu\frac{\partial I(r,\mu)}{\partial r}+\frac{1}{r}(1-\mu^2)\frac{\partial I(r,\mu)}{\partial \mu}+I(r,\mu)=g(r)+\frac{\omega}{2}\int_{-1}^{1}I(r,\mu')d\mu' \tag{5.3.1.a}$$

in $0<r<R, -1\leq\mu\leq 1$. For transparent boundary at $r = R$, with no externally incident radiation, the boundary condition is taken as

$$I(R,-\mu)=0 \qquad \mu\geq 0 \tag{5.3.1.b}$$

The geometry and coordinates of this spherical system is illustrated in figure 5.3.1. The source term is related to the temperature $T(r)$ in the medium by

$$g(r)=(1-\omega)\frac{\bar{n}^2\bar{\sigma}T^4(r)}{\pi} \tag{5.3.2}$$

where $\bar{n}$ is the refractive index of the medium and $\bar{\sigma}$ is the Stefan-Boltzmann constant.

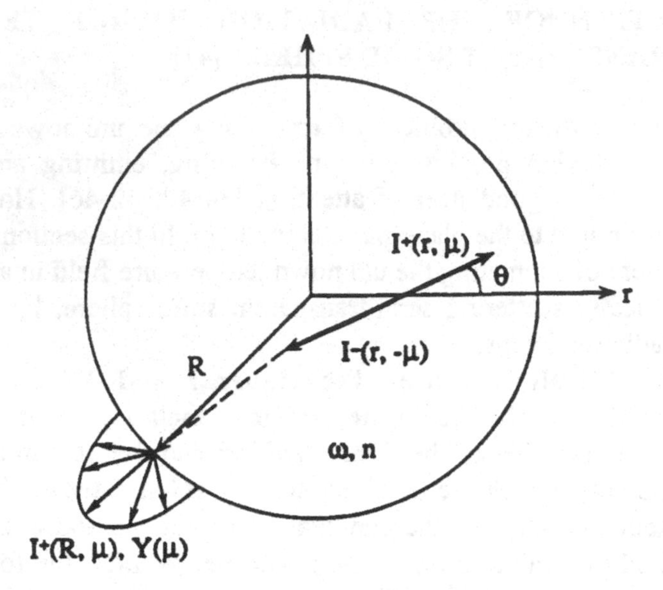

Figure 5.3.1 - Geometry, coordinates and measurement location.

When the source term $g(r)$, optical radius R, single scattering albedo ω and the boundary condition at $r = R$ are all specified, the problem defined above by equations (5.3.1) for determination of the radiation intensity $I(r,\mu)$ is called a *direct problem*. However, when the source term $g(r)$ is unknown and needs to be estimated from the knowledge of the measured exit intensities taken at the outer surface of the sphere, the problem becomes an *inverse problem*.

Pomraning and Siewert [47] developed the integral form of the above radiation problem in terms of incident radiation $I(r)$ as

$$rI(r) = \int_0^R x\left[\frac{\omega}{2}I(x) + g(x)\right]\left[E_1(|r-x|) - E_1(|r+x|)\right]dx \tag{5.3.3}$$

where $E_1(x)$ is the exponential integral function. The incident radiation, $I(r)$, is defined as

$$I(r) = \int_{-1}^{1} I(r,\mu)d\mu = \int_0^1 I^+(r,\mu)d\mu + \int_0^1 I^-(r,-\mu)d\mu \tag{5.3.4}$$

where $I^+(r,\mu)$ and $I^-(r,-\mu)$, for $\mu \geq 0$, are the forward and backward intensities, respectively.

Thynell and Özisik [30] solved this integral equation for $I(r)$, by representing $g(r)$ and $I(r)$ in a power series in the optical variable r as

$$g(r)=\sum_{n=0}^{N}P_n r^n \tag{5.3.5}$$

and

$$I(r)=\sum_{m=0}^{M}C_m r^m \tag{5.3.6}$$

where the expansion coefficients P_n are considered known in the case of the direct problem. The coefficients C_m are computed by using the Galerkin method; since the method is well documented [27-33], it will not be repeated here. However, we note that the accuracy of the solution of the resulting algebraic equations can be improved if the iterative improvement scheme suggested in reference [48] utilizing the L.U. decomposition is used. The scheme allows for many more terms to be included in the power series expansion given in equation (5.3.6). The iterative improvement scheme is found to be specially important when the system of equations tend to become ill-conditioned, as the optical radius R becomes large.

Once the coefficients C_m are determined, the incident radiation $I(r)$ is calculated from equation (5.3.6) and the angular distribution of the exit intensity $I^+(r,\mu)$ is computed from

$$I^+(r,\mu)=\frac{\omega}{2}\sum_{m=0}^{M}C_m\left[\lambda_1 X_m^+(r,\mu)+\lambda_2\,{}^*X_n^+(r,1)\right] \qquad \text{for } \mu\geq 0 \tag{5.3.7}$$
$$+\sum_{n=0}^{N}P_n\left[\lambda_1 X_n^+(r,\mu)+\lambda_2\,{}^*X_n^+(r,1)\right]$$

where P_n are the expansion coefficients specified by equation (5.3.5). Since we are interested only in the forward exit intensity at the outer radius, equation (5.3.7) will suffice for the solution of the direct problem with $r = R$. Here $X_i^+(r,\mu)$ and ${}^*X_i^+(r,1)$ are definite integrals defined by

$$X_i^+(r,\mu)=\int_{-r\mu}^{\left[R^2-r^2(1-\mu^2)\right]^{1/2}}\left[t^2+r^2(1-\mu^2)\right]^{i/2}\exp[-(t+r\mu)]dt \quad \mu<1 \tag{5.3.8.a}$$

and

$$^{*}X_{i}^{+}(r,1)=\int_{0}^{r}x^{i}\exp(-r+x)dx+\int_{0}^{R}x^{i}\exp(-r-x)dx \quad \mu=1 \tag{5.3.8.b}$$

for which explicit expressions are also available [49]. In addition, λ_i (i=1,2) are defined as

$$\lambda_1=1, \quad \lambda_2=0 \qquad \text{for } \mu<1, \tag{5.3.9.a}$$

$$\lambda_1=0, \quad \lambda_2=1 \qquad \text{for } \mu=1. \tag{5.3.9.b}$$

By substituting equations (5.3.8) and (5.3.9) along with the calculated values of C_m into equation (5.3.7), we can calculate the exit intensities $I^{+}(R,\mu)$ at any angle μ.

Inverse Problem

The problem defined by equations (5.3.1) with the source term $g(r)$ unknown and with measured exit intensities available is an *inverse problem*, which can be solved by minimizing the following objective function

$$S(\mathbf{P})=\int_{0}^{1}\left[I^{+}(R,\mu;\mathbf{P})-Y(\mu)\right]^{2}d\mu \tag{5.3.10}$$

where $Y(\mu)$ is the measured exit intensity and $I^{+}(R,\mu;\mathbf{P})$ is the estimated exit intensity at the outer surface, obtained by using the current estimate for the parameters $\mathbf{P}^{T}=\left[P_0,P_1,\cdots,P_N\right]$ of equation (5.3.5). A sufficiently large number of measurements is assumed available, so that it can be considered as continuous.

Sensitivity Problem

Equation (5.3.10) is minimized by differentiating $S(\mathbf{P})$ with respect to each of the unknown coefficients P_n, giving

$$\frac{\partial S}{\partial P_n}=\int_{0}^{1}2\left[I^{+}(R,\mu;\mathbf{P})-Y(\mu)\right]\frac{\partial I^{+}(R,\mu;\mathbf{P})}{\partial P_n}d\mu \tag{5.3.11}$$

The resulting equation for $\partial S/\partial P_n$ contains the sensitivity coefficients $\partial I^{+}/\partial P_n$, which are determined from the solution of the sensitivity problems, as now considered. Differentiating the direct problem given by equations (5.3.1) with respect to P_n generates the *sensitivity problems* for the determination of the *sensitivity coefficients* $\partial I^+ / \partial P_n$, $n=0,1,\ldots,N$, that is,

$$\mu \frac{\partial}{\partial r}\left(\frac{\partial I(r,\mu)}{\partial P_n}\right) + \frac{1}{r}(1-\mu^2)\frac{\partial}{\partial \mu}\left(\frac{\partial I(r,\mu)}{\partial P_n}\right) + \left(\frac{\partial I(r,\mu)}{\partial P_n}\right) = r^n + \frac{\omega}{2}\int_{-1}^{1}\left(\frac{\partial I(r,\mu')}{\partial P_n}\right)d\mu' \tag{5.3.12.a}$$

for $0 < r < R$, $-1 \le \mu \le 1$, and

$$\frac{\partial I(R,-\mu)}{\partial P_n} = 0 \qquad \text{for } \mu > 0 \tag{5.3.12.b}$$

The solution of equations (5.3.12) for each n, $n = 0,1,\ldots,N$, produces the sensitivity coefficient vector,

$$\left[\nabla I^+\right]^T = \left[\frac{\partial I^+}{\partial P_0}, \frac{\partial I^+}{\partial P_1}, \ldots, \frac{\partial I^+}{\partial P_N}\right] \tag{5.3.12.c}$$

The procedure for solving equations (5.3.12.a,b) is the same as that described previously for the solution for the direct problem, with $g(r)$ replaced by r^n.

It is noted that the sensitivity problem given by equations (5.3.12.a,b) is independent of the parameters P_n. Hence, the estimation problem is linear. The sensitivity coefficients are solved only once and need not be recalculated for each estimate of P_n.

The Solution with Technique II

As stated above, the inverse problem requires the minimization of the objective function, equation (5.3.10), by differentiating it with respect to P_n, $n = 0,1,2,\ldots,N$. In vector form, $\partial S / \partial P_n$ can be written as

$$[\nabla S(\mathbf{P})]^T = \left[\frac{\partial S}{\partial P_0}, \frac{\partial S}{\partial P_1}, \ldots \frac{\partial S}{\partial P_N}\right] \tag{5.3.13}$$

which can easily be computed with equation (5.3.11).

To estimate the unknown vector $\mathbf{P}$, we consider here the iterative procedure of Technique II given in the form:

$$\mathbf{P}^{k+1} = \mathbf{P}^k - \beta^k \mathbf{d}^k \qquad \text{for } k = 0,1,2,\cdots \tag{5.3.14}$$

where β^k is the *step size*, in going from iteration k to k+1. The *direction of descent* $\mathbf{d}^k$ is given by

$$\mathbf{d}^k = \nabla S(\mathbf{P}^k) + \gamma^k \mathbf{d}^{k-1} \qquad \text{for } k = 0,1,2,\cdots \tag{5.3.15}$$

where the conjugation coefficient γ^k is computed from the Fletcher-Reeves expression

$$\gamma^k = \frac{\left[\nabla S(\mathbf{P}^k)\right]^T \left[\nabla S(\mathbf{P}^k)\right]}{\left[\nabla S(\mathbf{P}^{k-1})\right]^T \left[\nabla S(\mathbf{P}^{k-1})\right]} \qquad \text{for } k=1,2,\ldots \quad \text{with } \gamma^0=0 \tag{5.3.16}$$

The step size β^k is determined by minimizing the objective function $S(\mathbf{P}^{k+1})$, that is

$$\min_{\beta^k} S(\mathbf{P}^{k+1}) = \min_{\beta^k} S(\mathbf{P}^k - \beta^k \mathbf{d}^k) \tag{5.3.17}$$

By using a Taylor series expansion, the following expression is obtained for the determination of β^k (see equation 2.2.16):

$$\beta^k = \frac{\int_{\mu=0}^{1} [I^+(R,\mu;\mathbf{P}^k) - Y(\mu)][\nabla I^+(R,\mu;\mathbf{P}^k)]^T \mathbf{d}^k d\mu}{\int_{\mu=0}^{1} \{[\nabla I^+(R,\mu;\mathbf{P}^k)]^T \mathbf{d}^k\}^2 d\mu} \tag{5.3.18}$$

Once $\mathbf{d}^k$ is calculated from Equation (5.3.15) and β^k from equation (5.3.18), the iterative process given by equation (5.3.14) is applied until a specified stopping criterion is satisfied.

If there is no measurement error, the conventional stopping criterion defined as

$$S\left(\mathbf{P}^{k+1}\right) < \varepsilon \tag{5.3.19}$$

where ε is a small positive number, can be used to terminate the iteration process. On the other hand, if the exit intensity measurements contain errors, the *discrepancy principle* is required to obtain the tolerance ε from equation (5.3.10) in the form

$$\varepsilon = \sigma^2 \tag{5.3.20}$$

so that the resulting solution is stable, where σ is the standard deviation of the measurement errors.

The computational algorithm for Technique II, as applied to the present inverse radiation problem, can be summarized as follows.

Assume $\mathbf{P}^k$ is known at the kth iteration. Then,

Step 1. Solve the sensitivity problem given by equations (5.3.12.a,b) and compute the sensitivity coefficient vector ∇I^+ given by equation (5.3.12.c).

Step 2. Given the current estimate of $\mathbf{P}^k$, compute the exit intensities $I^+(R,\mu;\mathbf{P}^k)$ for the outer radius R, from equation (5.3.7).

Step 3. Terminate the iterations when the stopping criterion given by equation (5.3.19) is satisfied. Continue otherwise.

Step 4. Knowing ∇I^+, $I^+(R,\mu;\mathbf{P}^k)$ and the measured exit intensities $Y(\mu)$, compute the gradient $\nabla S(\mathbf{P}^k)$ from equation (5.3.13).

Step 5. Knowing $\nabla S(\mathbf{P}^k)$, compute γ^k from equation (5.3.16) and then the direction of descent $\mathbf{d}^k$ from equation (5.3.15).

Step 6. Knowing ∇I^+, $I^+(R,\mu;\mathbf{P}^k)$, $Y(\mu)$ and $\mathbf{d}^k$, compute the step size β^k from equation (5.3.18).

Step 7. Knowing β^k and $\mathbf{d}^k$, estimate $\mathbf{P}^{k+1}$ from equation (5.3.14) and return to step 2.

The Solution with Technique I

Technique I, the Levenberg-Marquardt method, is also applied to the solution of the present inverse radiation problem, for the sake of comparison with Technique II. It also employs the direct and the sensitivity problems given above. The objective function is re-written in the following form for Technique I:

$$S(\mathbf{P}) = \sum_{i=1}^{I}\left[I_i^+(R,\mu,\mathbf{P}) - Y_i(\mu)\right]^2 \tag{5.3.21}$$

where I is the total number of measurements. To estimate the unknown parameter vector $\mathbf{P}$ by the *Levenberg-Marquardt method*, the following iterative procedure is used (see equation 2.1.13):

$$\mathbf{P}^{k+1} = \mathbf{P}^k + \left[\mathbf{J}^T\mathbf{J} + \mu^k\Omega^k\right]^{-1}\mathbf{J}^T\left[\mathbf{Y} - \mathbf{I}^+(R,\mu,\mathbf{P}^k)\right] \tag{5.3.22}$$

where μ^k is the damping parameter, and **J** is the sensitivity matrix defined by equation (2.1.7.b). The iterative method given by equation (5.3.22) should be continued until stopping criteria are met. The stopping criteria for the Levenberg-Marquardt method are given by equations (2.1.14.a-c).

The computational algorithm for the Levenberg-Marquardt method can be found in section 2-1.

Results

In order to illustrate the application of Techniques I and II to the solution of the present inverse problem, a set of simulated measured exit intensities was generated in order to estimate the unknown source term, $g(r)$, or the temperature $T(r)$, as defined by equation (5.3.2). To simulate the measured exit intensities, $Y(\mu)$, the direct problem was solved with a specified source term, $g(r)$ and exact exit intensities $I^+(R,\mu)$ were determined at 5-degree intervals from 0 to 90 degrees. The simulated exit intensities were then generated by adding random errors to such exact intensities, in a form similar to equation (2.5.2).

For the first test case, the source term was represented by a fifth degree polynomial in the optical variable r as

$$g(r) = 1 + 3r - 0.1r^2 + r^3 + 0.1r^4 - 4r^5 \qquad (5.3.23)$$

for $0 \le r \le 1$, $\omega = 0.3$. Figure 5.3.2.a shows the estimated value of the temperature distribution $T(r)$ as a function of the radial position r, for two different values of standard deviation, $\sigma = 0.03$ and $\sigma = 0.1$, by using the conjugate gradient method. Also included in the figure are the exact values of $T(r)$. Clearly, the accuracy of the estimation improves as the standard deviation of the measurement errors decreases. The accuracy of the estimate is better in the region near the outer surface of the sphere than towards the center. This is consistent with the fact that measured exit intensity receives less information from the interior regions of the sphere due to attenuation, than the region near the surface. Several different initial guesses were chosen as the starting point for equation (5.3.14) and it appeared that the solution was insensitive to the starting guess. Figure 5.3.2.b shows an estimate of the source function given by equation (5.3.23), but now using the Levenberg-Marquardt method. The estimated function obtained with the Levenberg-Marquardt method is clearly more accurate than that obtained with the conjugate gradient method. However, the starting guess for the parameters had to be chosen much closer to the exact values for the Levenberg-Marquardt method than for the conjugate gradient method. For example, the Levenberg-Marquardt method did not converge for a starting guess of $P_0 = P_1 = ... = P_5 = 0$, but the conjugate gradient method easily converged.

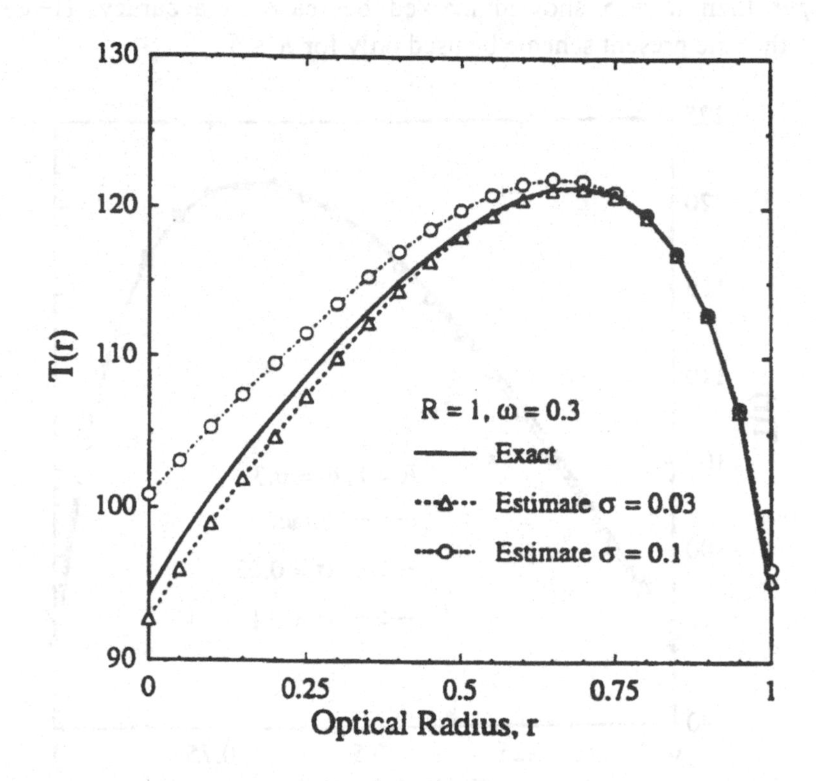

Figure 5.3.2.a - Temperature distribution for $R = 1$, $\omega = 0.3$ and $g(r) = 1 + 3r - 0.1r^2 + r^3 + 0.1r^4 - 4r^5$ using conjugate gradient method.

Next, we consider another source term expressed as a fifth degree polynomial in the optical variable r as

$$g(r) = 1 + 0.6r - 0.004r^2 + 0.008r^3 + 0.0016r^4 - 0.00128r^5, \tag{5.3.24}$$

for $0 \le r \le 5$, $\omega = 0.3$ and using measurement data of $\sigma = 0$ and $\sigma = 0.1$. The test source term was chosen such that the magnitude of the intensity would be of the same order as that of the case with $R = 1$, in figures 5.3.2.a,b. Figure 5.3.3.a illustrates the results obtained for such a test-case by using the conjugate gradient method. This figure shows that for $R = 5$ the solution deviates from the exact function even with errorless measurements. This is expected, since the majority of the information for the estimation comes from the radii closer to the outer radius. Although the agreement between exact and estimated temperatures is still quite good, the inverse problem becomes more sensitive to errors when a larger radius is used. As only one measurement location can be used at the outer radius, the problem is one of physics and not of the solution method. Figure 5.3.3.b presents the estimate of the source term given by equation 5.3.24 by using the Levenberg-Marquardt method. As was shown in the $R = 1$ case, the Levenberg-Marquardt method produced more accurate solutions, but once again the starting parameter guess had to be close to the exact solution. Inverse analysis for optical radii larger than $R = 5$ showed marked decrease in accuracy. Hence, it is suggested that the present scheme be used only for $R \leq 5$.

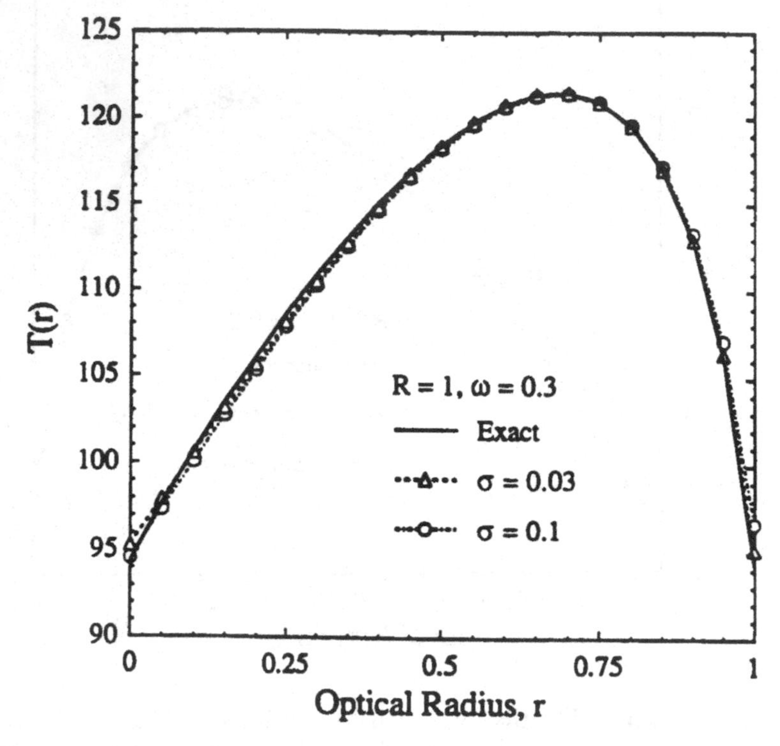

Figure 5.3.2.b - Temperature distribution for $R = 1$, $\omega = 0.3$ and $g(r) = 1 + 3r - 0.1r^2 + r^3 + 0.1r^4 - 4r^5$ using Levenberg-Marquardt method.

We also considered the estimation of the temperature source term for a small optical radius using the following fourth degree polynomial

$$g(r) = 1 + 2r + r^2 - 10{,}000r^4 \tag{5.3.25}$$

with $0 \leq r \leq 0.1$, $\omega = 0.3$ and using measurement data with $\sigma = 0.003$ and $\sigma = 0.006$. Figure 5.3.4.a shows that the estimation utilizing the conjugate gradient method is quite good. However, the error is greatest toward the center of the sphere even for small optical radii. Figure 5.3.4.b is the Levenberg-Marquardt solution. Again it was found to be more accurate, but needing an initial guess close to the exact solution.

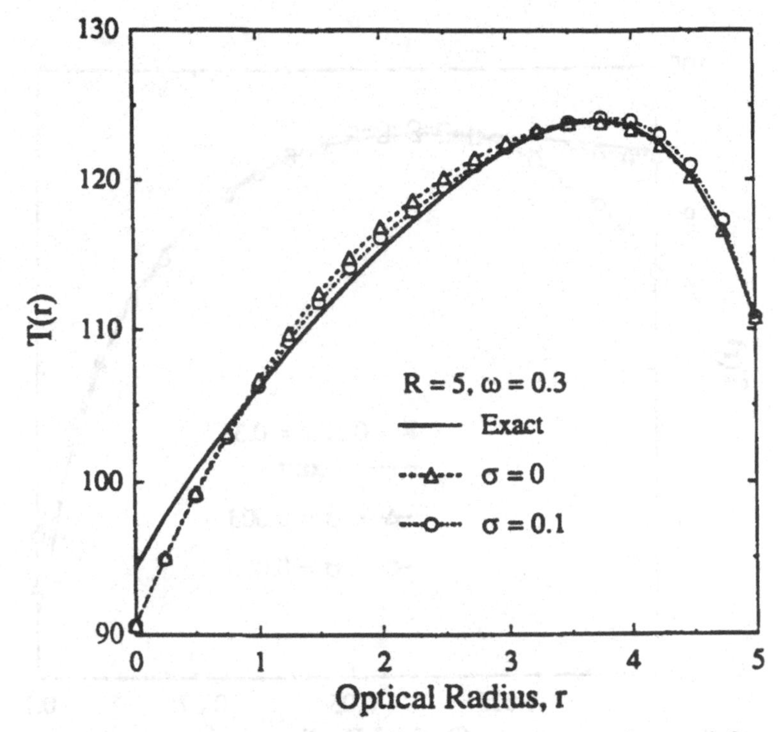

Figure 5.3.3.a - Temperature distribution for $R = 5$, $\omega = 0.3$ and $g(r) = 1 + 0.6r - 0.004r^2 + 0.008r^3 + 0.0016r^4 - 0.00128r^5$ using the conjugate gradient method.

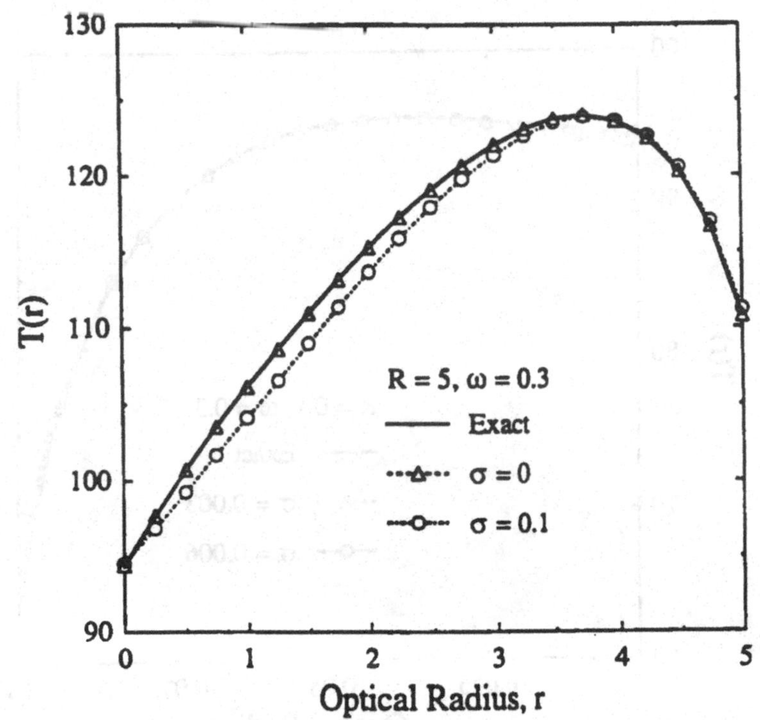

Figure 5.3.3.b - Temperature distribution for $R = 5$, $\omega = 0.3$ and $g(r) = 1 + 0.6r - 0.004r^2 + 0.008r^3 + 0.0016r^4 - 0.00128r^5$ using the Levenberg-Marquardt method.

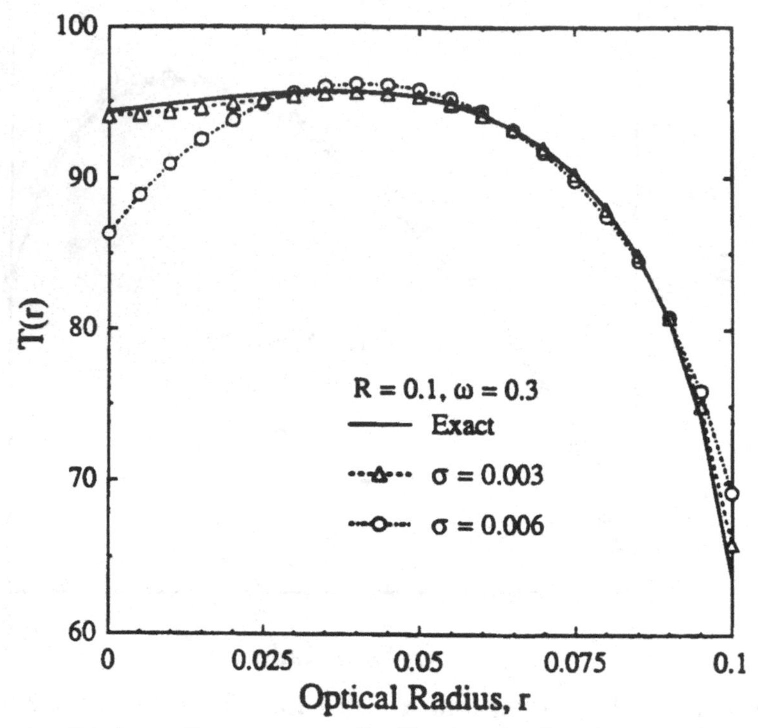

Figure 5.3.4.a - Temperature distribution for $R = 0.1$, $\omega = 0.3$ and $g(r) = 1 + 2r + r^2 - 10{,}000r^4$ using the conjugate gradient method.

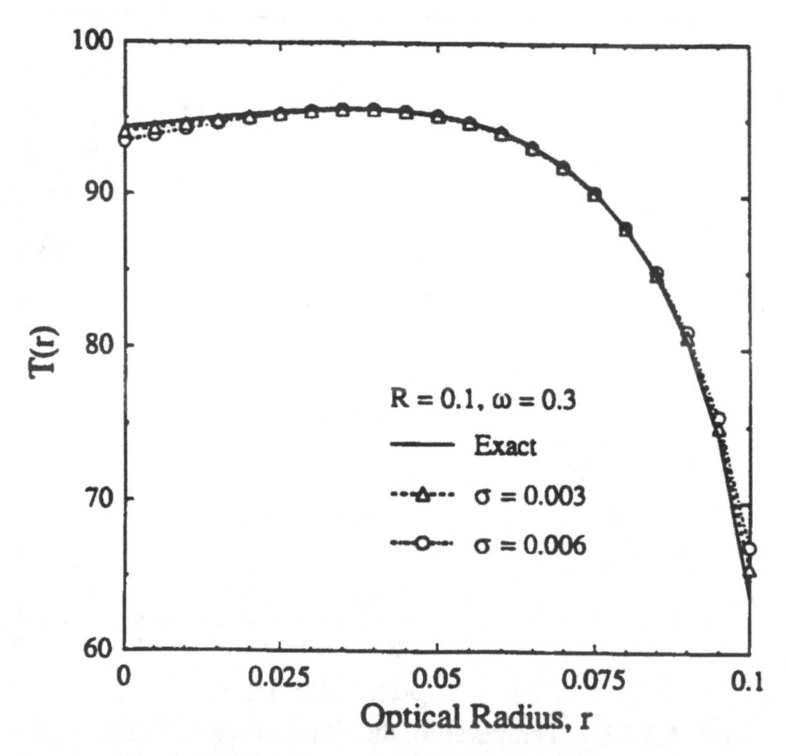

Figure 5.3.4.b - Temperature distribution for $R = 0.1$, $\omega = 0.3$ and $g(r) = 1 + 2r + r^2 - 10{,}000r^4$ using the Levenberg-Marquardt method.

Finally, we considered the variation of the temperature field with albedo, ω, for the source term given by equation (5.3.23). For the results presented below, we utilized initially the conjugate gradient method with a starting guess far from the exact solution and then, after several iterations, the current estimate was used as the starting guess for the Levenberg-Marquardt method. Results obtained with such an approach are shown in figure 5.3.5, for different values for the albedo ω. The use of the combination of Levenberg-Marquardt and conjugate gradient methods appears to yield more accurate results and allows one to perform the calculations with starting guesses far from the converged result. Figure 5.3.5 shows that, for a given source term, the temperature increases as albedo increases. This is expected, since the higher albedo scatters more energy back into the interior and, hence, produces a higher temperature.

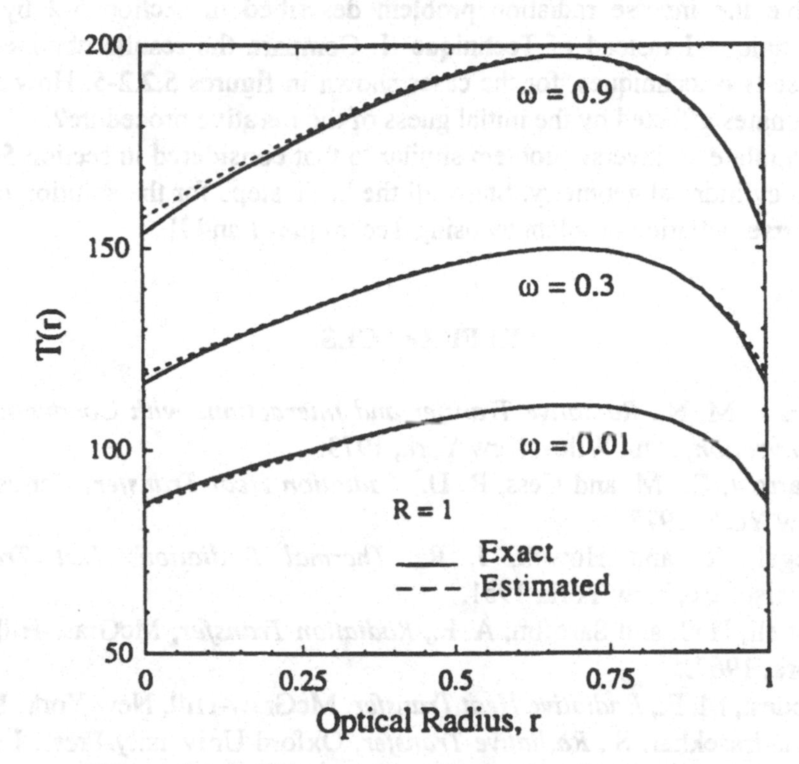

Figure 5.3.5 - Temperature distribution for $R = 1$ and $g(r) = 1 + 3r - 0.1r^2 + r^3 + 0.1r^4 - 4r^5$ for different values of ω utilizing combined conjugate gradient and Levenberg-Marquardt methods.

PROBLEMS

5-1 Derive the Equation of Radiative Heat Transfer for an absorbing, emitting, isotropically scattering plane-parallel gray medium.

5-2 Derive the Equation of Radiative Heat Transfer for an absorbing, emitting, isotropically scattering gray medium with spherical symmetry.

5-3 Derive the Equation of Radiative Heat Transfer for an absorbing, emitting, isotropically scattering gray medium with cylindrical symmetry.

5-4 Derive the sensitivity problem given by equations (5.1.4).

5-5 Derive equation (5.1.11.a) for the search step size in section 5-1.

5-6 Show all the basic steps for the solution of the inverse radiation problem described in section 5-1, by using Technique I.

5-7 Use Technique I to estimate the coefficients of the polynomial given by equation (5.1.13). Utilize in the inverse analysis the measurements of exit intensities at $\tau = 0$ and $\tau = \tau_0 = 1$, with standard-deviations $\sigma = 0$ and $\sigma = 0.03$. Examine the effects of the initial guess on the solution. How the results obtained with Technique I compare to those shown in figures 5.1.2,3, obtained with Technique II?

5-8 Show all the basic steps for the solution of the inverse radiation problem described in section 5-2, by using Technique I.

5-9 Solve the inverse radiation problem described in section 5-2 by using Technique I instead of Technique II. Compare the results obtained with these two techniques, for the cases shown in figures 5.2.2-5. How are the estimates affected by the initial guess of the iterative procedure?

5-10 Formulate an inverse problem similar to that considered in section 5-3, but in a cylindrical geometry. Show all the basic steps for the solution of such inverse radiation problem by using Techniques I and II.

REFERENCES

1. Özisik, M. N., *Radiative Transfer and Interactions with Conduction and Convection*, John Wiley, New York, 1973.
2. Sparrow, E. M. and Cess, R. D., *Radiation Heat Transfer*, Hemisphere, New York, 1978.
3. Siegel, R. and Howell, J. R., *Thermal Radiation Heat Transfer*, Hemisphere, New York, 1981.
4. Hottell, H.C. and Sarofim, A. F., *Radiation Transfer*, McGraw-Hill, New York, 1967.
5. Modest, M. F., *Radiative Heat Transfer*, McGraw-Hill, New York, 1993.
6. Chandrasekhar, S., *Radiative Transfer*, Oxford University Press, London, 1950, also Dover Publications, New York, 1960.
7. Viskanta, R. and Menguc, M. P., "Radiative Transfer in Combustion System", *Progr. Energy Combust. Sci.*, **13**, 97-160, 1987.
8. Hyde, D. J. and Truelove, J. S., "The Discrete Ordinates Approximation for Multidimensional Radiant Heat Transfer in Furnaces", *UKAEA Report* AERE-R 8502, AERE Harvel, 1977.
9. Fiveland, W. A., "Discrete Ordinates Solutions of the Radiative Transport Equation for Rectangular Enclosures", *J. Heat Transfer*, **106**, 699-706, 1984.
10. Fiveland, W. A., "Three-dimensional Radiative Heat Transfer Solutions by the Discrete Ordinates Method", *ASME HDT* 72, 9-18, 1987.
11. Truelove, J. S., "Discrete Ordinates Solution of the Radiation Transport Equation", *J. Heat Transfer*, **109**, 1048-1051, 1987.

12. Truelove, J. S., "Three-dimensional Radiation in Absorbing - Emitting - Scattering Media Using the Discrete-Ordinates Approximations", *Journal of Quantitative Spectroscopy Radiative Transfer*, **39**, 27-31, 1988.
13. Koch R., Wittig, W. and Viskanta, R., "Discrete Ordinates Quadrature Schemes for Multidimensional Radiative Transfer in Furnaces", *Journal of Quantitative Spectroscopy and Radiative Transfer*, **53**, 353-372, 1995.
14. Koch, R., Krebs, W., Wittig, S. and Viskanta, R., "A Parabolic Formulation of the Discrete Ordinates Method for the Treatment of Complex Geometries", *Proc. Radiative Transfer - 1*, Pinar Menguc (ed.), 43-61, 1995.
15. Tsai, J. R. and Özisik, M. N., "Transient, Combined Conduction and Radiation in an Absorbing, Emitting and Isotropically Scattering Solid Sphere", *Journal of Quantitative Spectroscopy Radiative Transfer*, **38**, 243-251, 1987.
16. Tsai, J. R. and Özisik, M. N., "Transient, Combined Conduction and Radiation in an Absorbing, Emitting, Isotropically Scattering Solid Cylinder", *J. Applied Physics*, **64**, 3820-3824, 1988.
17. Tsai, J. R. and Özisik, M. N., "Effects of Anisotropic Scattering of Radiation in Laminar Forced Convection in a Parallel Plate Duct", *Can. J. Chem. Engr*., **225**, 924-929, 1988.
18. Tsai, J. R. and Özisik, M. N., "Radiation in Cylindrical Symmetry with Anisotropic Scattering and Variable Properties", *Int. J. Heat Mass Transfer*, **33**, 2651-2658, 1990.
19. Li, H. Y., Özisik, M. N. and Tsai, J. R., "Two Dimensional Radiation in a Cylinder with Spatially Varying Albedo", *J. Thermophysics and Heat Transfer*, **6**, 180-182, 1992.
20. Tsai, J. R. and Özisik, M. N., "Radiation and Laminar Forced Convection of Non-Newtonian Fluid in a Circular Tube", *Int. J. Heat & Fluid Flow*, **10**, 361-365, 1989.
21. Jeans, J. H., "The Equations of Radiative Transfer of Energy", *Monthly Notices Royal Astromical Society*, **78**, 28-36, 1917.
22. Davison, B., *Neutron Transport Theory*, Oxford University Press, London, 1958.
23. Bayazitoglu, Y. and Hiyengi, J., "The Higher-Order Differential Equations of Radiative Transfer", *AIAA Journal*, 17, 424-431, 1973.
24. Menguc, M. P. and Viskanta, R., "Radiative Transfer in Three-Dimensional Rectangular Enclosures Containing Inhomogeneous, Anisotropically Scattering Media", *Journal of Quantitative Spectroscopy and Radiative Transfer*, **33**, 533-549, 1985.
25. Ou, S. C. S. and Liou, K. N., "Generalization of the Spherical Harmonics Method to Radiative Transfer in Multi-Dimensional Space", *Journal of Quantitative Spectroscopy and Radiative Transfer*, **28**, 271-288, 1982.
26. Benassi, M., Cotta, R. M., and Siewert, C. E., "The P_N Method for Radiative Transfer Problems with Reflective Boundary Conditions," *J. Quant. Spectrosc. Radiat. Transfer*, **30**, 547-553, 1983.

27. Özisik, M. N. and Yener, Y., "The Galerkin Method for Solving Radiation Transfer in a Plane-Parallel Participating Media", *J. Heat Transfer*, **104**, 316-322, 1982.
28. Cengel, Y. A. and Özisik, M. N., "The Use of Galerkin Method for Radiation Transfer in an Anisotropically Scattering Slab with Reflecting Boundaries", *Journal of Quantitative Spectroscopy Radiative Transfer*, **32**, 225-233, 1984.
29. Cengel, Y. A. and Özisik, M. N., "Solar Absorption in Solar Ponds", *Solar Energy*, **33**, 581-591, 1984.
30. Thynell, S. T. and Özisik, M. N., "Radiation Transfer in an Isotropically Scattering Homogeneous Solid Sphere", *Journal of Quantitative Spectroscopy and Radiative Transfer*, **33**, 319-330, 1985.
31. Cengel, Y. A. and Özisik, M. N., "Radiation Heat Transfer in an Anisotropically Scattering Plane-Parallel Medium with Space Dependent Albedo, $\omega(x)$". *22nd Natl, Heat Transfer Conference, Niagara Falls*, Aug. 5-8, 1984, ASME Paper # 84-HT-38.
32. Cengel, Y. A., Özisik, M. N. and Yener, Y., "Radiative Transfer in a Plane-Parallel Medium with Space Dependent Albedo, $\omega(x)$". *Int. J. Heat Mass Transfer*, **27**, 1919-1922, 1984.
33. Thynell, S. T. and Özisik, M. N., "Radiation Transfer in Isotropically Scattering Rectangular Enclosures", *J. Thermophysics and Heat Transfer*, **1**, 69-76, 1987.
34. Yeh, H. Y. and Roux, J. A., "Spectral Radiative Properties of Fiberglass Insulations", *J. Thermophysics and Heat Transfer*, **2**, 78-81, 1989.
35. Glicksman, L., Schuetz, M. and Sinotsky, S. "Radiation Heat Transfer in Foam Insulation", *Int. J. Heat Mass Transfer*, **30**, 187-197, 1987.
36. Kamiuto, K., Iwamoto, M., Sato, M. and Nishimura, T., "Radiation Extinction Coefficients of Packed-Sphere Systems", *J. Quant. Spectrosc. Radiat. Transfer*, **45**, 93-96, 1991.
37. Kamiuto, K., Iwamoto, Nishimura, T. and Sato, M., "Albedo and Asymmetry Factors of the Phase Function for Packed-Sphere Systems", *J. Quant. Spectrosc. Radiat. Transfer*, **46**, 309-316, 1991.
38. Sacadura, J. F. and Nicolao, V. P., "Spectral Radiative Properties Identification of Semi-Transparent Porous Media", *3rd UK National & 1st. European Conference Thermal Sciences*, Birmingham, UK, 717-723, 1992.
39. Li, H. Y. and Özisik, M. N., "Identification of Temperature Profile in an Absorbing, Emitting and Isotropically Scattering Medium by Inverse Analysis", *J. Heat Transfer*, **114**, 1060-1063, 1992.
40. Li, H. Y. and Özisik, M. N., "Inverse Radiation Problem for Simultaneous Estimation of Temperature Profile and Surface Reflectivity", *J. Termophysics and Heat Transfer*, **7**, 88-93, 1993.
41. Bokar, J. and Özisik, M. N., "An Inverse Problem for Estimation of Radiation Temperature Source Term in a Sphere", *Inverse Problems in Engineering*, **1**, 191-205, 1995.

42. Silva Neto, A. J. and Özisik, M. N., "An Inverse Analysis of Simultaneously Estimating Phase Function, Albedo and Optimal Thickness," *ASME/AIChe Conference*, San Diego, CA, Aug. 9-12, 1992.
43. Ho, C. H. and Özisik, M. N., "An Inverse Radiation Problem," *Int. J. Heat and Mass Transfer*, **32**, 335-341, 1989.
44. Li, H. Y. and Özisik, M. N., "Estimation of the Radiation Source Term with a Conjugate-Gradient Method of Inverse Analysis", *Journal of Quantitative Spectroscopy and Radiative Transfer*, **48**, 237-244, 1992.
45. Ruperti Jr., N. and Raynaud, M., "Estimation of the Incident Radiative Heat flux from Transient Temperature Measurements in a Semitransparent Slab", *Proc. 11th International Heat Transfer Conference, Vol. 7*, 325-330, South Korea, 1998.
46. Ruperti Jr., N., Raynaud, M. and Sacadura, J. F., "A Method for the Solution of the Coupled Inverse Heat Conduction-Radiation Problem", *ASME J. Heat Transfer*, **118**, 10-17, 1996.
47. Pomraning G. C. and Siewert, C. E., "On the Integral Form of the Equation of Transfer for a Homogeneous Sphere", *Journal of Quantitative Spectroscopy and Radiative Transfer*, **28**, 503-506, 1982.
48. Press, W. H., Flannery, B. F., Teukolsky, S. A. and Wetterling, W. T., *Numerical Recipes*, Cambridge University Press, New York, 1989.
49. Thynell, S. T. and Özisik, M. N., "Integral Involving an Exponential Integral Function and Exponentials Arising in the Solution of Radiation Transfer", *Journal of Quantitative Spectroscopy and Radiative Transfer*, **33**, 259-266, 1985.

Chapter 6

A GENERAL FORMULATION FOR INVERSE HEAT CONDUCTION

6-1 INTRODUCTION

As apparent from the material in the previous chapters, the solution of inverse problems involving different heat transfer modes generally requires the solution of their associated direct problems. Therefore, the ability of a method of solution of inverse problems to handle complex physical situations is closely related to the direct problem method of solution.

Several practical engineering applications involve geometries irregularly shaped, that is, geometries with boundaries not coinciding with surfaces of constant coordinates in the system where they are referred to. The traditional finite difference methods have computational simplicity when they are applied for the solution of problems involving a regular geometry, with uniformly distributed grids over the region. However, their major drawbacks include their inability to handle effectively the solution of problems over arbitrarily shaped complex geometries. When the geometry is irregular, difficulty arises from the boundary conditions because interpolation is needed between the boundaries and the interior points, in order to develop finite-difference expressions for nodes next to the boundaries. Such interpolations produce large errors in the vicinity of strong curvatures and sharp discontinuities. Therefore, it is difficult and inaccurate to solve problems with traditional finite difference methods over regions having irregular geometries.

Consider, for example, the annular region depicted in figure 6.1.1. It is impossible to discretize such a region in the Cartesian system of coordinates with constant grid spacing Δx and Δy, since the boundaries of the region do not coincide with surfaces of constant x or constant y. However, such difficulty can be easily overcome by using the polar system of coordinates (r,θ), instead of the Cartesian system (x,y). The polar system of coordinates (r,θ) is the natural one for the annular region, because its boundaries are surfaces of constant radius; hence, a uniform grid with constant increments Δr and $\Delta\theta$, in the r and θ directions, respectively, can be generated over the region. Such an example involving the annular region reveals important aspects that will be extended later for general regions: (i) The annular region, irregularly shaped in the Cartesian system of coordinates, was transformed (or mapped) into a rectangle in the polar system of coordinates, as shown in figure 6.1.2; (ii) Governing equations for the physical problem of interest for the annular region shall be written in terms of polar coordinates. Therefore, they can be discretized and solved over the regular region on the polar system of coordinates; (iii) Since the transformation from Cartesian to polar coordinates is one-to-one, the solution developed over the uniform grid on the polar system can be easily transformed backwards to the physical annular region in the Cartesian system of coordinates.

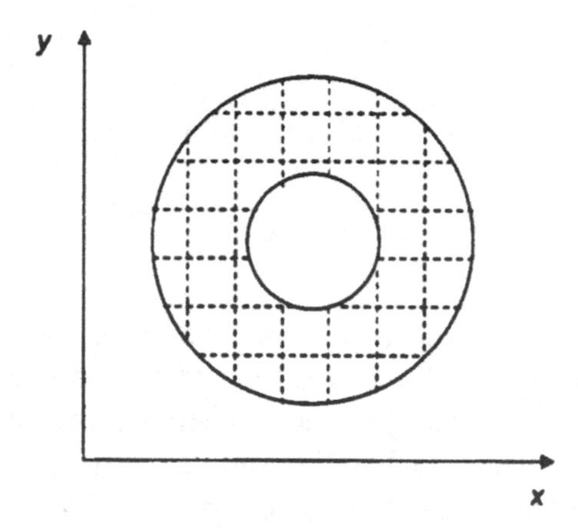

Figure 6.1.1 - Annular region in the Cartesian system of coordinates.

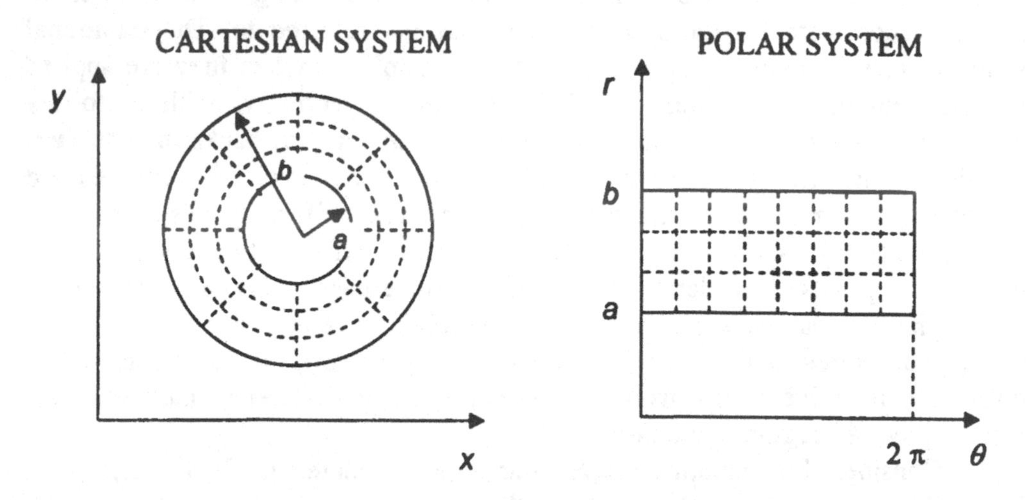

Figure 6.1.2 - Transformation of the annular region in the Cartesian system into a rectangle in the polar system of coordinates.

Many transformations are available in which the physical and computational coordinates (the Cartesian and polar coordinates, respectively, in the example above) are related with algebraic expressions. But such transformations are difficult to construct for general multidimensional cases. The coordinate transformation technique advanced by Thompson [1] alleviates such difficulties, because the transformation is obtained automatically from the numerical solution of partial differential equations. In this approach, a curvilinear mesh is generated over the physical domain, such that one member of each family of curvilinear coordinate lines is coincident with the boundary contour of the region. Therefore, it is also called *boundary fitted coordinates method*. The use of numerical grid generation has provided finite difference methods with the geometrical capabilities of treating irregular geometries of the finite element method, but maintaining their intrinsic simplicity of discretization.

To illustrate the basic concepts in the implementation of Thompson's technique of numerical grid generation, we consider a two-dimensional region, with (x,y) being the coordinates in the physical domain and (ξ,η) the generalized coordinates in the computational domain. The basic steps in Thompson's approach can be summarized as follows:

1. The transformation relations, for mapping the irregular region in the physical domain (x,y) into a regular region in the computational domain (ξ,η) (or vice versa), are determined automatically from the numerical solution of two elliptic partial differential equations of the Laplace or Poisson type. The parabolic and hyperbolic type differential equations have also been used for numerical grid generation; but elliptic equations are preferred because of their smoothing effect in spreading out the boundary slope irregularities. Customarily, the Cartesian coordinate system is used both in the (x,y) physical and (ξ,η) computational domains. It is also possible to use other coordinate systems, such as the (r,θ) polar coordinates in the physical domain and (η,ξ) polar coordinates in the computational domain. In either case, the irregular physical region is mapped into the computational domain as a regular region.
2. The partial differential equations governing the physical phenomena are transformed from the (x,y) independent variables of the physical domain to the (ξ,η) independent variables of the computational domain. Hence, traditional finite difference methods can be used to solve the governing equations in the computational domain.
3. Once the transformed field equations are solved in the computational domain, the solution is transformed from the (ξ,η) computational domain to the (x,y) physical domain, by using the transformation relations developed previously.

An extensive review of numerical grid generation is available in the book by Thompson et al. [1] and the application of this technique to the solution of various engineering problems can be found in references [2-12].

In this chapter, we solve the inverse heat conduction problem of estimating the transient heat flux, applied on part of the boundary of an irregular two-dimensional region, by using **Technique IV**, the conjugate gradient method with adjoint problem for function estimation. The irregular region in the physical domain (x,y) is transformed into a rectangle in the computational domain (ξ,η). The direct, sensitivity and adjoint problems, as well as the gradient equation, are formulated in terms of the generalized coordinates (ξ,η). Therefore, *the present formulation is general and can be applied to the solution of boundary inverse heat conduction problems over any region that can be mapped into a rectangle.*

Chapter 6 is organized as follows. An overview of coordinate transformation relations, required to transform the heat conduction equation and boundary conditions into the computational domain, is presented. We then discuss some basic ideas for mappings and present the boundary value problem of numerical grid generation. After developing the appropriate background, the auxiliary problems and expressions required for the solution of inverse problems with Technique IV over irregular geometries are derived in terms of generalized coordinates. The present approach is then illustrated with an inverse problem of practical engineering interest.

6-2 COORDINATE TRANSFORMATION RELATIONS

Consider a partial differential equation given in the (x,y) independent variables in the physical domain. We seek the transformation of this partial differential equation from the (x,y) to the (ξ,η) independent variables. The transformation from the (x,y) to the (ξ,η) variables can be expressed as

$$\xi \equiv \xi(x,y) \quad , \quad \eta \equiv \eta(x,y) \tag{6.2.1.a,b}$$

and the inverse transformation as

$$x \equiv x(\xi,\eta) \quad , \quad y \equiv y(\xi,\eta) \tag{6.2.2.a,b}$$

The transformation of governing equations requires relations for the transformation of various differential operators, such as the first derivative, gradient, Laplacian, etc. Therefore, in this section we present such relations for use as ready reference in later sections.

The Jacobian of the inverse transformation J is given by:

$$J = J\left(\frac{x,y}{\xi,\eta}\right) = \begin{vmatrix} x_\xi & y_\xi \\ x_\eta & y_\eta \end{vmatrix} = x_\xi y_\eta - x_\eta y_\xi \tag{6.2.3}$$

where the subscripts denote differentiation with respect to the variable considered, i.e.,

$$x_\xi = \frac{\partial x}{\partial \xi}, \quad y_\eta = \frac{\partial y}{\partial \eta}, \quad \text{etc.} \tag{6.2.4}$$

The Jacobian is required to be different from zero in order to obtain one-to-one transformations. This is accomplished by requiring that coordinate lines of the same family do not cross and lines of different families do not cross more than once.

The transformation relations can be developed by application of the chain rule of differentiation. Consider, for example, the first derivatives $\partial T/\partial x$ and $\partial T/\partial y$. By the chain rule of differentiation, we write

$$\frac{\partial T}{\partial x} = \xi_x \frac{\partial T}{\partial \xi} + \eta_x \frac{\partial T}{\partial \eta} \tag{6.2.5.a}$$

$$\frac{\partial T}{\partial y} = \xi_y \frac{\partial T}{\partial \xi} + \eta_y \frac{\partial T}{\partial \eta} \tag{6.2.5.b}$$

Interchanging x and ξ, as well as y and η we obtain

$$\frac{\partial T}{\partial \xi} = x_\xi \frac{\partial T}{\partial x} + y_\xi \frac{\partial T}{\partial y} \tag{6.2.6.a}$$

$$\frac{\partial T}{\partial \eta} = x_\eta \frac{\partial T}{\partial x} + y_\eta \frac{\partial T}{\partial y} \tag{6.2.6.b}$$

The solution of equations (6.2.6.a,b) for $\partial T/\partial x$ and $\partial T/\partial y$ with Cramer's rule gives the transformation relations for the first derivatives as

$$\frac{\partial T}{\partial x} = \frac{1}{J}\left(y_\eta \frac{\partial T}{\partial \xi} - y_\xi \frac{\partial T}{\partial \eta} \right) \tag{6.2.7.a}$$

$$\frac{\partial T}{\partial y} = \frac{1}{J}\left(-x_\eta \frac{\partial T}{\partial \xi} + x_\xi \frac{\partial T}{\partial \eta} \right) \tag{6.2.7.b}$$

A comparison of equations (6.2.5) and (6.2.7) gives

$$\xi_x = \frac{1}{J} y_\eta, \quad \xi_y = -\frac{1}{J} x_\eta \tag{6.2.8.a,b}$$

$$\eta_x = -\frac{1}{J} y_\xi, \quad \eta_y = \frac{1}{J} x_\xi \tag{6.2.8.c,d}$$

Example 6-1. Transform the continuity equation

$$\frac{\partial u}{\partial x}+\frac{\partial v}{\partial y}=0$$

from the (x,y) coordinates of the physical domain to the (ξ,η) coordinates of the computational domain.

Solution. The transformations of the first derivatives are given by equations (6.2.7.a,b). Then the transformation of the above continuity equation from the (x,y) to the (ξ,η) coordinates becomes

$$\frac{1}{J}\left(y_\eta\frac{\partial u}{\partial \xi}-y_\xi\frac{\partial u}{\partial \eta}\right)+\frac{1}{J}\left(-x_\eta\frac{\partial v}{\partial \xi}+x_\xi\frac{\partial v}{\partial \eta}\right)=0$$

The transformation of second derivatives can be obtained by utilizing the transformation relations for the first derivatives and the chain rule of differentiation. Thompson et al. [1] presented extensive relations for the transformation of the divergence, gradient, Laplacian, etc., for both conservative and non-conservative forms, from the Cartesian coordinates to general curvilinear coordinates. Here we present, for ready reference, some of these transformation relations from the (x,y), to the (ξ,η) coordinates in both the conservative and non-conservative forms. It is to be noted that the non-conservative forms can be obtained from the conservative forms, by expanding all derivatives and cancelling the identity terms.

Gradient

Consider the gradient of the scalar quantity T given in the form $\nabla T = T_x\,\mathbf{i}+T_y\,\mathbf{j}$, where i and j are the unit vectors in the x and y directions, respectively. The components T_x and T_y of the gradient vector can be written in the computational domain as:

Conservative form:

$$T_x=\frac{1}{J}\left[\left(y_\eta T\right)_\xi-\left(y_\xi T\right)_\eta\right] \tag{6.2.9.a}$$

$$T_y=\frac{1}{J}\left[-\left(x_\eta T\right)_\xi+\left(x_\xi T\right)_\eta\right] \tag{6.2.9.b}$$

Non-conservative form:

$$T_x = \frac{1}{J}\left(y_\eta T_\xi - y_\xi T_\eta\right) \tag{6.2.10.a}$$

$$T_y = \frac{1}{J}\left(-x_\eta T_\xi + x_\xi T_\eta\right) \tag{6.2.10.b}$$

where the Jacobian J is defined by equation (6.2.3). Note that when the product derivative terms in the conservative form are expanded, the identity terms cancel out and equations (6.2.9) reduce to the non-conservative form given by equations (6.2.10).

Divergence

We now consider the vector quantity **T**, that is,

$$\mathbf{T} = T_1\,\mathbf{i} + T_2\,\mathbf{j} \tag{6.2.11}$$

The divergence of **T** is written in the computational domain (ξ,η), in the conservative and non-conservative forms, respectively, by:

Conservative form:

$$\nabla\cdot\mathbf{T} = \frac{1}{J}\left[\left(y_\eta T_1 - x_\eta T_2\right)_\xi + \left(-y_\xi T_1 + x_\xi T_2\right)_\eta\right] \tag{6.2.12}$$

Non-conservative form:

$$\nabla\cdot\mathbf{T} = \frac{1}{J}\left[y_\eta\left(T_1\right)_\xi - x_\eta\left(T_2\right)_\xi - y_\xi\left(T_1\right)_\eta + x_\xi\left(T_2\right)_\eta\right] \tag{6.2.13}$$

Laplacian

We consider the Laplacian of a scalar quantity T in the physical domain (x,y), that is,

$$\nabla^2 T = \frac{\partial^2 T}{\partial x^2} + \frac{\partial^2 T}{\partial y^2} \tag{6.2.14}$$

This operator in the computational domain (ξ,η) is given in the conservative and non-conservative forms, respectively, by:

Conservative form:

$$J\nabla^2 T = \frac{\partial}{\partial \xi}\left\{\frac{1}{J} y_\eta\left[\left(y_\eta T\right)_\xi - \left(y_\xi T\right)_\eta\right] - \frac{1}{J} x_\eta\left[-\left(x_\eta T\right)_\xi + \left(x_\xi T\right)_\eta\right]\right\} +$$
$$+ \frac{\partial}{\partial \eta}\left\{-\frac{1}{J} y_\xi\left[\left(y_\eta T\right)_\xi - \left(y_\xi T\right)_\eta\right] + \frac{1}{J} x_\xi\left[-\left(x_\eta T\right)_\xi + \left(x_\xi T\right)_\eta\right]\right\}$$

(6.2.15)

Non-conservative form:

$$\nabla^2 T = \frac{1}{J^2}\left[\alpha T_{\xi\xi} - 2\beta T_{\xi\eta} + \gamma T_{\eta\eta}\right] + \left[\left(\nabla^2 \xi\right) T_\xi + \left(\nabla^2 \eta\right) T_\eta\right] \qquad (6.2.16)$$

where

$$\alpha = x_\eta^2 + y_\eta^2, \quad \beta = x_\xi x_\eta + y_\xi y_\eta, \quad \gamma = x_\xi^2 + y_\xi^2, \quad J = x_\xi y_\eta - x_\eta y_\xi \qquad (6.2.17.a\text{-}d)$$

Normal Derivatives

Conservative form:

The normal derivatives of T to the ξ-constant line along the normal $\mathbf{n}^{(3)}$ shown in figure 6.2.1 is given by

$$\frac{\partial T}{\partial \mathbf{n}^{(\xi)}} = \frac{1}{J\alpha^{1/2}}\left\{y_\eta\left[\left(y_\eta T\right)_\xi - \left(y_\xi T\right)_\eta\right] - x_\eta\left[-\left(x_\eta T\right)_\xi + \left(x_\xi T\right)_\eta\right]\right\} \qquad (6.2.18.a)$$

and to the η - constant line along the normal $\mathbf{n}^{(4)}$ shown in figure 6.2.1 is given by

$$\frac{\partial T}{\partial \mathbf{n}^{(\eta)}} = \frac{1}{J\gamma^{1/2}}\left\{-y_\xi\left[\left(y_\eta T\right)_\xi - \left(y_\xi T\right)_\eta\right] + x_\xi\left[-\left(x_\eta T\right)_\xi + \left(x_\xi T\right)_\eta\right]\right\} \qquad (6.2.18.b)$$

Non-conservative form:

The normal derivatives of T to the ξ-constant line along the normal $\mathbf{n}^{(3)}$ is given by

$$\frac{\partial T}{\partial \mathbf{n}^{(\xi)}} = \frac{1}{J\alpha^{1/2}}\left(\alpha T_\xi - \beta T_\eta\right) \qquad (6.2.19.a)$$

and to the η- constant line along the normal $\mathbf{n}^{(4)}$ is given by

$$\frac{\partial T}{\partial \mathbf{n}^{(\eta)}} = \frac{1}{J\gamma^{1/2}}\left(-\beta T_{\xi} + \gamma T_{\eta}\right) \tag{6.2.19.b}$$

where α, β, γ and J are defined by equations (6.2.17).

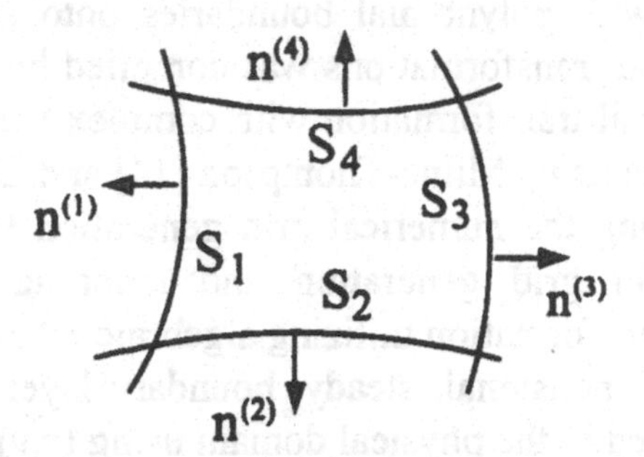

Figure 6.2.1 - Outward drawn unit normal vectors to ξ = constant and η = constant lines

The derivative along the normal vectors $\mathbf{n}^{(1)}$ and $\mathbf{n}^{(2)}$ are obtained by switching signs in equations (6.2.18.a,b), respectively, for the conservative form, or in equations (6.2.19.a,b), respectively, for the non-conservative form.

The reader should consult Thompson et al [1] for the transformation relations for other partial derivative operators such as for $\partial^2/\partial x \partial y$.

Example 6.2. Consider the two-dimensional transient heat conduction equation in the physical domain (x,y,t) given by

$$\frac{1}{\alpha_t}\frac{\partial T}{\partial t} = \frac{\partial^2 T}{\partial x^2} + \frac{\partial^2 T}{\partial y^2} \tag{6.2.20.a}$$

where α_t is the thermal diffusivity. Transform this equation from the (x,y,t) independent variables of the physical domain to the (ξ,η,t) independent variables of the computational domain.

Solution. By utilizing the non-conservative form of the Laplacian given by equation (6.2.16), we can write the heat conduction equation in terms of the generalized variables (ξ,η,t) as

$$\frac{1}{\alpha_t}\frac{\partial T}{\partial t} = \frac{1}{J^2}\left(\alpha T_{\xi\xi} - 2\beta T_{\xi\eta} + \gamma T_{\eta\eta}\right) + \left[\left(\nabla^2 \xi\right)T_{\xi} + \left(\nabla^2 \eta\right)T_{\eta}\right] \tag{6.2.20.b}$$

where α, β, γ and J are defined by equations (6.2.17).

6-3 SIMPLE TRANSFORMATIONS

A variety of approaches has been reported in the literature for the transformation of irregularly shaped regions into simple regular regions such as a square, rectangle, etc. The basic theory behind such transformations is quite old. For example, conformal transformation has been widely used in classical analysis. Schwarz-Christoffel transformation is well known for conformal mapping of regions with polynomial boundaries onto an upper-half plane. A dictionary of conformal transformations was compiled by Kober [13]. Details of application of conformal transformation with complex variable technique can be found in the standard texts by Milne-Thompson [14] and Churchill [15].

Before presenting the numerical grid generation technique, we illustrate the basic concepts in grid generation and mapping by considering one-dimensional simple transformation utilizing algebraic relations.

Consider two-dimensional, steady, boundary layer flow over a flat plate mathematically modeled in the physical domain using (x,y) Cartesian coordinates. To solve such flow problem with finite-differences, customarily a rectangular grid is constructed over the solution domain and the nodes are concentrated near the wall where the gradients are large, as illustrated in figure 6.3.1.a. A uniform grid is constructed in the x-direction, but a nonuniform grid is used in the y-direction. To alleviate the difficulties associated with the use of nonuniform grids, the problem can be transformed from the physical (x,y) domain to the computational (ξ,η) domain, with a coordinate transformation that will allow the use of uniform grids in both the ξ and η directions, as illustrated in figure 6.3.1.b.

A coordinate transformation that maps a nonuniform grid spacing in the y direction into a uniform grid spacing in the η direction, but allows the grid spacing in the x direction to remain unchanged, is given by Roberts [16] in the form

$$\xi = x \tag{6.3.1.a}$$

$$\eta = 1 - \frac{\ell n[A(y)]}{\ell n B} \tag{6.3.1.b}$$

where

$$A(y) = \frac{\beta + \left(1 - \frac{y}{h}\right)}{\beta - \left(1 - \frac{y}{h}\right)}, \qquad B = \frac{\beta + 1}{\beta - 1} \tag{6.3.2.a,b}$$

Here β is the ***stretching parameter***, which assumes values $1 < \beta < \infty$. As β approaches unity, more grid points are clustered near the wall in the physical domain. The inverse transform is given by

$$x = \xi \tag{6.3.3.a}$$

$$y = \frac{(\beta+1)-(\beta-1)B^{1-\eta}}{1+B^{1-\eta}}h \tag{6.3.3.b}$$

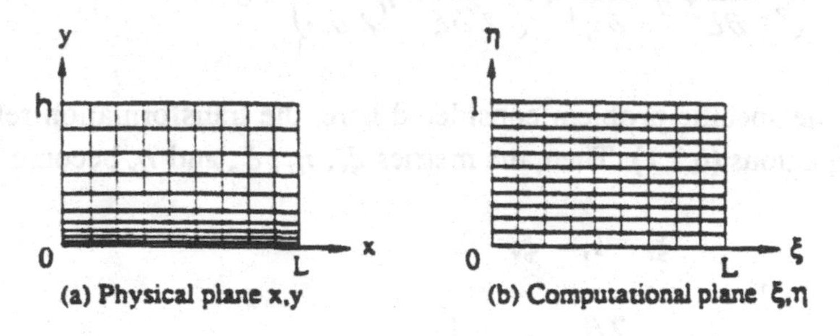

Figure 6.3.1 - One dimensional stretching transformation

To illustrate the grid concentration as $\beta \rightarrow 1$, we set, for example, $\eta = 0.4$ and calculate y for different values of β, as shown below.

β	1.5	1.1	1.01
y	0.327	0.205	0.0705

Once the relations for the coordinate transformation are established, the differential equations governing the physical phenomena must be transformed from the (x,y) independent variables of the physical domain to the (ξ,η) independent variables of the computational domain under the same transformation, since all numerical computations will be performed on the (ξ,η) computational domain. To illustrate the transformation of the governing partial differential equations, we consider, say, the continuity equation given by

$$\frac{\partial u}{\partial x} + \frac{\partial v}{\partial y} = 0 \tag{6.3.4}$$

The transformation of this equation from the (x,y) to the (ξ,η) variables under the general transformation defined by equations (6.2.1.a,b) was given in Example 6.1 by

$$\frac{1}{J}\left(y_{\eta}\frac{\partial u}{\partial \xi} - y_{\xi}\frac{\partial u}{\partial \eta}\right) + \frac{1}{J}\left(-x_{\eta}\frac{\partial v}{\partial \xi} + x_{\xi}\frac{\partial v}{\partial \eta}\right) = 0 \tag{6.3.5}$$

The computational derivatives y_η, y_ξ, x_η and x_ξ are expressed in terms of the metrics ξ_x, ξ_y, η_x and η_y, according to equations (6.2.8). Then, the transformed equation (6.3.5) takes the form

$$\left(\xi_x \frac{\partial u}{\partial \xi}+\eta_x \frac{\partial u}{\partial \eta}\right)+\left(\xi_y \frac{\partial v}{\partial \xi}+\eta_y \frac{\partial v}{\partial \eta}\right)=0 \tag{6.3.6}$$

For the specific problem considered here, the transformation relations are given by equations (6.3.1). Then, the metrics ξ_x, η_x, ξ_y, and η_y become

$$\xi_x = 1, \quad \xi_y = 0 \tag{6.3.7.a,b}$$

$$\eta_x = 0, \quad \eta_y = \frac{2\beta}{h\, \ell n B}\frac{1}{\beta^2-\left(1-\frac{y}{h}\right)^2} \tag{6.3.7.c,d}$$

By introducing equations (6.3.7) into equation (6.3.6), the transformed continuity equation takes the form

$$\frac{\partial u}{\partial \xi}+\eta_y \frac{\partial v}{\partial \eta}=0 \tag{6.3.8}$$

where η_y is defined by equation (6.3.7.d).

We note that the transformed continuity equation (6.3.8) retains its original general form, except for the coefficient η_y accompanying the $\partial v/\partial \eta$ term. Therefore, the transformed equation (6.3.8) is slightly more complicated than its original form given by equation (6.3.4); but it will be solved over a uniform grid both in the ξ and η directions in the computational domain, using the (ξ,η) rectangular coordinates. Clearly, the finite-difference solution in the (ξ,η) computational domain with a uniform grid is much easier and more accurate than solving the problem in the original physical domain with nonuniform grid. If the problem involves other partial differential equations, they also need to be transformed into the (ξ,η) computational domain in a similar manner.

Once the problem is solved in the computational domain, the results are transformed backwards into the physical domain from each (ξ,η) location to the corresponding (x,y) location, by using the inverse transformation given by equations (6.3.3).

Roberts [16] and other investigators have proposed numerous other simple stretching transformations. However, it is difficult to develop analytic transformations capable of clustering grids around arbitrary locations, whereas the numerical grid generation technique provides a unified approach for developing transformations capable of dealing with more general situations.

6-4 BASIC IDEAS IN NUMERICAL GRID GENERATION AND MAPPING

In finite difference solutions of partial differential equations over regions having regular shapes, such as a rectangle, cylinder or sphere, the discretization can be made to conform to the boundaries of the region. As a result, the boundary interpolation is avoided. For regions having an arbitrary irregular shape, this is not possible. One way to overcome such difficulty is to map the region, with a suitable transformation, into the computational domain where the geometry becomes regular, say, rectangular. The problem is then solved over the rectangular region with a square grid by using conventional finite-differences. The solution developed in the computational domain is then transformed backwards into the physical domain.

To illustrate the basic concepts in the mapping and development of curvilinear coordinates, we consider a two dimensional physical domain in the (x,y) Cartesian coordinates and a computational domain in the (ξ,η) Cartesian coordinates. The transformation between (x,y) and (ξ,η) coordinates should be such that the boundaries of the physical domain must be coincident with the curvilinear coordinates (ξ,η); thus there will be no need for boundary node interpolation.

Consider an irregular region ABCDA in the physical domain in the (x,y) Cartesian coordinates, as illustrated in figure 6.4.1.a. The region is called simply connected because it contains no obstacles in its interior. This region is to be mapped into a rectangle in the computational domain (ξ,η), in the following manner:

- Set η constant and let ξ to vary monotonically along the boundary segments AB and DC of the physical region, and
- Set ξ constant and let η to vary monotonically along the boundary segments AD and BC of the physical region.

Clearly, with such requirements on the values of ξ and η along the boundaries of the physical region, the segments AB and DC are mapped into the computational domain as horizontal lines, while the segments AD and BC are mapped into the computational domain as vertical lines, as illustrated in figure 6.4.1.b. Notice that each boundary segment of the irregular region in the physical domain is mapped into the sides of the rectangular region in the computational domain. Without loss of generality, we can choose $\Delta\xi = \Delta\eta = 1$ in the computational domain, so that M and N are the number of ξ and η grid lines in the region, respectively.

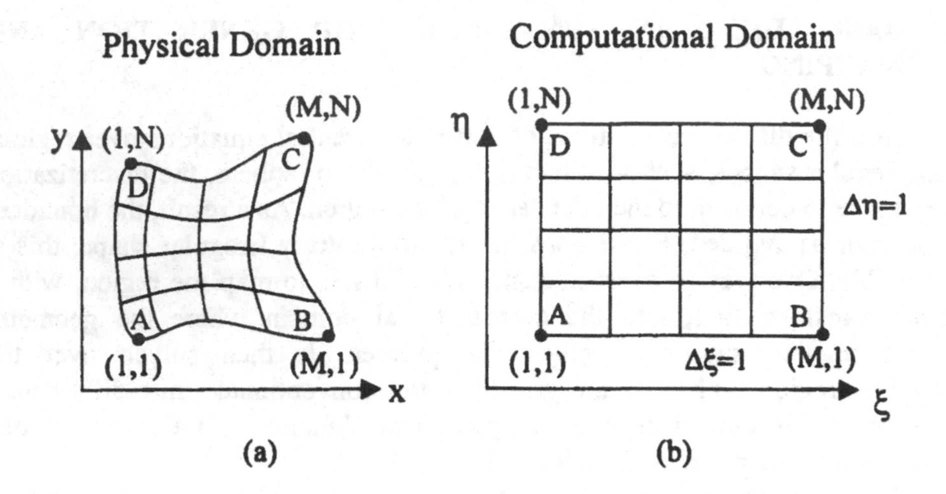

Figure 6.4.1 - Mapping an irregular simply connected region into the computational domain as a rectangle.

In the previous illustration of mapping, an irregular region in the physical domain is mapped as a rectangular region into the computational domain. Depending on the choice of the values of (ξ,η) along the boundary segments of the physical region, a variety of other acceptable configurations can be generated in the computational domain. To illustrate this matter, we consider an L-shaped irregular region ABCDEFA in the physical domain as shown in figure 6.4.2.a. One possibility is to map the region into an L-shaped regular region, as illustrated in figure 6.4.2.b. Another possibility is to map the L-shaped irregular region as a rectangle in the computational domain, as shown in figure 6.4.3.

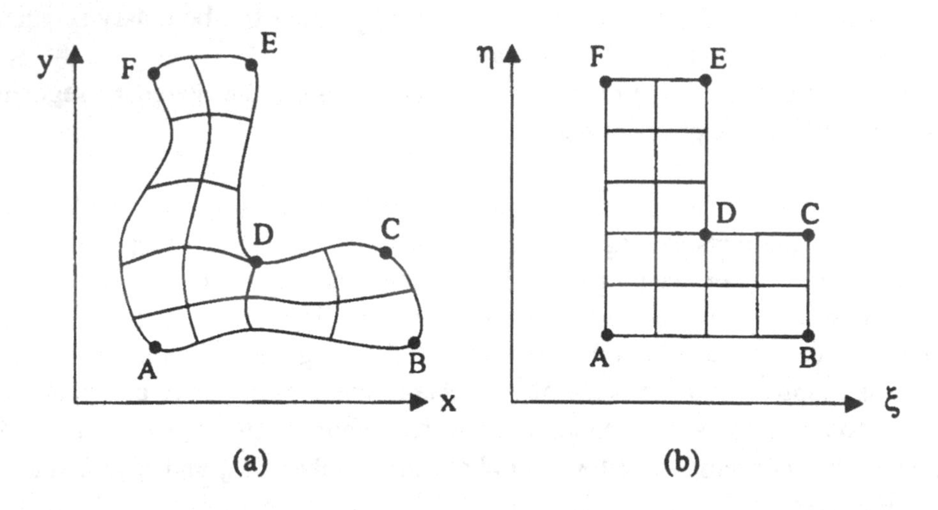

Figure 6.4.2 - Mapping of an L-shaped irregular region into an L-shaped regular region

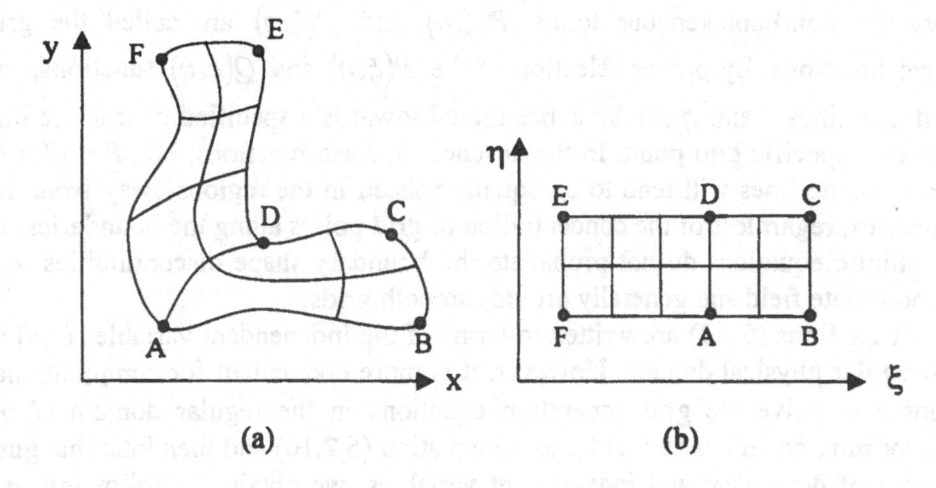

Figure 6.4.3 - Mapping of an L-shaped irregular region into a rectangle

The example presented in figures 6.4.2 and 6.4.3 shows that different possible mappings can be envisioned to transform the same irregular region in the physical domain into a regular region in the computational domain. The analyst must exercise his expertise in order to devise the most suitable mapping for each physical region of interest.

The mappings illustrated in figures 6.4.1-3 involved simply-connected regions in the physical domain. Similarly, irregular multiply-connected regions in the physical domain can be transformed into regular multiply-connected regions in the computational domain; or, alternatively, branch-cuts can be used so that the transformed region becomes a rectangle. The reader should consult references [1] and [12] for possible mappings involving multiply-connected regions, as well as simply-connected regions containing reentrant boundary surfaces.

6-5 BOUNDARY VALUE PROBLEM OF NUMERICAL GRID GENERATION

We present here the numerical grid generation and mapping technique advanced by Thompson [1], involving the solution of two elliptic partial differential equations in the form

$$\nabla^2 \xi = \frac{\partial^2 \xi}{\partial x^2} + \frac{\partial^2 \xi}{\partial y^2} = P(\xi, \eta) \tag{6.5.1.a}$$

$$\nabla^2 \eta = \frac{\partial^2 \eta}{\partial x^2} + \frac{\partial^2 \eta}{\partial y^2} = Q(\xi, \eta) \tag{6.5.1.b}$$

where the non-homogeneous terms $P(\xi,\eta)$ and $Q(\xi,\eta)$ are called the grid control functions. By proper selection of the $P(\xi,\eta)$ and $Q(\xi,\eta)$ functions, the coordinate lines ξ and η can be concentrated towards a specified coordinate line or about a specific grid point. In the absence of these functions, i.e., $P = Q = 0$, the coordinate lines will tend to be equally spaced in the regions away from the boundaries, regardless of the concentration of grid points along the boundaries. In fact, elliptic equations do not propagate the boundary shape discontinuities into the coordinate field and generally provide smooth grids.

Equations (6.5.1) are written in terms of the independent variables (x,y) in the irregular physical domain. However, it is more convenient for computational purposes to solve the grid generation equations in the regular domain (ξ,η). Transforming equations (6.5.1) by using equation (6.2.16) and then interchanging the roles of dependent and independent variables, we obtain the following two elliptic equations for the determination of the unknowns x and y, in terms of the independent variables ξ and η in the computational domain:

$$\alpha\frac{\partial^2 x}{\partial\xi^2}-2\beta\frac{\partial^2 x}{\partial\xi\,\partial\eta}+\gamma\frac{\partial^2 x}{\partial\eta^2}+J^2\left[P(\xi,\eta)\frac{\partial x}{\partial\xi}+Q(\xi,\eta)\frac{\partial x}{\partial\eta}\right]=0 \qquad \text{(6.5.2.a)}$$

$$\alpha\frac{\partial^2 y}{\partial\xi^2}-2\beta\frac{\partial^2 y}{\partial\xi\,\partial\eta}+\gamma\frac{\partial^2 y}{\partial\eta^2}+J^2\left[P(\xi,\eta)\frac{\partial y}{\partial\xi}+Q(\xi,\eta)\frac{\partial y}{\partial\eta}\right]=0 \qquad \text{(6.5.2.b)}$$

where the geometric coefficients α, β, γ and the Jacobian J are obtained from equations (6.2.17).

The mathematical problem defined by equations (6.5.2), subjected to appropriate boundary conditions, constitutes the *boundary value problem of numerical grid generation*. Generally, such a problem is solved by finite-differences, by utilizing either first-kind or second-kind boundary conditions, as described next.

Boundary Conditions

(i) Boundary Condition of the First Kind. In most applications, the values of the (x,y) coordinates of the boundaries of the physical domain are known, for each grid point (ξ,η). Then, the grid generation problem becomes one of solving equations (6.5.2) over the regular computational domain, with prescribed values of (x,y) at the boundaries.

(ii) Homogeneous Boundary Condition of the Second Kind: Orthogonality of Grid Lines. There are situations in which ξ (or η) grid lines are required to intersect some portion of the boundary segment in the physical domain at a specified angle, ϕ, as illustrated in figure 6.5.1. A commonly imposed condition is that ξ (or η) grid lines intersect some segment of the physical boundary normally, that is $\phi = \pi/2$. In fact, it has been shown that the numerical discretization error increases when the intersection angle departures from $\phi = \pi/2$ [1].

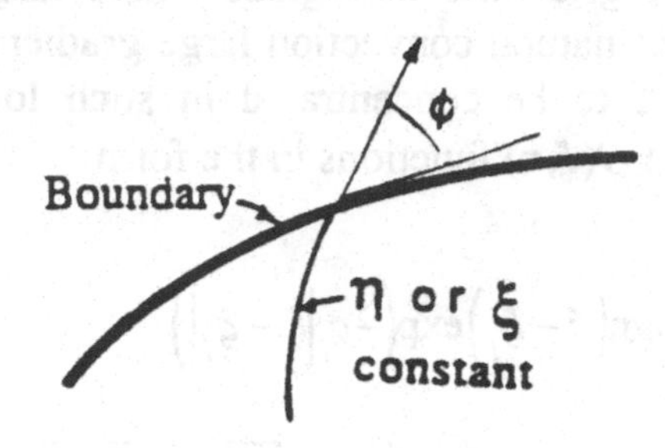

Figure 6.5.1 - The angle of intersection of grid lines.

To establish the mathematical expression to implement the requirement of orthogonality of grid lines, we consider the gradients of ξ and η defined respectively by

$$\nabla\xi = \xi_x \mathbf{i} + \xi_y \mathbf{j} \tag{6.5.3.a}$$

$$\nabla\eta = \eta_x \mathbf{i} + \eta_y \mathbf{j} \tag{6.5.3.b}$$

where **i** and **j** are the unit direction vectors. The dot product of $\nabla\xi$ and $\nabla\eta$, i.e.,

$$\nabla\xi \cdot \nabla\eta = \xi_x \eta_x + \xi_y \eta_y \tag{6.5.4}$$

represents the cosine of the angle ϕ. By introducting ξ_x, η_x, ξ_y and η_y from equations (6.2.8) into equation (6.5.4) we obtain

$$\nabla\xi \cdot \nabla\eta = -\frac{1}{J^2}\left(x_\xi x_\eta + y_\xi y_\eta\right) \tag{6.5.5}$$

In the case of orthogonality, we have $\phi = \pi/2$ or $\cos\phi = 0$. Then equation (6.5.5) reduces to

$$x_\xi x_\eta + y_\xi y_\eta = 0 \tag{6.5.6}$$

This is the criterion to be implemented in the computational domain, whenever the ξ (or η) constant grid lines are required to intersect the physical boundary normally.

Grid Control Functions

The user-specific grid control functions $P(\xi,\eta)$ and $Q(\xi,\eta)$ are useful to concentrate the interior grid lines in regions where large gradients occur. For example, in problems of natural convection large gradients occur near the walls, hence grid points need to be concentrated in such locations. Thompson [1] specified the $P(\xi,\eta)$ and $Q(\xi,\eta)$ functions in the form

$$\begin{aligned} P(\xi,\eta) = &-\sum_{i=1}^{n} a_i \operatorname{sign}\left(\xi-\xi_i\right)\exp\left(-c_i\left|\xi-\xi_i\right|\right) \\ &-\sum_{i=1}^{m} b_i \operatorname{sign}\left(\xi-\xi_i\right)\exp\left(-d_i\sqrt{\left(\xi-\xi_i\right)^2+\left(\eta-\eta_i\right)^2}\right) \end{aligned} \tag{6.5.7.a}$$

and

$$\begin{aligned} Q(\xi,\eta) = &-\sum_{i=1}^{n^*} a_i^* \operatorname{sign}\left(\eta-\eta_i\right)\exp\left(-c_i^*\left|\eta-\eta_i\right|\right) \\ &-\sum_{i=1}^{m^*} b_i^* \operatorname{sign}\left(\eta-\eta_i\right)\exp\left(-d_i^*\sqrt{\left(\xi-\xi_i\right)^2+\left(\eta-\eta_i\right)^2}\right) \end{aligned} \tag{6.5.7.b}$$

We note that the $P(\xi,\eta)$ and $Q(\xi,\eta)$ functions are similar, except that ξ and η are interchanged. Due to the form of the Poisson equations (6.5.1), the control function $P(\xi,\eta)$ acts on the attraction of ξ = constant lines, while the function $Q(\xi,\eta)$ acts on the attraction of η = constant lines. The physical significance of various terms in equations (6.5.7) are as follows.

In the first summation of equation (6.5.7.a), the amplitude a_i attracts ξ = constant lines towards the $\xi = \xi_i$ line; and in the second summation the amplitude b_i attracts ξ = constant lines towards the point (ξ_i,η_i). Figure 6.5.2 illustrates such effects. Similar effects are obtained on the η = constant lines with the $Q(\xi,\eta)$ grid control function.

The summation indexes n and m (or n^* and m^*) denote the number of line and point concentrations, respectively. The sign function, sign $(\xi-\xi_i)$, ensures that attraction of ξ lines occurs on both sides of the ξ_i line or (ξ_i, η_i) point. Without the sign function, the attraction occurs only on the side towards increasing ξ, with repulsion occurring on the other side. The coefficients c_i, c_i^* and d_i, d_i^* control the decay of attraction with the distance, while a_i, a_i^* and b_i, b_i^* give the amplitude of the attraction.

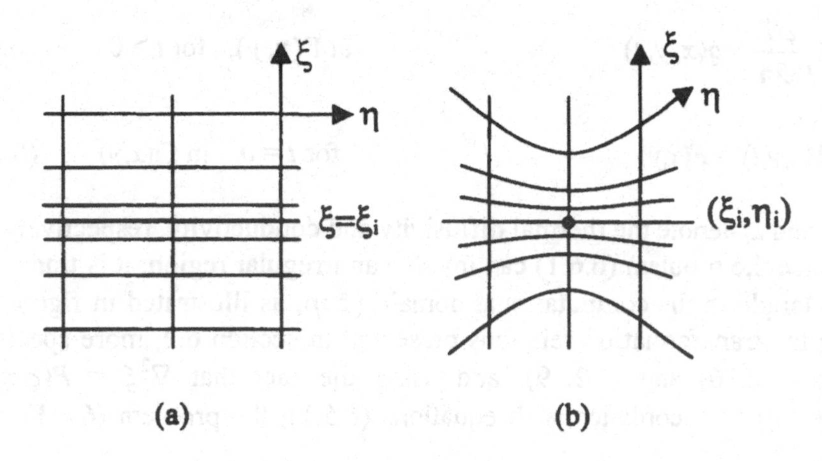

Figure 6.5.2 - The attraction of ξ = constant lines towards (a) the coordinate line $\xi = \xi_i$ and (b) the point (ξ_i, η_i).

For other grid control approaches, the reader is referred to references [1, 12].

6-6 A GENERALIZED COORDINATES APPROACH FOR INVERSE HEAT CONDUCTION [11]

In the previous sections of this chapter we presented the formulation and discussed aspects relevant for the numerical grid generation and domain transformation approach. We now use such an approach allied with **Technique IV**, for the estimation of the transient heat flux applied on part of the boundary of a general two-dimensional region.

The direct, inverse, sensitivity and adjoint problems, as well as the gradient equation, all required for the solution of inverse problems with Technique IV, are derived in terms of the generalized coordinates (ξ,η) in the computational domain, as described next.

Direct Problem

The physical problem considered here involves the linear heat conduction in a two-dimensional general region $\Omega(x,y)$, in the Cartesian coordinates system (x,y). The initial temperature distribution in the region is $F(x,y)$. For times $t > 0$, the boundary $\Gamma(x,y)$ of the region is subjected to a second kind boundary condition. The mathematical formulation of this problem is given by:

$$\frac{1}{\alpha_t}\frac{\partial}{\partial t}T(x,y,t) = \frac{\partial^2 T}{\partial x^2} + \frac{\partial^2 T}{\partial y^2} \qquad \text{in } \Omega(x,y), \text{ for } t > 0 \qquad (6.6.1.a)$$

$$k_t \frac{\partial T}{\partial \mathbf{n}} = q(x,y,t) \qquad \text{at } \Gamma(x,y), \quad \text{for } t>0 \qquad (6.6.1.b)$$

$$T(x,y,t) = F(x,y) \qquad \text{for } t=0, \quad \text{in } \Omega(x,y) \qquad (6.6.1.c)$$

where α_t and k_t denote the thermal diffusivity and conductivity, respectively.

Since the problem (6.6.1) can involve an irregular region, it is transformed into a rectangle in the computational domain (ξ,η), as illustrated in figure 6.4.1. By using the transformation relations presented in section 6.2, more specifically equations (6.2.16) and (6.2.19), and using the fact that $\nabla^2\xi = P(\xi,\eta)$ and $\nabla^2\eta = Q(\xi,\eta)$ in accordance with equations (6.5.1), the problem (6.6.1) can be written as:

$$\frac{1}{\alpha_t}\frac{\partial}{\partial t}T(\xi,\eta,t) = \frac{1}{J^2}\left(\alpha T_{\xi\xi} - 2\beta T_{\eta\xi} + \gamma T_{\eta\eta}\right) + \left(PT_\xi + QT_\eta\right) \qquad (6.6.2.a)$$

$$\text{in } 1<\xi<M, 1<\eta<N \ , \text{for } t>0$$

$$\frac{k_t}{J\sqrt{\alpha}}\left(\beta T_\eta - \alpha T_\xi\right) = q_1(t) \qquad \text{at } \xi = 1 \ , \ 1<\eta<N \ , \ \text{for } t>0 \qquad (6.6.2.b)$$

$$\frac{k_t}{J\sqrt{\gamma}}\left(\beta T_\xi - \gamma T_\eta\right) = q_2(t) \qquad \text{at } \eta = 1 \ , \ 1<\xi<M \ , \ \text{for } t>0 \qquad (6.6.2.c)$$

$$\frac{k_t}{J\sqrt{\alpha}}\left(\alpha T_\xi - \beta T_\eta\right) = q_3(t) \qquad \text{at } \xi = M \ , \ 1<\eta<N \ , \ \text{for } t>0 \qquad (6.6.2.d)$$

$$\frac{k_t}{J\sqrt{\gamma}}\left(\gamma T_\eta - \beta T_\xi\right) = q_4(t) \qquad \text{at } \eta = N \ , \ 1<\xi<M \ , \ \text{for } t>0 \qquad (6.6.2.e)$$

$$T(\xi,\eta,0) = F^*(\xi,\eta) \qquad \text{for } t=0 \ , \ \text{in } 1<\xi<M \ , \ 1<\eta<N \qquad (6.6.2.f)$$

where the subscripts ξ and η above denote partial derivatives and $F^*(\xi,\eta)$ is the initial condition $F(x,y)$ rewritten in terms of the independent variables ξ and η.

For the *Direct Problem*, the thermophysical properties α_t and k_t, the initial condition $F^*(\xi,\eta)$, the heat fluxes $q_1(t)$, $q_2(t)$, $q_3(t)$ and $q_4(t)$ applied on the boundary of the region, as well as the transformation from the physical domain into the computational domain, defined by α, β, γ, J, $P(\xi,\eta)$ and $Q(\xi,\eta)$, are considered known. The direct problem is concerned with the determination of the temperature field $T(\xi,\eta,t)$ in the region.

Inverse Problem

For the inverse problem considered here, the heat flux $q_3(t)$ at the boundary $\xi = M$ is regarded as unknown, while all the other quantities appearing in equations (6.6.2) are assumed to be known with sufficient degree of accuracy. The heat flux $q_3(t)$ is to be estimated by using the transient readings of NS temperature sensors located at the positions (ξ_m, η_m), $m = 1, ..., NS$, during the time interval $0 < t < t_f$. Such temperature measurements may contain random errors. The present inverse problem is solved as a function estimation approach by using Technique IV, that is, no information regarding the functional form of the unknown is considered available for the inverse analysis, except that it belongs to the Hilbert space L_2 in $0 < t < t_f$.

The ill-posed inverse problem stated above is re-formulated as a well-posed minimization problem. Hence, an estimation for the function $q_3(t)$ is obtained by minimizing the following functional

$$S[q_3(t)] = \frac{1}{2} \int_{t=0}^{t_f} \sum_{m=1}^{NS} \left[T(\xi_m, \eta_m, t; q_3) - Y_m(t)\right]^2 dt \tag{6.6.3}$$

where $Y_m(t)$ and $T(\xi_m, \eta_m, t; q_3)$ are the measured and estimated temperatures at the measurements positions (ξ_m, η_m), $m = 1, ..., NS$. The estimated temperatures are obtained from the solution of the direct problem (6.6.2) by using an estimate for the heat flux $q_3(t)$.

We note that the inverse problem, as stated above, can also be used for the estimation of the heat transfer coefficient at the boundary $\xi = M$, if the cooling fluid temperature is known. In fact, it has been shown that the computational time for the estimation of the heat flux and posterior estimation of the heat transfer coefficient, by using the estimated heat flux and known fluid temperature, is smaller than that for the solution of the inverse problem involving the heat transfer coefficient as unknown [17]. Such behavior is due to the fact that the functional given by equation (6.6.3) is quadratic and the inverse problem of estimating the boundary heat flux is linear. On the other hand, the estimation of the heat transfer coefficient involves the minimization of a non-quadratic functional.

In order to apply Technique IV for minimizing the functional given by equation (6.6.3), we need to develop the *sensitivity* and *adjoint problems*, as described next.

Sensitivity Problem

In order to develop the sensitivity problem, we assume that the temperature $T(\xi,\eta,t)$ undergoes a variation $\Delta T(\xi,\eta,t)$, when the unknown boundary heat flux $q_3(t)$ undergoes a variation $\Delta q_3(t)$. By substituting into the direct problem given by equations (6.6.2), $T(\xi,\eta,t)$ by $[T(\xi,\eta,t) + \Delta T(\xi,\eta,t)]$ and $q_3(t)$ by $[q_3(t) + \Delta q_3(t)]$, and subtracting from the resulting expressions the original direct problem, we obtain the following ***sensitivity problem*** for the determination of the ***sensitivity function*** $\Delta T(\xi,\eta,t)$:

$$\frac{1}{\alpha_t}\frac{\partial}{\partial t}\Delta T(\xi,\eta,t) = \frac{1}{J^2}\left(\alpha\Delta T_{\xi\xi} - 2\beta\Delta T_{\eta\xi} + \gamma\Delta T_{\eta\eta}\right) + \left(P\Delta T_\xi + Q\Delta T_\eta\right) \tag{6.6.4.a}$$

$$\text{in } 1<\xi<M, 1<\eta<N \text{ , for } t>0$$

$$\frac{k_t}{J\sqrt{\alpha}}\left(\beta\Delta T_\eta - \alpha\Delta T_\xi\right) = 0 \qquad \text{at } \xi=1 \text{ , } 1<\eta<N \text{ , for } t>0 \tag{6.6.4.b}$$

$$\frac{k_t}{J\sqrt{\gamma}}\left(\beta\Delta T_\xi - \gamma\Delta T_\eta\right) = 0 \qquad \text{at } \eta=1 \text{ , } 1<\xi<M \text{ , for } t>0 \tag{6.6.4.c}$$

$$\frac{k_t}{J\sqrt{\alpha}}\left(\alpha\Delta T_\xi - \beta\Delta T_\eta\right) = \Delta q_3(t) \quad \text{at } \xi=M \text{ , } 1<\eta<N \text{ , for } t>0 \tag{6.6.4.d}$$

$$\frac{k_t}{J\sqrt{\gamma}}\left(\gamma\Delta T_\eta - \beta\Delta T_\xi\right) = 0 \qquad \text{at } \eta=N \text{ , } 1<\xi<M \text{ , for } t>0 \tag{6.6.4.e}$$

$$\Delta T(\xi,\eta,0) = 0 \qquad \text{for } t=0 \text{ , in } 1<\xi<M \text{ , } 1<\eta<N \tag{6.6.4.f}$$

We note that the sensitivity problem is independent of the unknown heat flux $q_3(t)$. Hence the present estimation problem is linear.

Adjoint Problem

In order to develop the adjoint problem, we multiply the differential equation of the direct problem, equation (6.6.2.a) by the Lagrange multiplier $\lambda(\xi,\eta,t)$, integrate over the time and space domains and add the resulting expression to the functional (6.6.3). The following extended functional is obtained:

$$S[q_3(t)] = \frac{1}{2}\int_{t=0}^{t_f}\int_{\xi=1}^{M}\int_{\eta=1}^{N}\sum_{m=1}^{NS}[T(\xi,\eta,t;q_3) - Y_m(t)]^2\delta(\xi-\xi_m)\delta(\eta-\eta_m)d\eta\, d\xi\, dt +$$

$$\int_{t=0}^{t_f}\int_{\xi=1}^{M}\int_{\eta=1}^{N}\left\{\frac{1}{\alpha_t}\frac{\partial T}{\partial t} - \frac{1}{J^2}\left(\alpha T_{\xi\xi} - 2\beta T_{\eta\xi} + \gamma T_{\eta\eta}\right) - \left(PT_\xi + QT_\eta\right)\right\}\lambda(\xi,\eta,t)J\, d\eta\, d\xi\, dt \tag{6.6.5}$$

where $\delta(.)$ is the Dirac delta function.

We assume that the functional $S[q_3(t)]$ is perturbed by $\Delta S[q_3(t)]$ when the boundary heat flux $q_3(t)$ undergoes a variation $\Delta q_3(t)$. By substituting into equation (6.6.5), $T(\xi,\eta,t)$ by $[T(\xi,\eta,t)+\Delta T(\xi,\eta,t)]$ and $S[q_3(t)]$ by $\{S[q_3(t)+\Delta S[q_3(t)]\}$, and subtracting from the resulting expressions the original equation (6.6.5), we obtain the following expression for the variation of the extended functional:

$$\Delta S[q_3(t)] = \int_{t=0}^{t_f}\int_{\xi=1}^{M}\int_{\eta=1}^{N}\sum_{m=1}^{NS}\left[T(\xi,\eta,t;q_3) - Y_m(t)\right]\Delta T(\xi,\eta,t;q_3)$$
$$\delta(\eta-\eta_m)\,\delta(\xi-\xi_m)\,d\eta\,d\xi\,dt +$$
$$+\int_{t=0}^{t_f}\int_{\xi=1}^{M}\int_{\eta=1}^{N}\left\{\frac{1}{\alpha_t}\frac{\partial\Delta T}{\partial t} - \frac{1}{J^2}\left(\alpha\Delta T_{\xi\xi} - 2\beta\Delta T_{\eta\xi} + \gamma\Delta T_{\eta\eta}\right) - \left(P\Delta T_\xi + Q\Delta T_\eta\right)\right\}$$
$$\lambda(\xi,\eta,t)\;J\,d\eta\,d\xi\,dt \tag{6.6.6}$$

The second integral term in equation (6.6.6) is integrated by parts. By substituting the initial and boundary conditions of the sensitivity problem, equations (6.6.4.b-f), and then letting the terms containing $\Delta T(\xi,\eta,t)$ to vanish, we obtain after some lengthy but straightforward algebraic manipulations the following *adjoint problem* for the determination of the Lagrange multiplier $\lambda(\xi,\eta,t)$:

$$-\frac{J}{\alpha_t}\frac{\partial\lambda(\xi,\eta,t)}{\partial t} - \left(\frac{\alpha\lambda}{J}\right)_{\xi\xi} + 2\left(\frac{\beta\lambda}{J}\right)_{\eta\xi} - \left(\frac{\gamma\lambda}{J}\right)_{\eta\eta} + (PJ\lambda)_\xi + (QJ\lambda)_\eta +$$
$$+\sum_{m=1}^{NS}\left[T(\xi,\eta,t;q_3) - Y_m(t)\right]\delta(\eta-\eta_m)\,\delta(\xi-\xi_m) = 0$$
$$\text{in } 1<\xi<M,\, 1<\eta<N\ ,\ \text{for } t>0 \tag{6.6.7.a}$$

$$\frac{1}{J\sqrt{\alpha}}\left(\beta\lambda_\eta - \alpha\lambda_\xi\right) = 0 \qquad \text{at } \xi=1\ ,\ 1<\eta<N\ ,\ \text{for } t>0 \tag{6.6.7.b}$$

$$\frac{1}{J\sqrt{\gamma}}\left(\beta\lambda_\xi - \gamma\lambda_\eta\right) = 0 \qquad \text{at } \eta=1\ ,\ 1<\xi<M\ ,\ \text{for } t>0 \tag{6.6.7.c}$$

$$\frac{1}{J\sqrt{\alpha}}\left(\alpha\lambda_\xi - \beta\lambda_\eta\right) = 0 \qquad \text{at } \xi=M\ ,\ 1<\eta<N\ ,\ \text{for } t>0 \tag{6.6.7.d}$$

$$\frac{1}{J\sqrt{\gamma}}\left(\gamma\lambda_{\eta}-\beta\lambda_{\xi}\right)=0 \qquad \text{at } \eta=N\ ,\ 1<\xi<M\ ,\ \text{for } t>0 \tag{6.6.7.e}$$

$$\lambda\left(\xi,\eta,t_{f}\right)=0 \qquad \text{for } t=t_{f}\ ,\ \text{in } 1<\xi<M\ ,\ 1<\eta<N \tag{6.6.7.f}$$

Gradient Equation

In the limiting process described above for obtaining the adjoint problem, the following integral term is left:

$$\Delta S\left[q_{3}(t)\right]=\int_{t=0}^{t_{f}}\int_{\eta=1}^{N}-\left.\frac{\lambda(\xi,\eta,t)\sqrt{\alpha}}{k_{t}}\right|_{\xi=M}\Delta q_{3}(t)\,d\eta\,dt \tag{6.6.8}$$

By assuming that the unknown heat flux belongs to the Hilbert space L_2 in the time domain $0<t<t_f$, that is,

$$\int_{t=0}^{t_{f}}\left[q_{3}(t)\right]^{2}dt<\infty \tag{6.6.9.a}$$

we can write

$$\Delta S\left[q_{3}(t)\right]=\int_{t=0}^{t_{f}}\nabla S\left[q_{3}(t)\right]\Delta q_{3}(t)\,dt \tag{6.6.9.b}$$

Therefore, by comparing equations (6.6.8) and (6.6.9.b), we obtain the *gradient equation* for the functional as:

$$\nabla S\left[q_{3}(t)\right]=-\int_{\eta=1}^{N}\left.\frac{\lambda(\xi,\eta,t)\sqrt{\alpha}}{k_{t}}\right|_{\xi=M}d\eta \tag{6.6.10}$$

Iterative Procedure

The iterative procedure of the Technique IV, as applied to the estimation of the heat flux $q_3(t)$, can be written as

$$q_{3}^{k+1}(t)=q_{3}^{k}(t)-\beta^{k}d^{k}(t) \tag{6.6.11.a}$$

where β^k is the search step-size used to advance the estimation from iteration k to $k+1$.

The direction of descent $d^k(t)$ is given by:

$$d^k(t) = \nabla S\left[q_3^k(t)\right] + \gamma^k d^{k-1}(t) \tag{6.6.11.b}$$

The conjugation coefficient γ^k is obtained from the Fletcher-Reeves expression as:

$$\gamma^k = \frac{\int_{t=0}^{t_f} \left\{\nabla S\left[q_3^k(t)\right]\right\}^2 dt}{\int_{t=0}^{t_f} \left\{\nabla S\left[q_3^{k-1}(t)\right]\right\}^2 dt} \quad \text{for } k = 1,2,... \quad \text{with } \gamma^0 = 0 \tag{6.6.11.c}$$

The search step size β^k is obtained by minimizing the functional given by equation (6.6.3) with respect to β^k, in the same manner as described in Note 7 of Chapter 2. The following expression results:

$$\beta^k = \frac{\int_{t=0}^{t_f} \sum_{m=1}^{NS} \left[T\left(\xi_m, \eta_m, t; q_3^k\right) - Y_s(t)\right] \Delta T\left(\xi_m, \eta_m, t; d^k\right) dt}{\int_{t=0}^{t_f} \sum_{m=1}^{NS} \left[\Delta T\left(\xi_m, \eta_m, t; d^k\right)\right]^2 dt} \tag{6.6.12}$$

where $\Delta T(\xi_m, \eta_m, t; d^k)$ is the solution of the sensitivity problem given by equations (6.6.4) at the measurement point (ξ_m, η_m), obtained by setting $\Delta q_3^k(t) = d^k(t)$.

Stopping Criterion

The iterative procedure of the conjugate gradient method, given by equations (6.6.11-12), is applied to the estimation of $q_3(t)$ until a stopping criterion based on the *Discrepancy Principle* is satisfied. In such principle, as described in Chapter 2, we assume that the inverse problem solution is sufficiently accurate when the difference between estimated and measured temperatures is of the order of the standard deviation (σ) of the measurements. Thus, the value of the tolerance ε is obtained from equation (6.6.3) as

$$\varepsilon = \frac{1}{2} NS\,\sigma^2 t_f \tag{6.6.13}$$

The value of the functional (6.6.3) is then compared to the tolerance ε at each iteration. The iterative procedure is stopped when $S[q_3^{k+1}(t)]$ becomes smaller than ε.

Computational Algorithm

We suppose available an estimate $q_3^k(t)$ for the unknown heat flux $q_3(t)$ at iteration k. Thus:

Step 1: Solve the direct problem given by equations (6.6.2) to obtain the estimated temperatures $T(\xi,\eta,t)$.

Step 2: Check the stopping criterion given by the discrepancy principle with ε determined from equation (6.6.13). Continue if not satisfied.

Step 3: Solve the adjoint problem given by equations (6.6.7) to obtain the Langrange multiplier $\lambda(\xi,\eta,t)$.

Step 4: Compute the gradient of the functional $\nabla S\left[q_3^k(t)\right]$ from equation (6.6.10).

Step 5: Compute the conjugation coefficient γ^k from equation (6.6.11.c) and then the direction of descent $d^k(t)$ from equation (6.6.11.b).

Step 6: Solve the sensitivity problem given by equations (6.6.4) to obtain $\Delta T(\xi,\eta,t)$, by setting $\Delta q_3^k(t) = d^k(t)$.

Step 7: Compute the search step size β^k from equation (6.6.12).

Step 8: Compute the new estimate $q_3^{k+1}(t)$ from equation (6.6.11.a) and go to step 1.

Results

We illustrate below thę present approach for solving inverse problems based on generalized coordinates, with a practical example involving the cooling of an electronic component. Figure 6.6.1.a shows a module used for the cooling of thyristors [18, 19]. In such a module, a fluid in convective boiling is forced through channels to remove the heat released by the thyristor. The heat flux to the boiling fluid may vary depending on the two-phase flow regime and is to be estimated by using transient temperature measurements taken at appropriate locations inside the module. We consider for the analysis a single central channel with a half-circle cross section and take into account the symmetry of the channel. The geometry and relevant dimensions are shown in figure 6.6.1.b. The half-circle cross section is utilized because it permits more flexibility in the design of the condensing system [19]. The module is made of copper with dimensions $a = 5$mm, $t = 10$mm, $H = 5$mm, $d = 5$mm and $e = 15$mm.

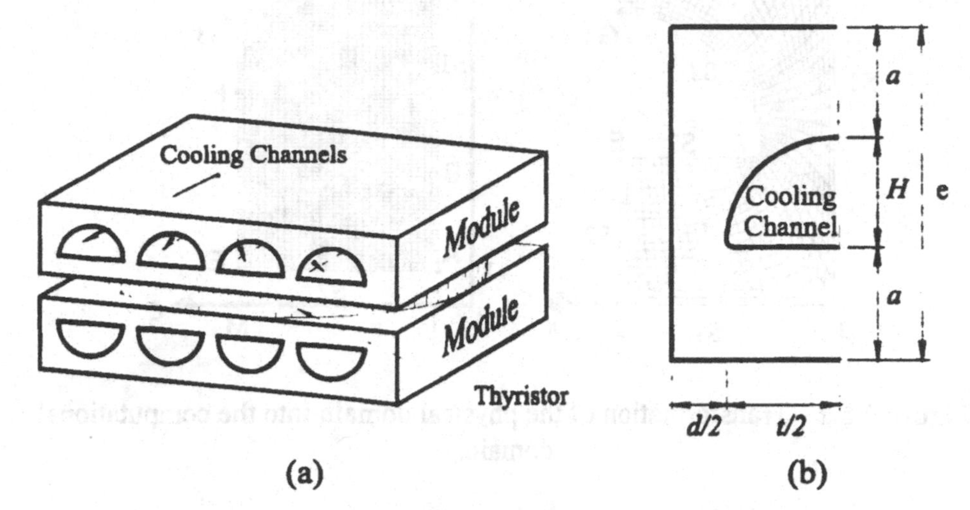

Figure 6.6.1 - (a) Module for the cooling of thyristors.
(b) Geometry in the physical domain with relevant dimensions.

The transformation of the irregular region in the physical domain into a rectangle in the computational domain is presented in figure 6.6.2. We note in this figure that the channel surface (E-F-G), with unknown boundary heat flux, is mapped into the boundary $\xi = M$ in the computational domain.

For simplicity in the analysis, we solve the present inverse problem in dimensionless form by introducing the following dimensionless groups

$$\bar{x} = \frac{x}{\ell} \qquad \bar{y} = \frac{y}{\ell} \qquad \bar{t} = \frac{\alpha_t t}{\ell^2} \qquad \bar{T}(\xi,\eta,t) = \frac{T(\xi,\eta,t) - T_0}{\frac{q_0 \ell}{k}} \qquad \bar{q}(t) = \frac{q(t)}{q_0} \tag{6.6.14.a-e}$$

where T_0, q_0 and ℓ are reference values for temperature, heat flux and length, respectively. The bars denote dimensionless variables in equations (6.6.14) and will be omitted hereafter.

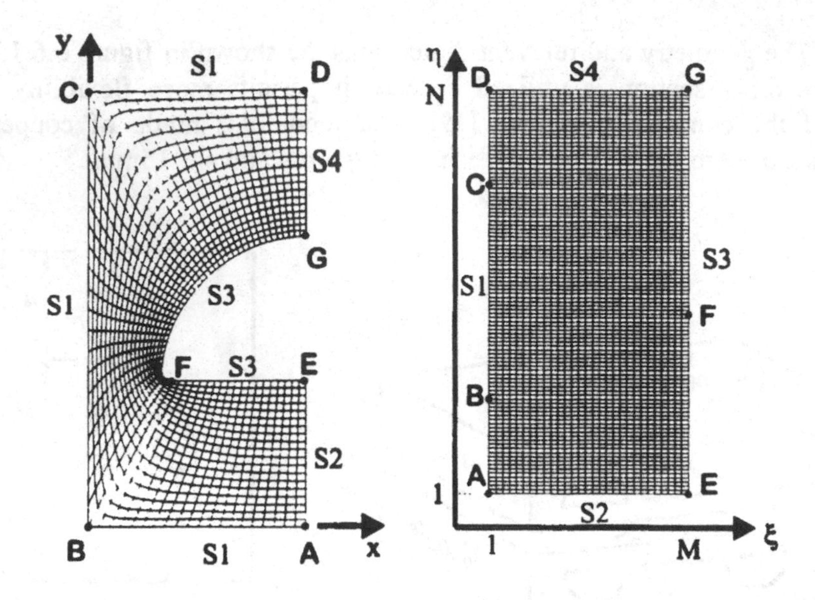

Figure 6.6.2 - Transformation of the physical domain into the computational domain.

For the results presented below, we assumed for the initial condition a uniform unitary temperature. A condition of symmetry was used for the boundaries $\eta = 1$ (A-E) and $\eta = N$ (D-G), while the boundary $\xi = 1$ (A-B-C-D) was supposed insulated. No generality is lost with this last assumption, since the heat flux at such boundary is considered known for the inverse analysis. The characteristic length was taken as $\ell = d/2 + t/2 = 7.5\text{mm}$, while the final dimensionless time was taken as 5.6, which corresponds to a dimensional time of 10 seconds. During this time interval, 50 measurements per sensor were considered available for the inverse analysis.

The direct, sensitivity and adjoint problems were solved with finite-differences by using the Alternating-Direction-Implicit (ADI) method [20, 21]. The resultant tri-diagonal systems were solved with a vector version of Thomas algorithm [22].

The domain shown in figure 6.6.2 was discretized with $M = 30$ and $N = 100$ points in the ξ and η directions, respectively. The time step was taken as 3.33×10^{-4}. Such time step and number of points were chosen based on a grid convergence analysis. The maximum difference between the results obtained with the above discretization for the direct problem and those obtained by doubling the number of points on both ξ and η directions, and using a time step 4 times smaller was less than 0.51%. The code for the direct problem was also validated by comparing the numerical results with known analytical solutions for cases involving regular geometries [23].

We used simulated measurements in order to assess the accuracy of the present approach of estimating the unknown boundary heat flux. Figures 6.6.3 and 6.6.4 present the results obtained for triangular and step variations for the heat flux respectively, and for two different levels of measurement errors, $\sigma = 0$

and $\sigma = 0.01\ T_{max}$, where T_{max} is the maximum measured temperature. Such results were obtained with the measurements of a single sensor located at the position A, as shown in figure 6.6.2. The agreement between exact and estimated functions, obtained with errorless measurements ($\sigma = 0$), is excellent for both functional forms tested. The triangular variation shown in figure 6.6.3 is exactly recovered and basically no smoothness is noticed in the corners. Also, very little oscillations are observed in the neighborhood of the discontinuities in figure 6.6.4. Similarly, the results obtained with measurements containing random errors ($\sigma = 0.01\ T_{max}$) are in very good agreement with the exact functional forms.

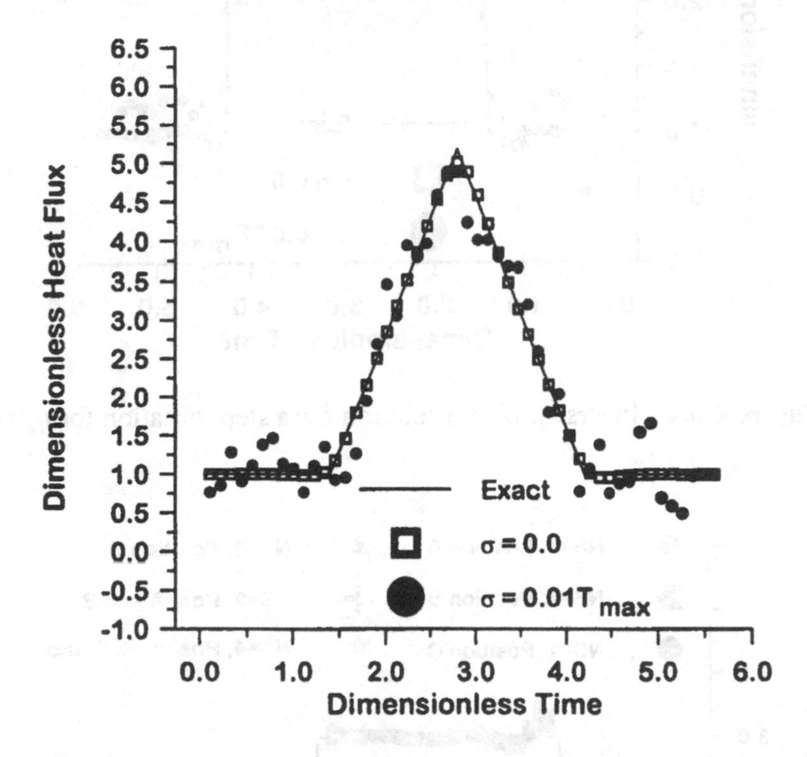

Figure 6.6.3 - Inverse problem solution for a triangular variation for $q_3(t)$

The effects of number and location of the sensors on the inverse problem solution were examined. Different configurations were tested, including: a single sensor (NS = 1) located at the position A, B, C, or D; two sensors (NS = 2) located at positions A and B; and four sensors (NS = 4) located at positions A, B, C and D. Figure 6.6.5 presents the solutions obtained with such configurations for the step variation of $q_3(t)$, by considering errorless measurements ($\sigma = 0$). The inverse problem solution appears to be insensitive to the location and number of sensors for the configurations tested. This is probably due to the reduced dimensions of the module studied. Similar behavior was also observed with the triangular variation, as well as with measurements containing random errors.

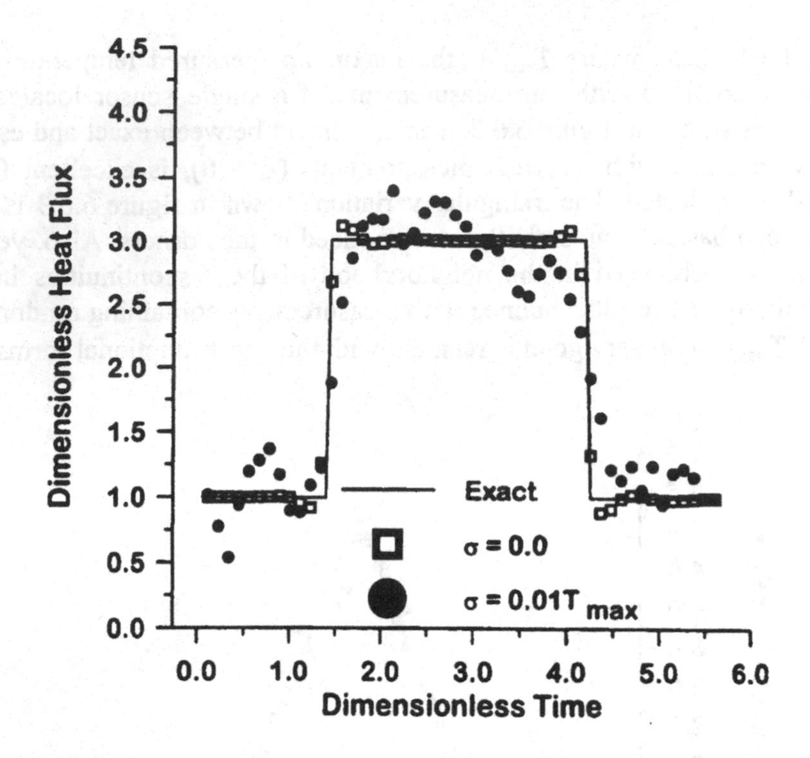

Figure 6.6.4 - Inverse problem solution for a step variation for $q_3(t)$

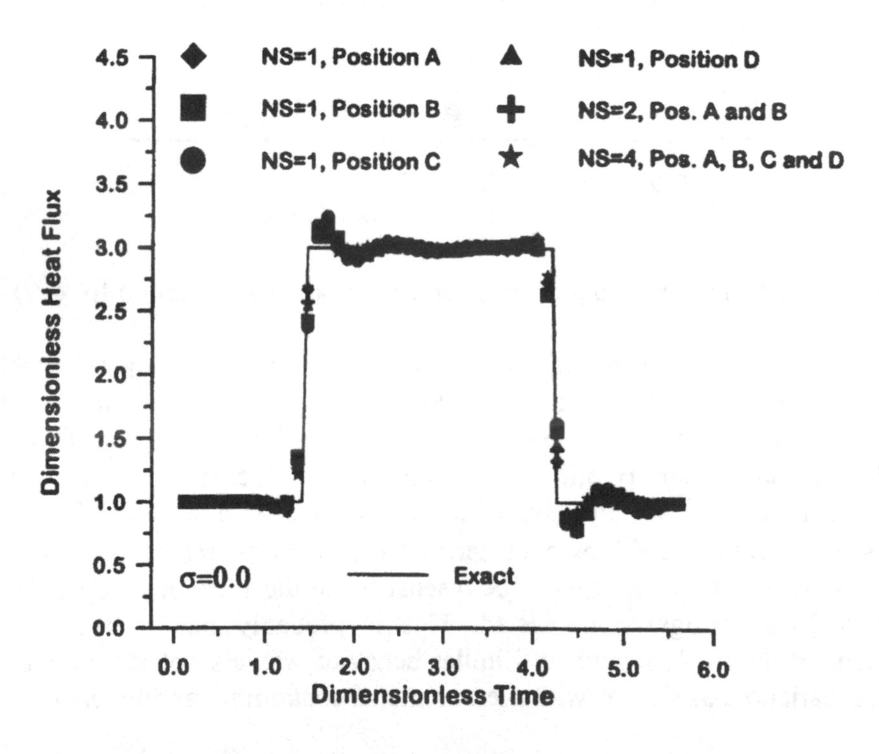

Figure 6.6.5 - Inverse problem solution for different sensor configurations

Although the solution of boundary inverse problems in irregular geometries as presented above is general, the preceding analysis of the number and location of sensors depends on the geometry under study, and should be performed for each case of interest. The present approach can be extended with few modifications to the analysis of problems involving multiply-connected regions.

Other numerical techniques, such as finite elements [17,24-27] and boundary elements [28-31], have also been applied to the solution of inverse problems involving irregular regions.

PROBLEMS

6-1 Derive the expressions given by equations (6.2.9,10), for the gradient in terms of the generalized coordinates (ξ,η).

6-2 Derive the expressions given by equations (6.2.12,13), for the divergence in terms of the generalized coordinates (ξ,η).

6-3 Derive the expressions given by equations (6.2.15,16), for the Laplacian in terms of the generalized coordinates (ξ,η).

6-4 For a square region in the physical domain with sides of unitary length, so that $h=L=1$ in figure 6.3.1, consider the discretization with 11 points in the x and y directions. Plot the grid lines on the region in the physical domain for different values of β, say, $\beta=1.5$, 1.1 and 1.01, by using the transformation given by equations (6.3.3.a,b).

6-5 Derive the elliptic grid generation equations (6.5.2.a,b).

6-6 Write a computer program for grid generation, by using the elliptic scheme given by equations (6.5.2.a,b) with first kind boundary conditions. Use such a program to generate grids on the region presented in figure 6.6.1.b. Examine the effects of the control functions $P(\xi,\eta)$ and $Q(\xi,\eta)$ on the grids generated.

6-7 Modify the program developed in problem 6-6 in order to allow the use of homogeneous second kind (orthogonality) boundary conditions.

6-8 Derive the sensitivity problem given by equations (6.6.4).

6-9 Derive the adjoint problem given by equations (6.6.7).

6-10 Consider the following heat conduction problem in a general two-dimensional region $\Omega(x,y)$:

$$\frac{1}{\alpha_t}\frac{\partial T(x,y,t)}{\partial t}=\frac{\partial^2 T}{\partial x^2}+\frac{\partial^2 T}{\partial y^2}+\frac{g(x,y,t)}{k_t} \quad \text{in } \Omega(x,y), \text{ for } t>0$$

$$\frac{\partial T}{\partial \mathbf{n}}=0 \quad \text{at } \Gamma(x,y), \text{ for } t>0$$

$$T(x,y,0)=0 \quad \text{in } \Omega(x,y), \text{ for } t=0$$

where $\Gamma(x,y)$ is the boundary of $\Omega(x,y)$.

By using the approach described in this chapter, derive all the basic steps of Technique IV in terms of generalized coordinates for the solution of the inverse problem of estimating the timewise variation of the heat source term $g(x,y,t)$. Assume the spatial distribution of $g(x,y,t)$ as known for the analysis.

6-11 Repeat problem 6.10 for the estimation of both the timewise and spacewise variations of $g(x,y,t)$.

6-12 In the heat conduction problem given by equations (6.6.2), assume that the heat flux at the boundary $\xi = M$ is a function of time as well as of the spatial position. Therefore, equation (6.6.2.d) needs to be replaced by

$$\frac{k_t}{J\sqrt{\alpha}}\left(\alpha T_\xi - \beta T_\eta\right) = q_3(\eta,t) \quad \text{at } \xi = M, \; 1 < \eta < N \;, \text{ for } t > 0$$

Derive all the basic steps of Technique IV in terms of generalized coordinates, for the estimation of the unknown function $q_3(\eta,t)$.

REFERENCES

1. Thompson, J. F., Varsi, U. A. and Mastin, C., *Numerical Grid Generation Foundations and Aplications*, North-Holland, Elsevier Science Publishers, Amsterdam, Netherlands, 1985.
2. Goldman, A. and Kao, Y. C., "Numerical Solution to a Two-dimensional Conduction Problem Using Rectangular and Cylindrical Body-Fitted Coordinate Systems", *J. Heat Transfer*, **103**, 753-758, 1981.
3. Ushikawa, S. and Takeda, R., "Use of Boundary-fitted Coordinates Transformation for Unsteady Heat Conduction Problems in Multiconnected Regions with Arbitrarily Shaped Boundary", *J. Heat Transfer*, **107**, 494-498, 1985.
4. Eiseman, P. R., "Automatic Algebraic Coordinate Generation", in *Numerical Grid Generation*, J. F. Thompson (ed.), North-Holland, Amsterdam, 447-463, 1982.
5. Guceri, S. I., "Finite Difference Methods in Polymer Processing", in *Fundamentals of Computer Modelling for Polymer Processing*, C. L. Tucker (ed.), Hanser, München, 1988.
6. Coulter, J. P. and Guceri, S. I., "Laminar and Turbulent Natural Convection within Irregularly Shaped Enclosures", *Numerical Heat Transfer*, **12**, 211-227, 1987.
7. Subbiah, S., Trafford, D. L. and Guceri, S. I., "Non-Isothermal Flow of Polymers Into Two-Dimensional Thin Cavity Molds: A Numerical Grid Generation Approach", *Int. J. Heat Mass Transfer*, **32**, 415-434, 1989.
8. Elshamy, M. M., Özisik, M. N. and Coulter J. P., "Correlation for Natural Convection Between Confocal Horizontal Elliptical Cylinders", *Numerical Heat Transfer*, **18**, 95-112, 1990.

9. Elshamy, M. M., Özisik, M. N., "Numerical Study of Laminar Natural Convection from a Plane to its Cylindrical Enclosure", *J. Solar Energy Engineering*, **113**, 194-199, 1991.
10. Orlande, H. R. B. and Özisik, M. N., "Transient Thermal Constriction Resistance in a Finite Heat Flux Tube", *J. Heat Transfer*, **117**, 748-751, 1995.
11. Alencar Jr., J. P., Orlande, H. R. B. and Özisik, M. N., "A Generalized Coordinates Approach for the Solution of Inverse Heat Conduction Problems", *Proceedings of the 11th International Heat Transfer Conference, Vol. 7*, 53-58, Kyongju, South Korea, 1998.
12. Özisik, M. N., *Finite Difference Methods in Heat Transfer*, CRC Press, 1994.
13. Kober, H., *Dictionary of Conformal Representation*, 2nd ed., Dover Publications, New York, 1957.
14. Milne-Thompson, L. M., *Theoretical Hydrodynamics*, 2nd ed., MacMillan, New York, 1950.
15. Churchill, R. V., *Introduction to Complex Variables*, McGraw-Hill, New York, 1948.
16. Roberts, G. O., "Computational Meshes for Boundary Layer Problems", Proc. Second Int. Conf. Num. Methods Fluid Dyn., *Lecture Notes in Physics*, Springer-Verlag, New York, **8**, 717-777, 1971.
17. Truffart, B., Jarny, Y. and Delaunay, D., "A General Optimization Algorithm to Solve 2-D Boundary Inverse Heat Conduction Problems Using Finite Elements", *First International Conference on Inverse Problems in Engineering: Theory and Practice*, Palm Coast, FL, 53-60, 1993.
18. Scaringe, R. P., *Forced Vaporization Cooling of HVDC Thyristor Valves*, Electric Power Research Institute, Report EL-2710, 1982.
19. Cavalcanti, E. S. C., Cruz, F. R. L., Orlande, H. R. B., "A Forced Convective Boiling Module for the Cooling of Thyristors", *X Brazilian Congress of Mechanical Engineering*, Rio de Janeiro, 375-378, 1989.
20. Peaceman, D. W. and Rachford, H. H., "The Numerical Solution for Parabolic and Elliptic Differential Equations", *J. Soc. Ind. Appl. Math.*, **3**, 787-793, 1955.
21. Anderson, D. A., Tannehill, J. C. and Pletcher, R. H., *Computational Fluid Mechanics and Heat Transfer*, Hemisphere, New York, 1984.
22. Ortega, J. M., *Introduction to Parallel and Vector Solution of Linear System*, Plenum Press, New York, 1988.
23. Özisik, M. N., *Heat Conduction 2nd edition*, Wiley, New York, 1993.
24. Osman, A. M., Dowding, K. J. and Beck, J. V., "Numerical Solution of the General Two-dimensional Inverse Heat Conduction Problem", *ASME J. Heat Transfer*, **119**, 38-44, 1997.
25. Zabaras, N. and Ngugen, T. H., "Control of the Freezing Interface Morphology in Solidification Processes in the Presence of Natural Convection", *Int. J. Num. Meth. Eng.* , **38**, 1555-1578, 1995.

26. Zabaras, N. and Yang, G., "A Functional Optimization Formulation and Implementation of an Inverse Natural Convection Problem", *Comput. Methods Appl. Mech. Engrg.*, 144, 245-274, 1997.
27. Yang, G. and Zabaras, N., "An Adjoint Method for the Inverse Design of Solidification Processes with Natural Convection", *Int. J. Num Meth. Eng.*, **42**, 1121-1144, 1998.
28. Huang, C.H. and Tsai, C.C., "An Inverse Heat Conduction Problem of Estimating Boundary Fluxes in an Irregular Domain with Conjugate Gradient Method", *Heat and Mass Transfer*, **34**, 47-54, 1998.
29. Huang, C.H. and Tsai, C.C., "A Shape Identification Problem in Estimating Time-Dependent Irregular Boundary Configurations", *HTD-Vol.340 ASME National Heat Transfer Conference vol. 2*, 41-48, Dulikravich, G.S. and Woodburry, K. (eds.), 1997.
30. Huang, C.H. and Chiang, C.C., "Shape Identification Problem in Estimating Geometry of Multiple Cavities", *AIAA J. Therm. and Heat Transfer*, **12**, 270-277,1998.
31. Dulikravich, G.S. and Martin, T.J., "Inverse Shape and Boundary Condition Problems and Optimization in Heat Conduction", *Chapter 10 in Advances in Numerical Heat Transfer,* **1**, 381-426, Minkowycz, W.J. and Sparrow, E.M. (eds.), Taylor and Francis, 1996.

Index